Dedication

This Almanac is dedicated to all those who have been pushed off the land, who have lost land to banks, to fires, to hurricanes, to drought, to colonization, corporatization, and eminent domain.

It is dedicated to all who have fought to protect the land and the diversity of life it supports, who have resisted leaving and instead tended the land, who have put land in trust, who have reclaimed and recovered and tilled, responsibly.

And it is dedicated to the Oakland Institute and all those who steward the land now and in the future—to those who will plant trees, support mycorrhizal networks, share space and ideas, save seeds, and collaborate within and across land-based communities.

This Almanac Belongs To:

____

THE NEW FARMER'S ALMANAC

2019

THE GREATER "WE"

GREENHORNS

THE NEW FARMER'S ALMANAC VOLUME IV
THE GREATER "WE"

Published by the Greenhorns
greenhorns.org

GREENHORNS

Printed in the USA by McNaughton & Gunn.
ISBN 978-0-9863205-2-1

EDITOR IN CHIEF
SEVERINE VON TSCHARNER FLEMING

LEAD EDITOR
BRIANA OLSON

VISUAL EDITOR AND DESIGNER
KATIE EBERLE

COPY EDITORS AND PROOFREADERS
CHRISTOPHER CHURCH
SUE DUVALL
ALLI MALONEY

THANK YOU

A first-order thank you to Ben Franklin, who prototyped the form on this continent.

To all our writers and contributors—what a lot of love!

To Elisabeth Keller and Steven Bonsey—thank you for supporting our work.

Thank you to Rick and Megan Prelinger for setting us up for such a thrilling intersectional curriculum, for passing down those boxes of ephemera from the top shelf of your library (prelingerlibrary.org), and casually wondering aloud how easy it would be to do a contemporary version. Your open-handed wondering makes so many of us rise to the challenge in service to the deep future, and in a community relation with our historical archives. Thank you to Alessandro D'Ansembourg, who believed and helped so much to make the second version possible, getting us over the hump toward an ongoingness.

A thank you to the tireless diligence and pursuit of quality held by our evolving almanac team, including Briana Olson, Katie Eberle, Nina Pick, Charlie Macquarie, Audrey Berman, Hallie Chen, Tess Diamond, Francesca Capone, and Nicole Lavelle. These core almanackers, and the peripheral almanackers who make this scatterplot blob of Greenhorns cultural productions actually happen, beautifully and full of heart.

Thank you to Thomas Baldwin for his marvelous love, companionship, and a sustained inquiry across topics and geographies, intersections, and overland flights—direct data, dog-eared atlases, song lines, road snacks, and siestas. What a teammate! —SvTF

Thanks to all who have sent us their writing and art, and to Severine, for sparking curiosity and inspiration. Thanks to the Greenhorns team and our friends at the Agrarian Trust, especially Emma O'Leary, Jean Willoughby, Soraya Farivar, Travis Singley, Christopher Church, and Ian McSweeney, and outgoing editors Charlie Macquarie and Nina Pick; to Alli Maloney, for donating many hours of thoughtful editing; to Katie Eberle, for late-hour flexibility, and making this volume beautiful; and to Sarah Wentzel-Fisher, for introducing me to the Greenhorns in the first place. —BRO

CONTENTS

Index of Contributors

RACHEL ALEXANDROU
Forager, artist, propagandist
Alna, Maine
rachelalexandrou.wordpress.com

JASON BENTON
Gardener, goatherd
Prospect, Pennsylvania

KIRSTEN BONANZA
Seed keeper, life coach
Clinton Hollow, New York
kirstenbonanza.com

BRETT F. BRALEY
Writer, baker, blogger
Pennsylvania

KERI BRANDT OFF
Professor
Del Norte, Colorado
offfamilyranch.com

LIZ BRINDLEY
Artist, educator, farmer
Santa Fe, New Mexico
printsandplantspress.com

KAREN A. CANTOR
Filmmaker
Santa Fe, New Mexico
returndocumentary.com

WILLY CARLETON
Historian, farmer, editor
New Mexico

RENATA CHRISTEN
Researcher, crop diversity enthusiast
Amsterdam, The Netherlands
localwhey.tumblr.com

KIRK CRIPPENS
Artist, photographer
San Francisco Bay Area, California
kirkcrippens.com

DOUGLASS DECANDIA
Gardener
South Salem, New York
diggersdig.blogspot.com

GLORIA DECATER
Farmer, educator
Covelo, California
livepower.org

WILLIAM EDWARD BURGHARDT DU BOIS
Scholar, author, civil rights and peace activist, editor
W.E.B. Du Bois Memorial Centre for
Pan African Culture
Accra, Ghana

MICHELLE J. FELICETTI
Broad range creative
Windsor, Vermont
@the_rhodes_leap

JANINE FITZGERALD
Professor of sociology
Bayfield, Colorado
Crowbar Creek Ranch

MALORY FOSTER
Wannabe homesteader, farmer co-op starter
Gulfport, Florida
floridagrownherbs.com

LUCIA GAIA
Artist, community organizer
Brooklyn, New York
luciagaia.com

JOHN D. GALUSKA
Urban farmer, food activist, educator
Bloomington, Indiana

LINCOLN GEIGER
Farmer, educator, visionary
Temple, New Hampshire

JESSICA GORDON-NEMBHARD
Professor, economist
CUNY New York

CHRISTIE GREEN
Landscape designer, writer
Santa Fe, New Mexico
beradicle.com

MARC GRIGNON
Thinker, writer
Menominee Territory, Wisconsin

CLAUDIA K. GUNSCH
Environmental engineer
Durham, North Carolina
gunsch.pratt.duke.edu

REV. GRACE G. HACKNEY
Founding director, Life Around the Table
Orange County, North Carolina
lifearoundthetable.org

KATHERINE HALE
Naturalist, gardener, writer
Durham, North Carolina

LAYTON HANSEN
Illustrator
Albuquerque, New Mexico

FRANCES ELLEN WATKINS HARPER
Poet, writer, abolitionist
Philadelphia, Pennsylvania

ALLISON ADELLE HEDGE COKE
Human being
US/Canada
allisonhedgecoke.com

SAM HEDGES
Vegetable farmer, writer
Albuquerque, New Mexico

CHRISTINE HEINRICHS
Writer
Cambria, California
christineheinrichs.com

ELIZABETH HENDERSON
Organic farmer, writer
Rochester, New York
thepryingmantis.wordpress.com

MARIE HOFF
Grazier
Potter Valley, California
@stargrazers

ELIZABETH HOOVER
Professor, gardener, writer
Providence, Rhode Island
gardenwarriorsgoodseeds.com

SAVI HORNE
Executive director at Land Loss Prevention Project
Durham, North Carolina
landloss.org

ADAM HUGGINS
Plant person, podcaster
Vancouver, British Columbia
futureecologies.net

SARAH HUGHSON
Gardener, artist
Albuquerque, New Mexico
lulabell409@gmail.com

JAMIE HUNYOR
Writer, farm worker
Northern California

JOSIE ISELIN
Artist, author, seaweed enthusiast
San Francisco, California
josieiselin.com

FORREST JACKSON
Merchant seaman
Seattle, Washington
wayforasailor.com

Index of Contributors

LUCIA JAZAYERI
Writer
Boston, Massachusetts
cloverfoodlab.com

CASH JOHNSON
Pot farmer, songwriter, musician
Humboldt County, California
camocowboys.com

JAMES WELDON JOHNSON
Poet, writer, civil rights activist
Brooklyn, New York

GRANT JONES
Landscape healer, farmer bard
Oroville, Washington
jonesandjones.com

CONNOR JONES
Permaculture teacher and designer
Ojai Permaculture + East End Eden farm
ojaipermaculture.com

ROBERT KARP
Writer, consultant, social entrepreneur
Milwaukee, Wisconsin
robertkarp.net

AZURÉ KEAHI
Land artist, mother
Troy, New York
moku.farm

JOHANNA M. KEEFE
Visionary, writer, holistic educator, healer
Greater Boston Area, Massachusetts
growingsuccesstories.org

KEEFE KEELEY
Agroforester
Driftless, Wisconsin
savannainstitute.org

BRAD KIK
Co-director of Crosshatch Center
for Art and Ecology
Bellaire, Michigan
crosshatch.org

RUTHIE KING
Rancher, educator
Willits, California
adaptiveagriculture.org

PETER KIRN
Farmer

BINYAMIN KLEMPNER
Farmer, writer
Tiberias, Israel
galilsoil.com

NIKKI LASTRETO & SWAMI CHAITANYA
Cannabis cultivators and educators, founders of
Swami Select
The Emerald Triangle, California
swamiselect.com

GRETCHEN LEMAISTRE
Artist, photographer
San Francisco Bay Area, California
gretchenlemaistre.com

GINNY MCCLURE
Artist
Central Minnesota

FRANCIS MCGILL
Poet, husband, educator
Bristol, Vermont

URSULA MACFARLANE
Art that bends reality
New Zealand
spacedart.weebly.com

GABRIELLA MARKS
Photographer
New Mexico
gabriellamarks.com

IAN MCSWEENEY
Visionary, practitioner, creator
Weare, New Hampshire
agrariantrust.org

STEPHANIE MILLS
Bioregionalist, author
Northwest Lower Michigan
smillswriter.com

RAY RAY MITRANO
Social practice artist
Rochester, New York
rayraymitrano.com

NIKKI MOKRZYCKI
Artist, designer
Cleveland, Ohio
nikmoz.com

ÁINE MULLAN
Photographer, illustrator
Hobart, Tasmania
ainemullan.com

CASEY O'NEILL
Farmer
HappyDay Farms, Mendocino County, California
happydayfarmscsa.com

DANJO PALUSKA
Postcard aficionado
12 Dunning St, Brunswick, ME, 04011

SHAWN PORTER
Musician, poet, bioneer
Murray, Arkansas
facebook.com/staymore

ALEXANDER POSEY
Poet, writer, satirist, politician
Creek Nation

VANESSA RADITZ
Farmer, educator, organizer, dreamer
Bordentown, New Jersey
bit.ly/queerecoproject

ALAYNA RASILE-DIGRINDAKIS
Textile innovator
Bozeman, Montana
absorka.com and maywestmilkweed.club

ANTHONY REYES
Farm manager at the Homeless Garden Project
Santa Cruz, California
homelessgardenproject.org

CAMERON RHUDY
Staff attorney at Sustainable Economies Law Center
Sacramento, California
theselc.org

SAM RYERSON
Cowboy
New Mexico

M.T. SAMUEL
Youth worker, poet
Alaska and Wisconsin

CARL SANDBURG
Poet, biographer, radical
Remembrance Rock, Galesburg, Illinois

KATHLEEN SHANNON
Farmer, educator
the Rockies

VINCENT SHERIDAN
Artist
Dublin, Ireland
vincentsheridan.com

Index of Contributors

JULIA SHIPLEY
Word and homestead wrangler
Northeast Kingdom, Vermont
writingonthefarm.com

JOSH SLOTNICK
Farmer, Teacher
umt.edu

CHARLOTTE X.C. SULLIVAN
Citizen designer
Bristol, Vermont
charlottexcsullivan.com and maywestmilkweed.club

ROXANNE SWENTZELL
Artist, permaculturist
Santa Clara Pueblo, New Mexico
roxanneswentzell.net

NEIL THAPAR
Food and farm attorney
Oakland, California
twitter.com/neilthapar

DON TIPPING
Farmer, seedsman, teacher
Williams, Oregon
siskiyouseeds.com and sevenseedsfarm.com

FRANK UTPATEL
Artist, illustrator
Wisconsin

SAVANNAH J. VOLKOFF
PhD candidate
Durham, North Carolina

MAUREEN WALRATH
Interdisciplinary artist, basket weaver, farmer, and student midwife
Seasonally nomadic, living among the willow
woventhresholds.com

SOPHIE WANZER
Student
Littleton, Massachusetts

RACHEL WEAVER
Field philosopher
Denton, Texas

ROWEN WHITE
Seedkeeper, writer, teacher
Akwesasne and Northern California
sierraseeds.org

ELICIA WHITTLESEY
Sustainable agriculture program manager for the Old Fort at Hesperus
La Plata Mountain watersheds
singthewinterwren.wordpress.com

TOM WILLEY
Farmer emeritus
T&D Willey Farms
Madera, California
tdwilleyfarms.com

JEAN WILLOUGHBY
Author, filmmaker
USA
bit.ly/jeanwilloughby

ZACH WITHERS
Pig farmer
San Antonito, New Mexico

JUSTIN YOUNG
Teacher
Amarillo, Texas

JENNY ZHAO
Digital marketing director, photographer
Boston, Massachusetts
instagram.com/rogerandjennyphotography

Almanaque

EDITOR'S NOTE

BRIANA OLSON, LEAD EDITOR

Have high hopes in the future of all humanity.
—PHRA PRAYUDH PAYUTTO

A longtime wanderer, I belong to the many in that I've probably raised less than 0.001 percent of what I have up to now consumed. But as I began to gather the poems and essays and drawings that you now hold in your hands, I found myself with a 10-by-12-foot plot of healthy earth and a long season of undiluted sunshine. For the first time in my life, I pushed beans into the earth and watched with amazement as shoots sprouted from the soil and vines climbed up the iron trellis my partner had put up for this purpose.

Our neighbors, seasoned composters and canners and prize-winning picklers whose garden is thrice or more the size of ours, were impressed with the variety we'd planted in such a small space. We grew Filipino eggplants and French radishes and chiles and greens for every persuasion. Because of that, and despite our failures and losses (curly top virus in half the tomatoes, wilt in the tomatillos, the peas dried up, corn abandoned), we were able to supply ninety-some percent of our produce, to eat day after summer day what we'd raised.

If I was struck with wonder by the simple act of germination, I was just as awed with the commitment, the consistency and love that sustaining the life of the garden required.

Assembling this almanac has been a parallel education. Some of the lessons here are practical—Save seeds! Diversify, but don't take on too much! Plant milkweed!—and some are aspirational. What is pragmatic advice for one grower is a distant goal, an ideal, for the next. Some of the lessons are of hardship, and some are stories of collaboration. Some reflect on our American pasts—the pasts we share, and the pasts that divide us—and sketch blueprints for a more just future. Others dive back in planetary time. All speak to the well-grounded belief that beyond the gift of life itself, what is magical about the earth is its exquisite variation.

At its core, this collection is guided by the same principles that have guided my humble

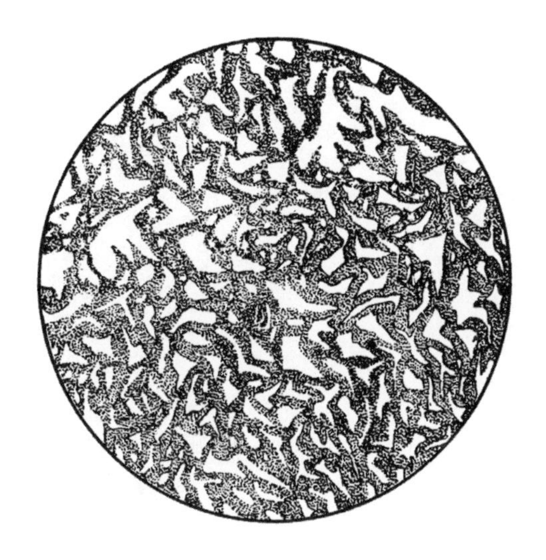

garden. First, start *somewhere*. Small is good. Here, you won't get lost in a thousand acres of corn being raised for ethanol. You won't find a hundred iterations of the myth of the American farmer; instead, meet actual farmers, with actual back pain, ruminating on myth and ideal while they rotate their crops.

Our contributors are small-scale growers, cowboys and hunters and activists, researchers, permaculturists, rural and urban idealists—listeners and advocates—photographers and doodlers—and among them are more women than men. Their stories are not reports from an idealized past. Farming is as hard, even harder, as it ever was. It's no wonder, given the onslaught of worst-case scenario reports on climate change and economic hardship, coupled with the experience of ongoing drought, record-breaking fires, farmer neighbors given in to debt and despair, that so many humans dismiss the future of the land as a lost cause. We inhabit the age of apocalypse, where constant screenings of the world's end provide an entertaining, if anxiety-ridden, distraction from the slow violence being done to the living world and its people even as the heroes on the screen find their way to the last safe place—a place that is often rural and green, welcoming to the metaphor of seed.

Seeds themselves, however small, are objects that exist in the physical realm. They do not last forever. To grow, to remain viable, they must be planted and fed. They adapt to place—to soil and climate—just as humans have adapted to the earth. Too often, when scientists and journalists speak to the people of changing practices to limit climate change, they speak in general terms: abstractions, statistics, and macro-level mapping that seem to take solutions out of our hands.

In these pages, contrarily, you will encounter people thinking with their feet on the ground. People who take industrial waste and repurpose it. People convincing the wealthy to place their land in trusts, or reclaiming space—figurative and literal—for the black and brown and indigenous peoples whose knowledge and labor are the root of that wealth. People who grow trees, and people who forage. People who, if you live on the coast or in the city, might have been reduced in your mind to the red meadows stretching across one and another infographic, mere pieces of political machinery. This almanac is for them, and all you others, working the fields, the rivers and seas. It's also for you, hungry urban dweller, fellow nomad, contemplating your escape from the cubicle—even if just for the time it takes to read an essay or story by someone who feeds you.

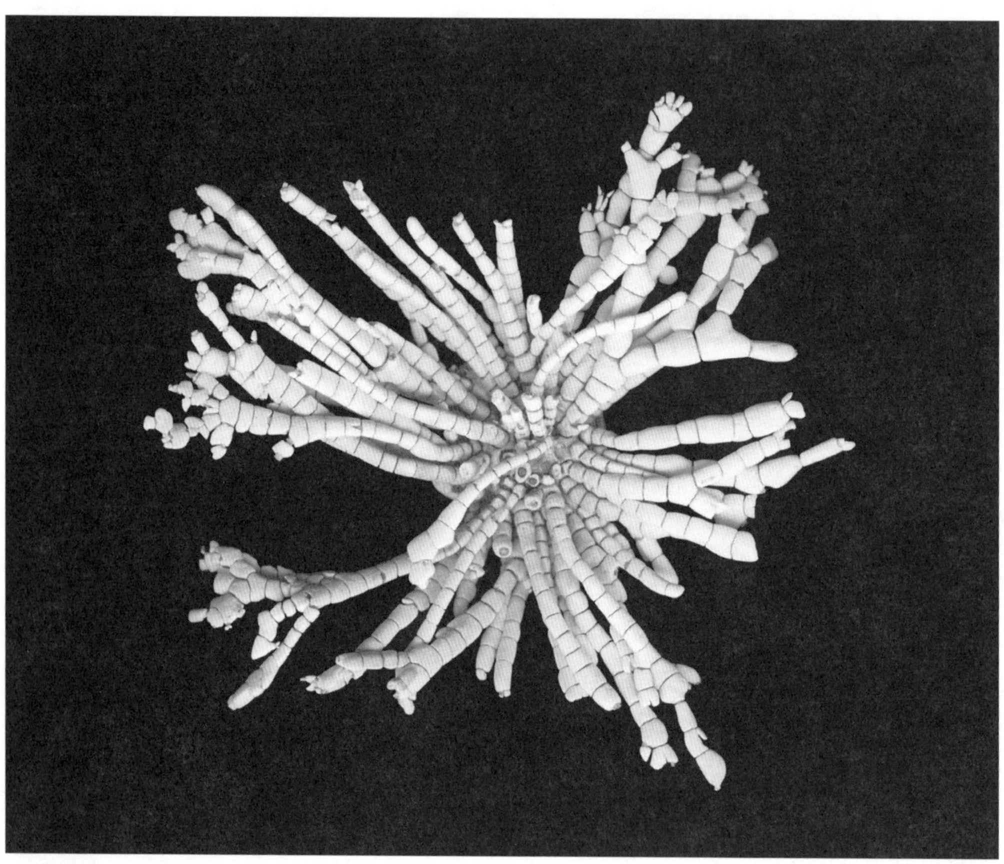

Josie Iselin | Coralline Algae

INTRODUCTION

SEVERINE VON TSCHARNER FLEMING, EDITOR IN CHIEF

@smithereenfarm
@thegreenhorns
@reversinghall

NOTABLE NEWS

Groundskeeper Dewayne "Lee" Johnson was awarded $289 million in damages by Monsanto—now Bayer—for illness sustained as a result of routine workplace use of their herbicide, Roundup. He is terminally ill with cancer.

UPDATE

After ten years running Greenhorns, I finally was able to buy and start my own farm. Come visit Smithereen Farm and the Reversing Hall in Downeast Maine. August is blueberry season, June is the Seaweed Symposium, and there's a full calendar of workshops and happenings. Pick berries, make your own jam, dig clams, and stay awhile in our seaside campsite with timber frame kitchen.

Due to love and farming, this essay is shorter than I would have liked.

IDENTITY OF EXTRACTION

In August, when the tide is right, boys in Downeast Maine switch from blueberries to clams, leaving their rakes in the field and coming back with a joint. It pays more.

The abandoned motel at the Pennamaquan River says "Best clams by a dam site." Which probably isn't true; the dam stops the river's sediment and slows down the coursing flow of anadromous fish and fertility up and down the river.

Down on the shore in the cooler months, it's older men you see clamming, bent down in the muck at the tail-end of a wet hole trail. They all say it's not as good as it once was. The extreme tides expose extra beach, and often bring on big weather as well. Gossip in town says that aching backs lead to painkillers, then illegal opioids. But when you get down to it, practically all of the natural-resource jobs seem culturally coded for painful overexertion. Plundering hurts the plunderer, as well as the ecosystem it takes too much from. From logging and shipbuilding, to dory and schooner fishing, to cannery work[1]—even the extra-large size of hay bales seems like testimony of this maximalist culture of human exertion. A colonialist, producer-identity or ethic encoded in the very fabric of human opinion, as well as the mode of economic exchange. Our predecessors hollowed and degraded the quarries, forests, and fisheries of their source material for wealth—plundering and rushing to get it all, the biggest armful possible. A Chinese company recently bought the paper mill in Baileyville, and they are doubling production. Our woods, it seems, are perfect for toilet paper.

The dam straddles a river, alongside a rotted-out alewife ladder, state-built in the 1980s. The ladder allows some fish to climb up around the dam, although the modification seems to advantage the eagles, as much as the fish. Sometimes during a run, you'll see twenty-three lazy raptors, turkey vultures, eagles, and osprey hunching with their tummies out, preening the fish grease off their front plumage. Birds can binge, but humans can dismantle and reconstitute the river-system.

Wealth of land, wealth of history—my new home is a place blessed with navigational relevance and lucky winds. Washington County was the site of the first naval battle of the War of Independence waged by then-colonies that would become the United States. A British convoy arrived to fetch lumber for their forts and armament of Boston Harbor, and were violently rebuffed by a motley crew led by Jeremiah O'Brien, plans hatched in fields and taverns to protect their liberty, and their trees, from forced removal by the king. These early colonists had hewn towns, pastures, and settlements from exposed rocky coastline lumberland where the indigenous Passamaquoddy people gathered and fished. Traders and merchants got news from Boston in the time it took the wind to blow downeast; the land itself gave them confidence to fight. That American revolution was won in the minds of the the new republic far before the British fought to suppress it. How do we, now, hold in our minds a similarly fierce—if peaceful—commitment? How will we dismantle the dam?

MACHINE-MANAGED LANDSCAPE

Living on a damaged planet, so says Donna Haraway, and is a great responsibility. The responsibility to respond, react, repair, refurbish. She challenges us to stay with the trouble, to see the trouble, and simultaneously imagine beyond it. To embody and exert a kind of distributed, and visionary human kindness on a violently colonized landscape. She challenges us to live-into the possibility, to see past the computers and toxic leviathan sprawl, and bring forth what is calling to be done.

So much evidence of that trouble presents itself on the land, and so much is visible and obvious from the road. I took the train from Kansas westward after a talk at the Land Institute and learned that armadillos (native to South America and migrating after the collision of continents) carry leprosy, which they acquired through contact with Old World Europeans who passed the same passage as they came overland into the Southwest.[2] In Dodge City, cheerful eighty-year-old volunteers tend the coffee pot at 4 a.m., doing their part to keep the Amtrak numbers up to preserve federal funding. Shrunken towns allude to a future of farming without farmers. Seventy-year-old-men in ironed shirts pump diesel into their millions of dollars of equipment, bouncing up high in their tractor-cabs, beaming up data to a wired-in satellite link. They dodge streams and gullies, smooth steering with routine nuance, teaching machines how to farm their land.

"A number of years back," historian Tom Giessel tells me, "we had a 270 horsepower tractor we'd purchased, used, but less than a year old. It sat in the barn over winter; in the spring we were getting ready to do something. Engine started just fine, but it wouldn't move. It wouldn't go forward or back. Called the dealership, so they would send a tech over. They don't call them mechanics anymore, they call them techs now. He hooked up a laptop computer and it needed new software installed. It was rendered useless in a few months. Without them coming out and doing those updates, that tractor was worthless. When you buy a piece of equipment, this nice equipment, they tell you that you bought the equipment but you didn't buy the technology … so really you don't own anything. They have full control. If you don't pay a bill, they can shut you off, literally, with a touch of a button, not even a button, but the touch of a screen. That's really really scary, especially when you are spending $300,000, $500,000, $700,000 for a piece of equipment. Really, you don't own it. Somebody else has the 'keys.' If everyone realized what they can do,[3] you'd shiver like a dog shitting razor blades."

Wilted towns radiate out from the upright columnar center of the grain bins. Great greying towers, surrounded by trailer hulks, drainage humps, and ramps—alongside warehouse bins full of blue, pink, and purple dust—amendments for sale, commodities for sale. Animated by diligent women with punch cards and IBM computers at the desk inside. It used to be that granges and farmers unions controlled these co-ops, exerting some control over the prices paid, lobbying effectually for "price parity," and taking charge of their destiny by coordinating and administrating the grain dumped through the metal grate onto their scales, belted aloft then pumped out to market. These days, speculators get their data straight from the tractors—grain harvest poundage, moisture content, the exact GPS angle of the combine fed directly from front-mounted computers—directly to the cloud.[4] Grain prices are lower now, in relative dollar value, than they were in the farm crisis years of the 1970s and 1980s, when one-third of family farms went out of business. To recompense for Chinese trade war retaliation on US exports abroad, it was announced that twelve billion dollars in direct payments was to be made to producers of hogs, sorghum, dairy, cotton, pork, wheat, soy, and corn.[5] The so-called "soybean bailout" was put forward by Sonny Perdue, US Secretary of Agriculture, and the Trump administration proposed a production payment of $1.65 per bushel for soy, $8 per head for hogs, and a measly 12 cents per hundredweight of dairy. Producers driven by farm policy into an overproduction frenzy now have to apply for yet another payment to keep from bankruptcy. Sales of soybeans to China are down 94 percent from last year. According to the *New York Times*, only $838 million had been paid out by November 2019.[6] The harvest is in and the farmers are still waiting for their checks. This is not an emancipatory mathematics.

Tom and I went out to visit Fort Larned, now a National Historic Site, a marvelously preserved golden sandstone quadrangle of proud buildings that adjoins the old Santa Fe Trail. These early mercenaries enjoyed decorating with buffalo pelts, rawhide fringe, bright indigo buntings, and thin-legged frontier furniture. The station was preserved intact because it was adopted by the stockman's association. A fittingly barbed intertwining, the very emblem of the imperialist US military and its supply chain of producers. The cavalry was stationed at this frontier post in 1867 to protect trade and settlers, to distribute food aid to the "friendly Indians,"[7] and to facilitate removal of the indigenous peoples from the grazing lands to be occupied by the bovine herds or plowed up.

Bonanza has always speculated, profiteered, and exploited the wealth, wisdom, bodies, and land of the subsistence homesteaders.[8] From the now-gullied track of the Santa Fe Trail, human skeletons are exposed as the decades erode hasty burials, exposing the pathos of US rural policy, insidiously binding the big business of export, the big military protection of the emancipation narrative, and the notion of "manifest destiny" employed to recruit the pioneers and workers. Today's large- and larger-scale operations follow directly in this lineage of concentration and consolidation: those who remain on the land and operate at the "trainload scale" are galloping at utmost human speed, riding tractors for twenty-hour shifts to get it all planted, sprayed, hauled in. How laughable for Trump to call these farmers "resilient."

Up to their eyeballs in high-liability details, engineering, weather foibles, and equipment repair, these gray-haired men making the grain crops are not villains, nor the enemy. Just another feedstock to the ever-merging juggernaut of the agri-pharma-chemi-clouds-syn-innovation leviathan. We ought to be strategizing how to structure a dignified retirement plan with reinvestment to diversify farm practices and restore soil health and ecological function. The current system is literally putting these operators underground in debt, for a food system that destroys the land for money.[9]

CASCADING DISPLACEMENT

May the lord, if there is one, save my soul, if I have one.

—VOLTAIRE

Should you wish it, there's a noble brick house for sale. A four-bedroom home on five acres, built in 1890 on a prominent avenue in a small town in Ohio, Michigan, or Indiana for $250,000; includes elm tree. For that price, you could buy half a damp vinyl-clad duplex in Gloucester, Massachusetts, or half a one-bedroom studio in Williamsburg, Brooklyn, or two hundred acres along a river with big barn and farmhouse in New Brunswick, Canada.

Why this misalignment, we wonder? Those lovely fellows working with data visualization will show you in graphic three-block bars the legacies of red-lining and speculation in Chicago: how the wealth, jobs, service economy, commerce, and overblown speculation concentrate themselves in corridors of growth. It's easy to be cynical when it comes to hypergentrification and the self-serving "creative placemaking" efforts aimed at recruiting a talent base for technology companies. Portland, Denver, Nashville—the loft-living, high-rent urbanista set have got their habitat-formula provided by big developers, with amped-up parking garages, hot yoga, craft and paper shops, and impulse displays full of mylar-pouch superfoods; these hipster-life hyperzones represent displacement and disinvestment of traditional neighborhoods and the human fabric of community. More frustratingly, they insulate creative, ambitious young people who we farmers could be working with. Feeding and creating economy within their own Instagram-pastiche bubble, it resembles the homogeneity of the blue-haired retirees clustered in their gated districts in Florida, San Diego, and South Africa.

A DOUBLE-EDGED FJORD!

If only US land culture, and real estate, were more practical—more like Southwest Airlines than the Virgin Galactic space shuttle. Geographies associated with economies of extraction and accumulation, like dynastic wealth, create overinflated real estate values that enable their beneficiaries a disproportionate potency in both resettlement and displacement.[10] Wealth is concentrated on the coasts, the cultural coasts, the economic coasts—and yes, the actual coastline! The coastlines of the U.S., settled as they were by sailing ships whose convenient ports and harbors became the hubs of colonial commerce, culture, settlement, and architecture—these are also the mouths of rivers, important for trading upstream, and for supporting offshore fisheries. The happenstance of geography meant that our cultural identity as a young and growing nation was born next to the sea, and still our intellectual headquarters are clustered on the east and west coastlines. It is now another happenstance of climate unpredictability (caused by soil destruction inland, oil and gas drilling on public lands and in public waters, etc.) that while shifts in weather will be felt across the continent, their most dramatic expressions take place in these river, harbor, and coastal zones, where the combination of overbuilt headwaters, manipulated and degraded hydrology, coastal weather patterns, and high populations make for more disastrous weather outcomes (e.g., Katrina, Irene, Sandy, Harvey, Florence). Wildfires express a different but similar causality in that the poorly managed woodlands packed in close to recreational and high real estate values set up a similar crosshatch.

These places were settled for good resource reasons—tradeable rivers, loggable uplands, rich productive harbors. While growth here has allowed the "cashing out" of boomers and empty nesters, these areas are also disproportionately vulnerable to ecological consequence. Beware, hipsters, you are paying too much for places that are not stable habitat!

I point to this demographic, geographic, economic, and dynamic trend not to call out an injustice against the young, or yet another harangue about explosive speculation in real estate; rather this analysis points to a pattern on the land that our young farmer strategies must address.[11] My direct observation of the landscape shows me that the current continuing prosperity cycle is driving "lifestyli-fi-cation" of semirural towns as well. Not just in the Hudson Valley or Oregonian marijuana zones, it is extending in a exurban circle around many major cities. This means it isn't just old farmers, or absentee landowners, or land trusts we must interact with to gain land access. It means there are opportunities to work with middle-age people who've made good equity on the inflation of their urban housing stock and are now in a position to make partnerships in rural areas. They could build a silly gated lifestyle farm, but they could also leverage their equity to create infrastructure for a community based on responsible land stewardship that cultivates healthy food and connection for generations to come. Retired people who already have a fully operational toolshed in their basement are not the problem; they are potential partners who need their security lightbulbs changed, and who are often more than willing to lend a table saw.

Given our national failure to enact progressive land use policy, or to set forth an agenda of regional food sovereignty—it falls on us, as farmers, advocates, and new economic workers, to understand and react to these outcomes and this context. Just as we react to and straddle the colonial legacies and injuries, social and racial wealth gap, and distortion and subsidy of exports and monoculture, so too must we take into account the future we are born into working with.[12]

RECONSTITUTING OURSELVES: WHICH WE ARE WE?

I feel the need to address a rhetorical framework, quotidian to our daily lives: violent sucky we's. The inferred Trumpian "we," as pervasive as misogyny and black mold. This inferred we is not a we I want in my mouth. *We the dominant, the still included, the not yet oppressed. We the unchallenged male ego of global envy. We with the freedom to cringe or pity or disdain those to whom the current tragedy is occuring. We whose credit cards are swiped without an angry sound.*[13] *We who are not afraid to carry guns. We who choose our warlords.*

The reflex flip-side we, the superioristic moderate left we, is, I'd argue, also toxic.

White urban wired educated First World we?

People like us we?

We who do not participate directly in violence that is committed in systematic ways to keep us comfortable and who therefore do not acknowledge or associate with the violence?

I refute such we-ing. I suggest we replace it with: We the implicated, we the complicit. We the orphans of oppressive biocide, we the granddaughters of extraction. We, neighbors and creatures nearby the supply chain. We, the most certainly impacted by the chemistry in our watersheds, and therefore in immutable alliance with one another.

Which brings me to "they"—the ill-defined, the other. The befuddler, the misunderstander, the bureaucrat, the corporate, the confounding other who has opposed the preceding, exclusive we.

I wonder what role TED Talks have played in this habit: the script that says, "These days, people are ..." (revisionist, enclosed, forgetful, desensitized to violence, disconnected from their sexuality) and provides then a succinct critique. I wish this TED way of being were a bit more embodied, connected to a commitment of action.

Name your subject! Who is doing the action? Who holds the cards, who made the rules, who is affected, who is driving, who is alongside—an amorphous, blaming "they" won't cut it, not in these times of concentrating power and gangster politics. Coalition building, subtle sidestepping, subversion, subterfuge—all these tactics require a precise read on who exactly is causing the action, and how. Contributing your own clarity of analysis is step one; committing to participate in the often muddy process beyond commentary is step two.

Step one: A gentle and insistent naming of the subject, specifically in a way that clarifies and requires discipline—not vilifying or "they"-ing ruthlessly and without account.

For instance: "They fucked up the bridge" becomes "The municipality of Scottsburg, Oregon, ran into a conflict with the state highway department over the recent construction project which leveraged federal funds from the stimulus package to enact major changes to the flow of traffic through the historic center. Those charged with planning profoundly misunderstood and sadly compromised the beloved bridge at the center of town." Especially in front of children, showing the discipline to identify the actors, factors, decisions, jurisdictions, and processes that lie behind governance decisions is the first step to reclaiming our agency in a democracy.

I challenge myself, and I challenge you. Name the subject. Name the object. Learn what happened, and—step two—use that learning to inform and design proposals for a saner way. This is the world we are in, and the times we have together.

WHAT DOES THE LAND WANT?

How are we tracking the overlaid injuries to our underlying ecologies, when we have tuned our bodies to the signals of market gossip and a predictable teeter-tot of partisan emotion? Where is this landscape moving? If it's degraded, what type of ecosystem is working to establish or reestablish itself? Can you see the patchy emergence of a further-future system, the destiny climate of your own particular place? Can you sense what is coming? Can you see a role for yourself, your crystal pattern of relationships, talents, and kinds of power? Can you find ways to relate to, and find consonance with, this land's destiny or potential destiny?

This feels like a humbler, less partisan, more responsive approach—less about identity, vanity, oppositionality, or rhetoric. More about sensing, tasting, touching, tuning into the genetic expressions of multitudinous lifeforms, and aligning with the life goals of the place you inhabit.

The next time I'm presented with the option to match my italic angulature to the coded political sentiments of my conversation-mate, I pledge to deflect towards stiffer stuff: *Work that wants to be done.* Primary research. Fact-checking. Due diligence. Pointed and proactive conversation with someone in a position to make change. I commit to an evidence-based framework of intervention, information gathered through direct observation and organoleptic testing. In direct retort to the Facebook-KGB fake-news hyperbolic shrieking, I will commit to staying grounded in place as I respond like a Zen journalist. I pledge that I will withhold vague liberal bitchery or reactionary emotional othering. When I disagree, I pledge to learn from my conversation and to follow up afterward with helpful links and resources. I pledge to be concrete and constructive.

TOWARD A SEAWEED COMMONS

Called by a mysterious seaweed songline, I followed the curiosity commons to the end of a peninsula in Maine, where I now live and farm. Here, I am surrounded by a stubbly, second-growth, slightly elfin, self-willed subboreal spruce, fir, alder, and birch woodland. Cobscook Bay, a backsplash to the Bay of Fundy, has twenty-two-foot tides—and so the withdrawing ocean leaves naked behind a dancing, bouncy, striated intertidal territory, a place of intense bioproductivity and interspecies intrigues. Seaweed evolved in this meeting point of land and sea—suspended for millennia in the wild tumult of an all-connected ocean. The seaweed zone, apart from having birthed photosynthetic processes later adapted by land plants, and credited by paleo-meteorologists with creating the oxygen that shifted our atmosphere to enable mammalian life, does more carbon sequestration than all the earth's grasslands. The metabolic understory of the ocean, these simple organisms specialize in nutrient filtering and buffering of wave action—a nursery of aquatic life. Converting sunlight into useable energy at the bottom of the food chain, hundreds and hundreds of species are protected, fed, and hunted in the intertidal zone, hovered over by fronds of love and grace, the micro- and macroalgae.

Seaweed beds exposed at low tide risk desiccation in the sunlight, and so they release iodide particles which drift up into the sky, creating that characteristic "bright maritime cloudbank"—a particularly stable, sheep-wooly cloud form that holds moisture (another coastal phenomenon) and plays a role in carrying that coastal moisture inland.[14]

It is here that I am committing to a place-based we, a we of small-scale seaweeding and wild-crafting. Here on this remote rural coastline I can pursue direct-action research aimed at discovering balance and stewardship on land and sea. What drew me here was the high quality habitat for edible macroalgae, and the relative shelter of Cobscook Bay—an advantage for a young boatswoman. Here I can harvest the wild rose petals, St. John's wort flowers, raspberry, and clover from the meadows, balsam fir, chaga, and spruce tips from the forest, and operate both small-scale multitrophic aquaculture and a small wild harvest of edible seaweeds. It is here I can participate with my own body, my own senses in direct relationship with a wild ecosystem.

In such a small rural place connected to natural resource economies, there lies the prospect of a community management framework that improves on state regulation, one that more closely resembles the commons described by Elinor Ostrum.[15] It is here we can move from theory to practice and perform an inquiry-based owner-operated fishery that learns from its marine ecology and history of extraction, that looks across jurisdictions and national boundaries to learn the outcomes of policy decisions and confronts the behaviors of overexploitation, ready to discover best practices with the layers of beings and bodies in columnar water whose various filtrations have coincided harmoniously for eons. How can I grow oysters plus kelps that are healthy and fetch a good price with low-impact, no-input practices? How can I limit my presence in the water so as not to displace the eider ducks, migrating birds, seals, and otters, instead using my lease sites to blockade the cove against draggers hunting urchins? Should you wish to participate in the learning and honest discovery of the boundaries and thematics of human-scale mariculture,[16] I invite you to join us in June 2019 for the second Wild and Domestic Algae Workshop, a two-day forum at the Reversing Hall in Downeast Maine.

Meanwhile, offshore—the Feds have a very different scheme. In November 2017, the US Department of Energy released a set of grants targeting creation of twenty-two million acres of mechanically harvested offshore kelp farms, with an eye to biofuel production.[17] These grants were issued to university researchers in a format called "teaming grants" whose purpose is to conjoin disparate areas of expertise (marine biology, engineering, robotics, biochemistry) in the manner of the Manhattan Project. Picture the unholy coalition's powerpoint meetings, cross-disciplinary presentations, the high-dollar RFPs and merging skill sets from biotech, venture capital, and engineers who've spent a career building off-coast oil drilling platforms. Imagine how such a coalition might think and act in a massive grant–funded, hyper-hyped "green revolution" of the seas.[18] The Farm Bureau says, "We feed the world"; the mega-aquaculture operatives say, "The ocean must feed us."

I believe there is a role for conservation

aquaculture, for small-scale aquaculture, and for mitigation installations, as in the Port of San Diego, created to absorb the excess nitrogen running off from our agrichemically leaky farm systems, from our sewers and our cornfields. I believe for this to proceed properly, we'll need a lot of literacy and public conversation about ecological design and governance in the public trust. For example, while some propose planting "mimic" kelp forests off the Pacific Coast to buffer waves and shelter creatures, the general public are not yet in a position to know enough to form opinions. The task therefore is to help build a more general marine literacy. Much good can come of this! Seaweeds can metabolize excessive nitrogen and bind carbon at the same time, a valuable service. These seaweeds could be carefully harvested and brought on shore to remediate soil health on highway margins, for they are composed of polysaccharides digestible by, and stimulating of, healthy soil microbes. Yes, seaweed is potentially an incredible tool in land restoration, in harbor restoration, as an amendment to young trees and hedgerow habitats, as a remineralizer of degraded, mined, or herbicided lands.[19] Such aquaculture is not the same as a massive corporate offshore biofuels plantation, and the public needs to understand this difference.[20] In my view, this is what is at stake in the seaweed commons—that we make it a comedy and not a tragedy.

Imagine fire and charcoal and middens; imagine small diversions of water and stone. Imagine the long line and lineage of interaction between human desire and the destiny of plant and animal and shellfish communities. Hold in your mind the ancient genetic tongue-tastings, recessive traits, respectful harvest. Contrast that with the 19th-century colonial swaps of plant systems across continents, the creation of plantations, the assertive breeding of foodstocks, fuel stocks, solvents, lubricants, fibers—the violent rearrangement and coercion of landscapes, human bodies, and machines. How now shall we relate to the seaweed underworld? I invite you to watch this space as a new generation of seaweed harvesters and small scale farmers discover our governance, our culture, our sharing of direct observation and historical thinking. Join us as we convene an emerging seaweed stewardship commons, as we learn and articulate our ethics, our holistic goals. Join us as allies in marine conservation, gastronomy, and land-bound environmentalism to protect the aquatic commons.

NOTES

1. At the height of the boom there were twenty-seven canneries around the shores of Cobscook Bay. Mackerel and herring built these towns.
2. "A New Customary Commons," youtube.com/watch?v=c8ukCyMZdtQ. The Spanish explorer Coronado came through Kansas in 1541, inspired by rumors of horse-sized fish and golden chalices. He found neither. A few kinds of armadillos produce four identical offspring from one egg, a phenomenon called polyembryony. This is a trait otherwise only seen in aquatic invertebrates, parasitic wasps, and tapeworms.
3. Much of the current right-to-repair fracas began with the lowly tractor. More specifically, it started when John Deere decided to ban anything but "authorized repairs," inadvertently turning countless ordinary citizens into technology policy activists. A lengthy End-User License Agreement (goo.gl/NHG411) the company required customers to sign back in 2016 forbids the lion's share of repair or modification of tractors customers thought they owned, simultaneously banning these consumers from suing over "crop loss, lost profits, loss of goodwill, loss of use of equipment … arising from the performance or non-performance of any aspect of the software."
4. Descarteslab.org, which emerged from Los Alamos Labs and is backed heavily by Morgan Stanley, gathers crop, weather, and climate-change data to help speculators predict price trends ahead of the crops' arrival to the

grain bins.

5. Natalina Sents, "Perdue Says Trump Is 'Delivering on a Promise' with $12 Billion in Farm Aid," *Successful Farming*, July 30, 2018, goo.gl/c41hJR.
6. Alan Rappeport, "A $12 Billion Program to Help Farmers Stung by Trump's Trade War Has Aided Few," *New York Times*, November 19, 2018, nytimes.com/2018/11/19/us/politics/farming-trump-trade-war.html.
7. Flour and oil rations to make fry bread—a.k.a. commodity rations, often discussed in the process of decolonizing diets to ground nutrition in native or "first" foods.
8. To animate this era, I recommend the wonderful film called *Heaven's Gate* with Kris Kristofferson, a big-budget masterpiece and flop at the box office. The left-leaning sheriff of Johnson County, Wyoming, confronts the powers that be and their private hired guns, who arrive in the county to execute one hundred thirteen alleged cattle rustlers—which basically amounts to the entire male leadership of the young county, and includes the madam of the whorehouse who takes rustled cattle in payment. Needless to say, the homesteaders gear up and fight back. The colluding calvary come in at the last minute to rescue the goons, on the verge of being finished off by women and old people in a prolonged shootout. Before the guns get going, it's an incredible orientation to the politics of the grazing wars and the role of US officialdom in sanctioning violence and privileging the greed of out-of-state interests.
9. In 2018 the median farm income in the United States was -$1,630. Siena Chrisman, "Is the Second Farm Crisis Upon Us?," *Civil Eats*, September 10, 2018, civileats.com/2018/09/10/is-the-second-farm-crisis-upon-us/.
10. Chuck Collins, *Is Inequality in America Irreversible?* (Cambridge, UK: Polity, 2018).
11. In the coming years of coastal abrasion, wealth may tend to seek safer places and so continue the trend of cascading displacement. Let's normalize a broader conversation about "real estate" and address directly the underlying land access, housing access, speculation, inflation, and inequality in a shared analysis. If we make the right alliances with city watersheds, conservationists, smart growth advocates, and adjacent interest groups—might we co-conceive a framework to redress these mistaken land-use incentives?
12. A brilliant free film: *We Power Our Future with the Breastmilk of Volcanoes*, 2016, vimeo.com/165412717. While you are at it, see the inspirational *All Watched Over by Machines of Love and Grace*, 2013, vimeo.com/groups/96331/videos/80799353.
13. Those barely discernible microaggressions of the machines.
14. For more on cloud-forming algae, read: Centre National de la Recherche Scientifique, "Iodide Build-up in Brown Algae Influences the Coastal Climate," May 6, 2008, 2.cnrs.fr/en/1198.htm.
15. Elinor Ostrom, in *Governing the Commons: The Evolution of Institutions for Collective Action* (Cambridge, UK: Cambridge University Press, 1990), identifies eight principles for establishing the proper management of a common good:
   - clearly defined borders: clear definition of the contents of the common good and effective exclusion of external actors who have no right to participate
   - rules concerning the appropriation and supply of common resources that are adapted to the local context
   - collective choice methods that allow most members to participate in the decision-making process
   - effective monitoring by members or by individuals who must be accountable to members
   - a spectrum of progressive sanctions for those who violate community rules
   - conflict resolution mechanisms that are cheap and easy to access
   - self-determination of the community recognized by high-level authorities
   - in the case of large common assets, organization in the form of multiple levels of nested businesses, with smaller common goods at the basic level.
16. In Hawaii, Molokai school kids are using "invasive" spiny seaweed for biochar—the carbon structure that can host effective microorganisms. See also the indigenous clam gardens of the Pacific Northwest, studied by Judith Williams. Their gentle, handmade stone terraces extend the lower intertidal zone to increase habitat for butter clams—indigenous mariculture.
17. The original text of the Department of Energy grant: "Department of Energy Announces 18 New Projects to Accelerate Production of Macroalgae for Energy and Other Uses," September 19, 2017, goo.gl/uuCVXt.
18. Yes, they are pursuing the breeding of super-kelps.
19. Remember that Roundup, sprayed on 250 million acres of land in the U.S., was developed as a chelator, which means it binds minerals. One major user of marine kelp is the dairy industry, which feeds cows corn, silage, and alfalfa that is often sprayed or grown on sprayed land. The cows benefit from the added minerals, but does the chemical chelator glyphosate actually demineralize the cows? If you know, please send me articles to read. Scientists at the Marine Lab in Oban, Scotland, are concerned about these proposed super-kelp farms because kelp's first cousin is the oomycete, first cousin of the potato blight. Farmed kelp can and does transmit disease to wild populations. Kelp are adapted to difficult habitat, and have a relatively narrow set of genetics, making them especially vulnerable.
20. Abrahm Lustgarten, "Palm Oil Was Supposed to Help Save the Planet. Instead It Unleashed a Catastrophe," *New York Times*, November 20, 2018, nytimes.com/2018/11/20/magazine/palm-oil-borneo-climate-catastrophe.html.

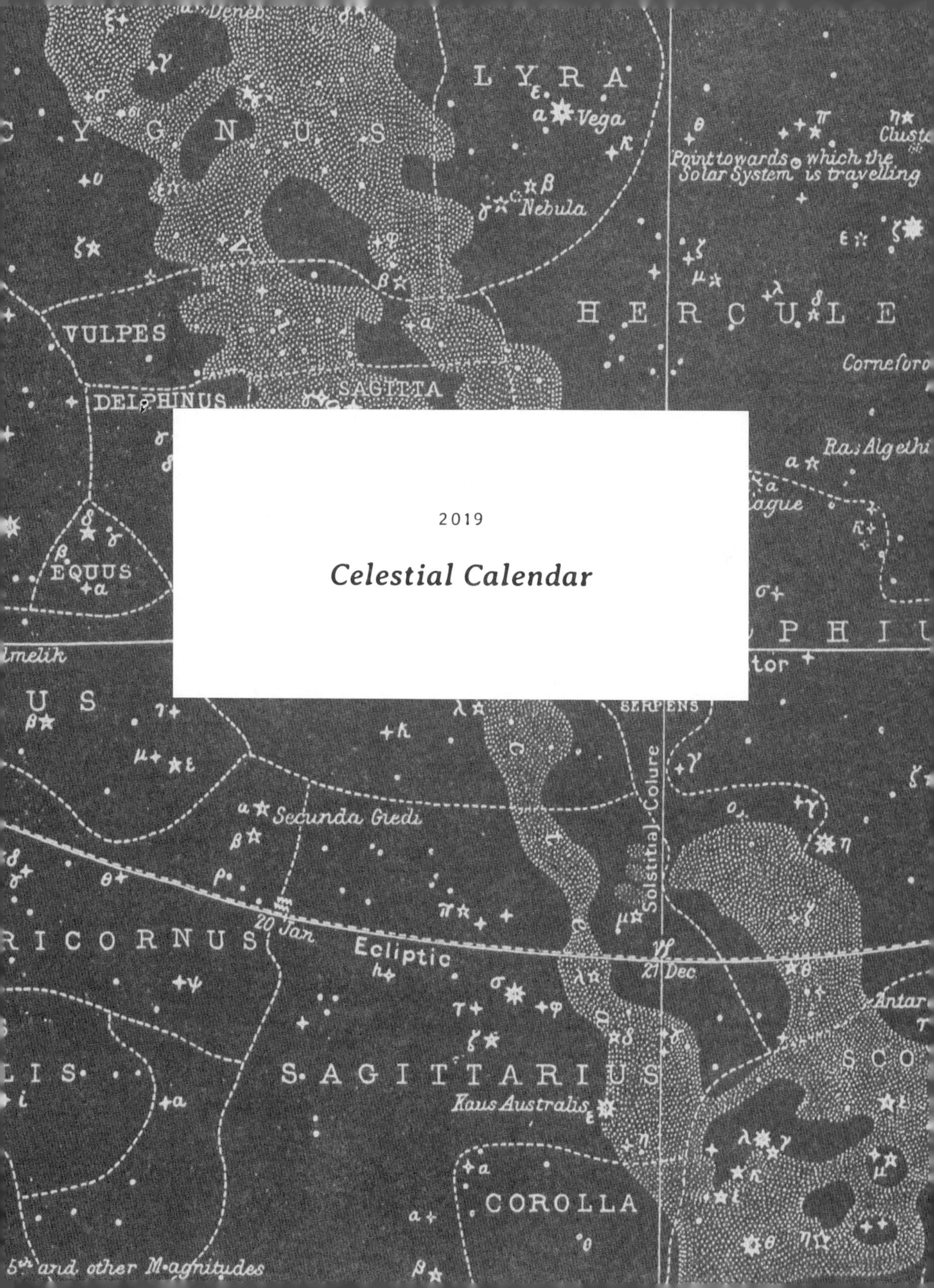

2019

Celestial Calendar

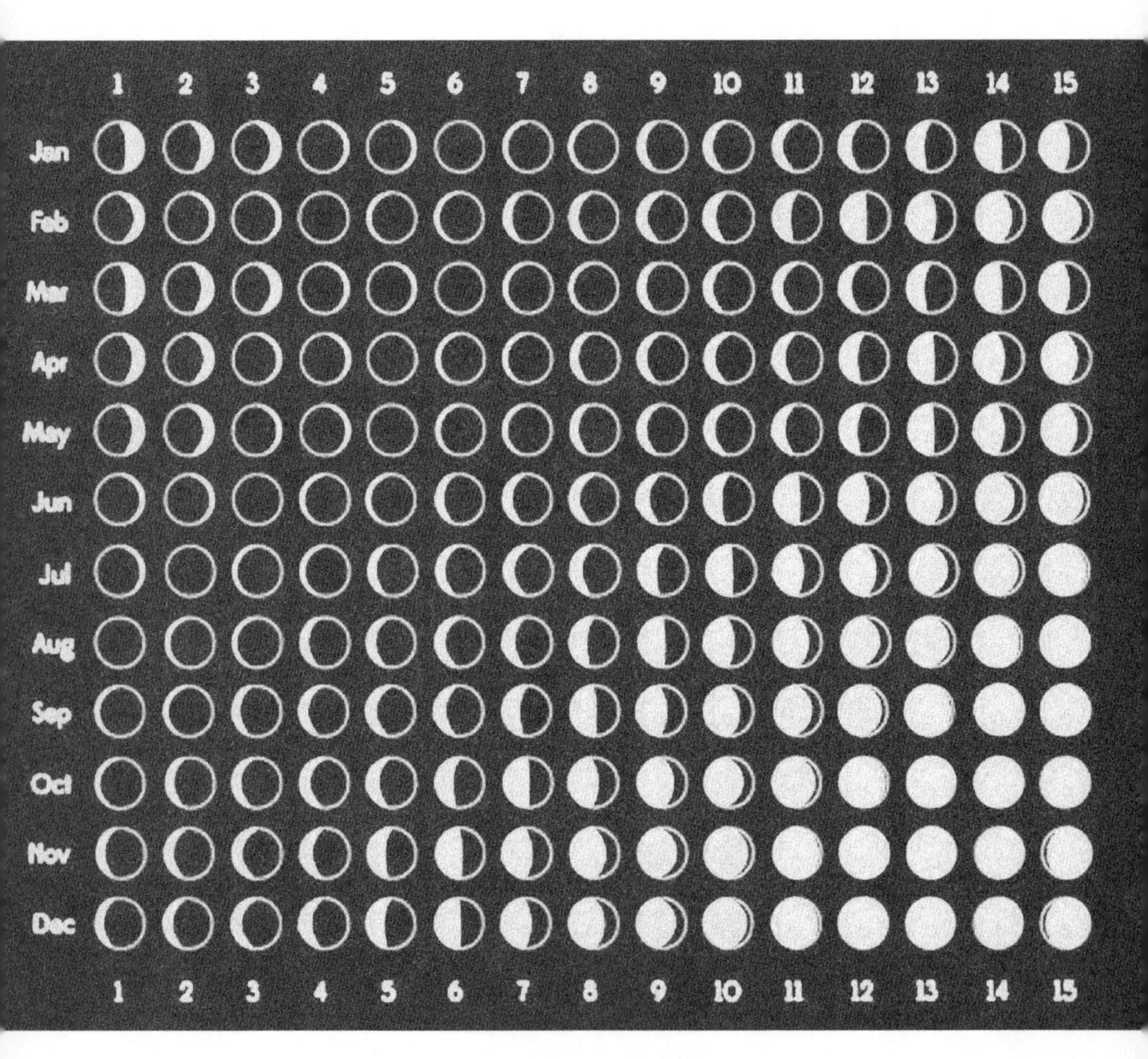

PHASES OF THE MOON

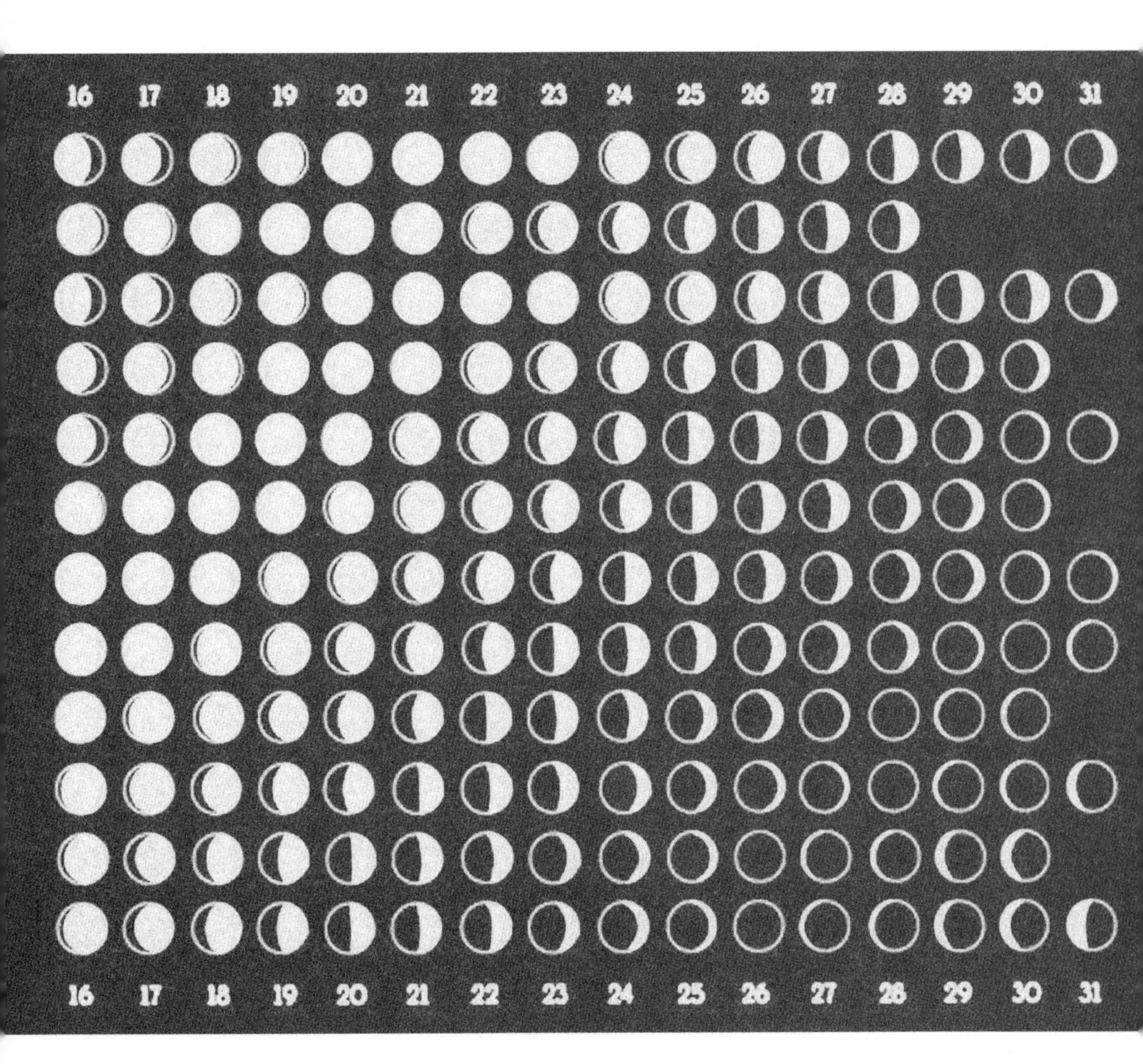

FOR THE YEAR 2019

DATA EXCERPTED FROM THE U.S. NAVAL OBSERVATORY ASTRONOMICAL ALMANAC 2019

A NOTE ON TIME The astronomical data are expressed in the scale of universal time (UT); this is also known as Greenwich Mean Time (GMT) and is the standard time of the Greenwich meridian (0° longitude). A time in UT may be converted to local mean time by the addition of east longitude (or subtraction of west longitude), where the longitude of the place is expressed in time-measure at the rate of one hour for every 15°.

PRINCIPAL PHENOMENA

PHENOMENON	DATE	TIME
Perihelion	2019 Jan 03	05:20
Equinox	2019 Mar 20	21:58
Solstice	2019 Jun 21	15:54
Aphelion	2019 Jul 04	22:11
Equinox	2019 Sep 23	07:50
Solstice	2019 Dec 22	04:19

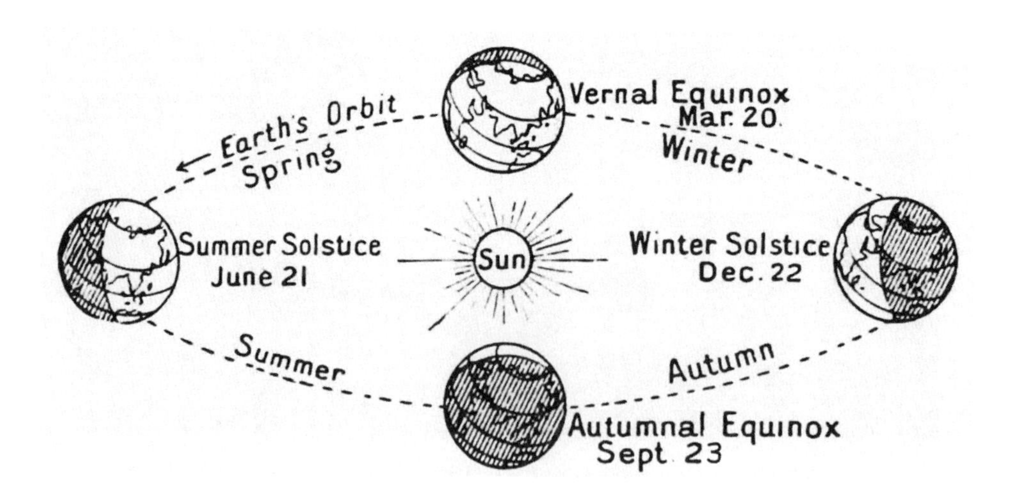

VISIBILITY OF PLANETS IN MORNING AND EVENING TWILIGHT

	MORNING	EVENING
VENUS	January 1–July 8	September 20–December 31
MARS	October 17–December 31	January 1–July 18
JUPITER	January 1 – June 10	June 19–December 15
SATURN	January 19–July 9	July 9–December 27

TOTAL ECLIPSE OF THE MOON JANUARY 21
TOTAL ECLIPSE OF THE SUN JULY 2

The Earth in Space.

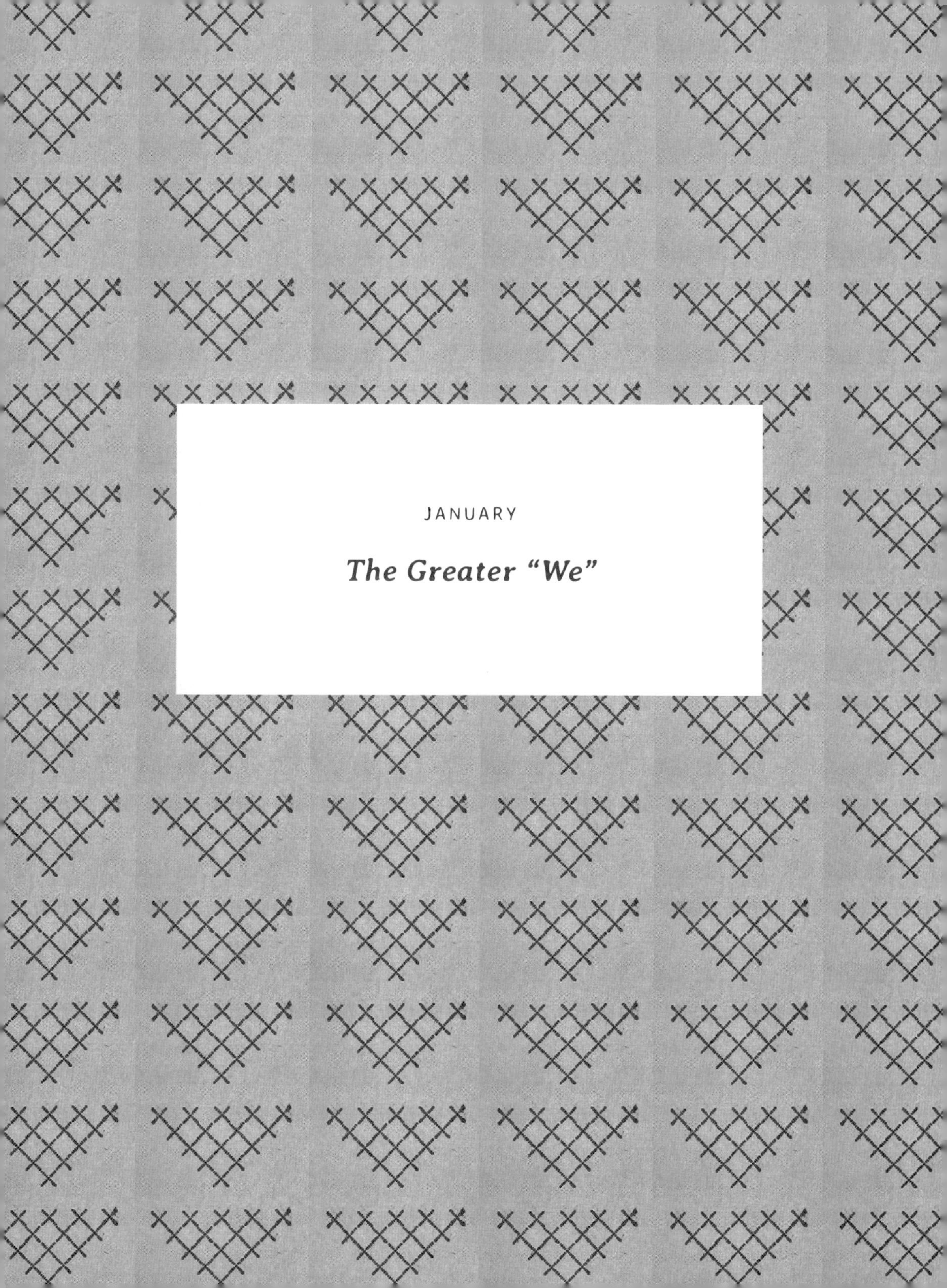

JANUARY

The Greater "We"

A RUDERAL REUNION

Ancestral Tools for Self-Remediation

AZURÉ KEAHI

No one raised me to a be a farmer, to find life's solace converting vacant lots into paradise.

When I struggle with words, I take comfort in knowing I can trace my lineage by drawing lines in the soil, guiding seeds to the ends of their trenches. From there, life grows—and my ancestors smile through veins of emerging cotyledons. I nourish with what I don't know. My genealogy feeds me, and from that, I will never go hungry.

Back home, many people are all mixed up—racially and otherwise. I call myself a person "off-color," of the "grey race," or a product of the plantations. I have channeled the skills of my sugar-caned predecessors. From the horrors of monoculture, a reparative polyculture of people has risen. I am part of it. Though I am proud of my Hawaiian-Chinese-Portuguese-Filipino-German descent, I suffer from a shattered identity. For thirty-two years, I have mastered code-switching, melted into invisibility, labored to articulate my rainbow. Lost in those alleys between racial colors, I search for some enlightening pot of gold.

My earliest memories of race-based shame are of my nagging fixation on blood quantum. As a part-Hawaiian, born and raised in Hawaii, the issue has tortured me. Studying the history of Hawaii in school, my mind was primed with stories of all the wrongdoings brought upon the land and its people. As a result, I yearned to separate myself from colonizing ideals and behaviors. To me, that meant being as Hawaiian as my *kupuna*. I wanted people to see me as Hawaiian. I wanted to be eligible for Hawaiian Home Lands.[1] I wanted to be a real Hawaiian who dances hula, speaks the language, knows how to wrap a *lau lau*, grows her own *kalo*, who is just the right shade of brown. But what is a real Hawaiian? And how could I ever be one if my blood was of so many others? I have harbored an immeasurable amount of shame from my mixed race status—for I know, I am not just Hawaiian. The bloodline was diluted long ago. Humans were humans—they shared earth, resources, meals, and each other. Skin color was a costume.

Often, I am plagued with the question: How do I properly represent all of my ethnicities? My skin is never brown or white enough, my eyes never slanted or round enough, my nose never flat or wide enough. In my mind, someone is always sizing me up, waiting for a chance to test my ties. I am on guard to prove my authenticity, waiting for someone to call fraud.

It is clear to me now that I never wanted to choose sides or deny my white blood. I simply wanted to be a hero to my people, free of colonizing habits and mentality. It took cross-coast displacement—living in places both highbrow and economically-distressed—to bring me closer to the end of my rainbow. Consoled by the landscape that has been pillaged, forgotten, and reclaimed, I embrace the familiarity of its circumstances.

In upstate New York, I have learned the challenge of cultivating diversity, to mend an ancestral tapestry wrought with agricultural and cultural exploits. Pushed from my homeland due to the high cost of living, I jumped from metropolis to metropolis only to settle in the seemingly forgotten landscape of Troy, New York. This is my gift produced by the past. Without the colonizers of my tropical homeland coupled with that of my present, post-industrial residence, land access would be far-fetched. In Troy, the Dutch colonizers came in, purged the Mohicans, built their manors, created slums, leaving many empty tracts of land. Here, I confront my reflection in this detritus of opportunity. To whom do I owe this blessing? Genetically, being part-colonizer, part-coolie, part-indigenous, part, part, part—what part do I play?

I fill every void with productivity as homage to both my ancestors and successors, reclaiming and empowering territory that has survived the fluctuation of time and values. With the help of my community, my newfound extended family, we have transformed previously vacant lots into edible, medicinal landscapes. These grounds have experienced the gifts of indigenous wisdom, the surge of the colonizer's monetary wealth at the former's expense, and the resulting state of poverty. Villages were eradicated, manors were built then bulldozed, and we are left with a clean slate to start to rehabilitate. Now, we build a new kind of village, planting fruit trees for both our predecessors and successors. Their harvests will speak for us beyond our lifetime and the understory of medicinal crops and fodder for pollinators will help to feed and heal future generations.

Just as poisoned terrain remediates through the welcoming of invasive, pioneer species that consume and process toxicity to their advantage, I have acknowledged my role as one that must inspire others to collectively adapt and heal from generations of trauma. I am an outsider to my new home—a displaced indigenous, mixed-race person—here to carry the common thread of my ancestors through the eye of the future. Coincidently, my work is focused in a neighborhood whose population of black, white, and brown peoples has been disenfranchised by the same systems of oppression that have inspired me to rise up. Like my ancestors of many races who met on the plantations, together, this new community and I strive to create a new language, a creole formed from how we blend our differences in skin color, culture, and life experience. We communicate through bites of fruit, the vibrance of summer blooms, and our most-prized recipes for collard greens. I strive to be an ally, remaining cognizant of my role as a newcomer and sensitive to those who are here and came before me. Through this heightened sense of awareness, I strive to work *with* this diverse community and not merely within it.

Generations back, my ancestors were brought together by Hawaii's sugar plantations, lured by tales of improvement. From places like China, the Philippines, Portugal, and Germany, they ventured to this mysterious island in the middle of the Pacific Ocean, knowing very little. Similarly, I was tempted and steered by this waterfront wasteland along the Hudson River. Unexpected access to land and a simpler way of life awakened skills and values that seemed to be whispered by the spirits of my progenitors. They spoke through the soils of vacant lots and the mix of so-called "weeds" that chose to thrive there. Caring for these forgotten plots acts as a test for me to be a more resourceful citizen, to rely on that innate wisdom that cannot be taught, to pay attention and question myself and my motives. There is an air of magic in the work that I do. It is a blessing of accident, illuminated by skills stored in my marrow. Displacement has forced me to confront my multi-colored role in society's landscape. Like a seed secretly carried to new lands to germinate for future generations, I am here to sprout, spread, and feed beyond my mortality.

NOTES

1. From the Department of Hawaiian Homelands: "The Department of Hawaiian Home Lands is governed by the Hawaiian Homes Commission Act of 1920, enacted by the US Congress to protect and improve the lives of native Hawaiians. The act created a Hawaiian Homes Commission to administer certain public lands, called Hawaiian home lands, for homesteads. Native Hawaiians are defined as individuals having at least fifty percent Hawaiian blood."

LESSONS FOUND IN MOVING HOME

BRETT F. BRALEY

We bought our farm when I was twenty-five. Five acres, nothing big. It sits in the middle of two mountains. Hills, really. It rains a lot, gets cold easily here. Stays cold longer. It was hard to drive the first winter. I still had my rear-wheel drive Mercedes and couldn't quite make it up the road. People stopped to help and I cried in a stranger's truck. That's Pennsylvania for you, and a perfect example of how I relate to the world I'm now in. People are there to help, even when I feel vulnerable and embarrassed.

I'm not good at asking the right questions, so I let others take the wheel. I sense the divide between myself and strangers, likely more than others do. I know my shortcomings, and juxtaposed against the kindness of strangers, I initially felt them all the more. But that's started to change the more time I've spent discovering myself on our farm in Pennsylvania, less constrained here than I was in California.

My partner Nolan and I had been living out West for about five years, and we needed a change. I think the dream of being other people never took root for me. I missed my family. I missed greenery. I missed expansiveness and lightning bugs and snowfall. California felt stagnant. Where we lived, in San Diego, holidays were unemotional rituals, just reasons for a day off from work. Christmas decorations stood plastic and cheap against palm trees. We ate Thanksgiving dinner in t-shirts. All the seasons melted into one, a prolonged summer that left me suffocating in traffic jams, desert vistas, and a bastardized version of twentieth-century manifest destiny.

It wasn't for us. In many ways, we weren't equipped to move back to Pennsylvania, but we eventually did.

Lived apart to save money. Split the dogs up between my parents' house and his.

Fought sometimes. Struggled to find out what it meant to be home again.

Bought a farm a year later.

When we got chickens, none died, and I took that as a positive. Our three dogs have changed here, coming out of their shells and growing up a bit more, and we take that as a positive, too. Each has felt out a niche in the house, and taken on the roles they want, unfettered by rules or space or neighbors, as it had been in California. Elsa, a rough collie mix, barks at the squirrels, the chickens, the cardinals on the tree, and exhausts herself in the process. Milo, our grumpy-faced pug mix, follows suit, but comes to rest at our feet when he tires out. And Murphy, our sweet old man, gentle and calm, smells the rain on the wind and stretches out on warm days. They made the transition look easy; they gave me purpose and allowed me to make mistakes once in a while. But they've made it feel like home, too.

The transition wasn't easy. I'm still learning and making mistakes and learning again. Since moving back, I have learned to mow a lawn, treat a chicken for mites, bake bread, and garden. I've learned the names of the librarians in town and have been been taking French lessons. We painted a room and bought a truck. These were all things to do to fill time, to test our boundaries, but they did not define us as farmers. I think the thing that defines us is our approach to raising animals.

I am not an owner to any animal on our land; I am a steward to their happiness. We facilitate their lives with our own resources. We give them the boundaries, food, and land to live their lives as they want. We don't get mad if they're doing what they think is best. We do not subscribe to the moral dichotomy of a good or bad dog. Neither a good or bad chicken exists on our land. Instead, through protection, safety, food, and activity, we hope that their environment will best develop their personalities. They will guide themselves to complement one another and the land.

The dogs do not know tricks because we never saw the point. Instead, we've engaged with them as often as possible and in wholly human ways—through conversations and choices. We leave the back door open so they can decide when

to come in and out. Each animal is special and the less restraints we give them, the more they express themselves.

I've noticed this in the chickens, too. We recognize each hen by its coloring and when two look similar in the dark, we can usually tell them apart by the way they act. One of our brown hens is broody and nips when I collect eggs. Our only golden hen is fearless and follows me around the barnyard, and within feet of the barking dogs. All ten girls were raised under the same conditions, and each is just slightly different than the other. I love them for it. I love them because here, they are not afraid. I love them for what I have been able to provide to them. I love that they are not workhorses and do not need to live in fear or a cramped cage. Their only job is to live as comfortably as they can with us. If we get eggs, we'll sell them. If we don't, we don't mind. And when they stop laying, the same principles will apply. There is no hierarchy to happiness here.

And while the dogs are loud and bossy, we know it's because we never tell them no. But we've never seen a reason to. They'll tire themselves out, we think. They'll come inside when they're ready. When they do, we're always waiting for them to jump into bed with us. When we're gone for more than a few hours, they never let us out of sight once we return. That's how I know what we're doing is working, because they love us not for the rules we've given, but for the freedom we allow.

I am no different than these animals. Since moving back to Pennsylvania, since buying a small farm, I, too, do not have a set of rules to follow. I left that all back in California. Instead, I spend my days baking. I spend my days writing. I spend my days trying to be better. I spend my days in the happy clamor of a life shared with animals, with mud on my palms and eggs in my fridge. I am not so different than the animals I've raised, because I've given them the tools to their own happiness. I'm sharing this land with them. I've grown up in the same conditions and I, too, have turned out to be more myself. And I am grateful to be a steward to their lives, to share the land with them and be a co-owner of it. And I am grateful we moved back to Pennsylvania to experience it all—I just had to leave California to find it out.

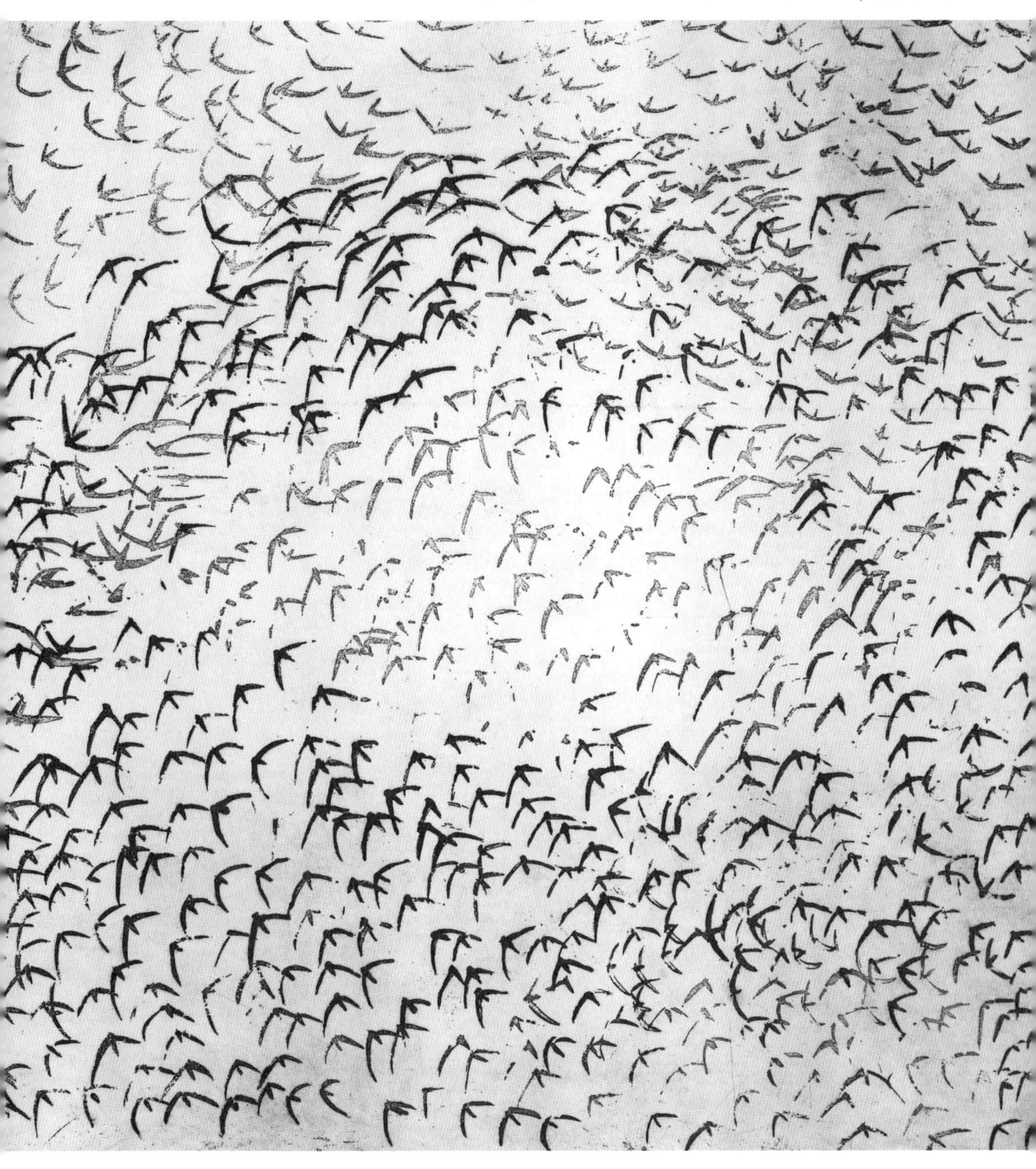

Vincent Sheridan | Migration

WE CLICHÉ

JOSH SLOTNICK

She knew the kids' names
"other side of the truck," "wait here," "get that deer out"
She'd keep up with me on the tractor on the way to the field around the corner
pinning her ears back, throwing those front legs out
she'd go horizontal in the air
pounding paws at speed along the ditch roadside
sleek in the wind, and smiling

Now I ratchet back the rpms on the way over
so she doesn't fall too far behind
Yesterday driving to the field, as we have always done
I caught the eye of a kid
zipping past
kayaks strapped to his Subaru
headed to the end of the road river access
He nodded
lifted the corner of his smile a bit, like he knew
I tried to imagine what he saw

An old-timer farmer in a frayed ball cap on a tractor with a white-faced border collie
loping alongside

Have we become a cliché
shallow, easy to read
like an image of
a sunset on the beach, a homeless guy pushing a shopping cart
or
a cool shades dude with kayaks strapped to his Subaru

While I wonder at these things
our story becomes richer, and our purchase on this life
grows thinner
Regardless
she knows better than to throw mental energy in such a direction
At the field gate
she smiles up at me when I lift the fence chain
in awe of our great good fortune, to be here, now, with each other
once more
doing as we should

"BILL," THE BUTCHER'S DOG.

ONE ACRE OBSERVATIONS

JUSTIN YOUNG

I have been "farming" for about six years. I write "farming" instead of farming because the land I work is really a large garden with animals on it. I also write "farming" because I don't make my living from the crops I grow. My wife and I get paid to teach people how to "farm" and run a one-acre education site.

When I began this work, I believed that the world of organic growing was black and white. There were the right ways to do things (my way) and the wrong ways to do things (not my way). It was an incredibly exciting time.

With that excitement came expansion. Not in physical space—our boundaries were set—but in what we did within that limited space. We began to replicate and initiate all kinds of beautiful and amazing ideas and practices in order to better steward our land and educate others on a greater variety of ways to produce food.

First came animals. Poultry and rabbits grazed in mobile tractors. Dairy goats were tethered out when they would tolerate it, and when they wouldn't, they composted our endless supply of picked weeds. Then came a high tunnel for season extension, and aquaponics for raising tilapia and catfish. The greenhouse had microgreens and worms and soldier flies, and a tree nursery in the off season. Each year, we made all our own compost from over fifteen thousand pounds of food waste. In spring, we would start over one hundred seventy-five plant varieties from seed, cold stratifying, scarifying, soaking, covering or not covering accordingly. We caught rainwater, kept bees, and grew medicinal herbs. We taught workshops on everything from gardening and cooking to raising meat rabbits, and also started and supported over a dozen community gardens. Somewhere around this time, I began to look at all we had created and saw that it was good.

But even using the best and purest practices that we knew, we'd built a house of cards, a system with almost no resilience. Vulnerabilities in our delicate system became exposed and broke down. Crop rotation became virtually impossible, which resulted in an increase of pest population, disease, and soil depletion. Harvesting such diverse plants became extremely labor intensive. Rows planted with five to ten different varieties were confusing and, honestly, tiresome for volunteers, leaving the task to staff, which pulled us from other jobs. Animals reproduced, and suddenly, pens were too small and more needed to be built. Structures deteriorated, breakers tripped, and things broke. We were scrambling as best as we could, but the biggest vulnerability of our system had been revealed: our time.

The initial shock of this would have been humbling, if my pride would have allowed me to accept it. Instead, we blamed our issues on having received too little funding, or the difficulty of

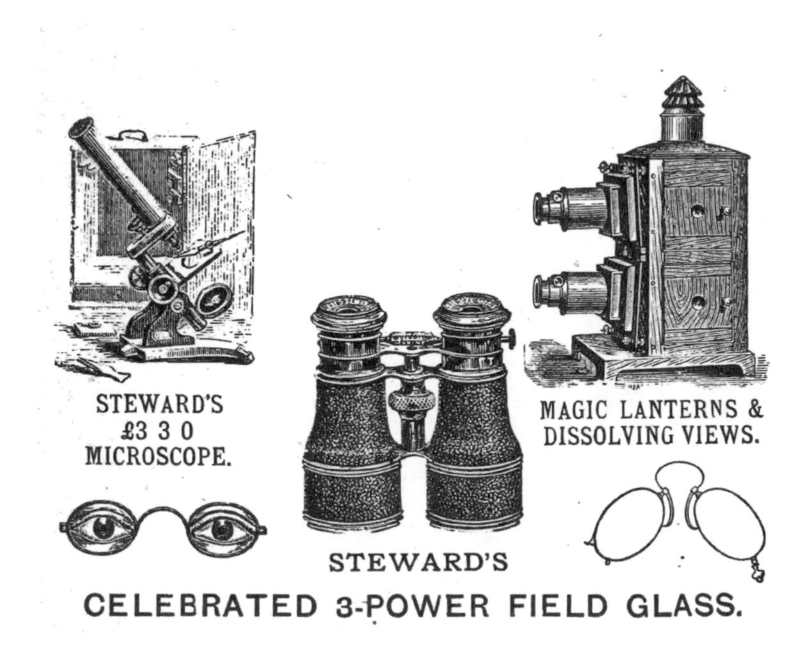

growing food crops in the Texas Panhandle. The truth is that I fell for a lie and it led me to compromise, and those compromises brought down the integrity of the land.

All the expansion and diversification did have to do with excitement, but it also had to do with image. Being a nonprofit, I thought we needed to prove our relevance. Social media and magazines that inspired me also challenged our relevance. My drive to secure our place led us to a place of unsustainability. I was continuously spent, trying to hold it all together, which only led to exhaustion and further compromise.

Through all of this, I've learned the importance of observing. I am trying to observe what our little piece of Earth can and should support with a realistic amount of management, and an understanding of how it can be most productive. I am trying to observe where its resilience is, and see where it is lacking. For years, our program was based on how things looked—especially to other people. But observing takes all the senses. It goes beyond a right or wrong way of doing things. It goes beyond formulas. If I had turned my observation inward years ago, I would have recognized the early signs of exhaustion in myself and the land that I worked. I would have noticed the subtle changes. I would have seen my own limitations.

There are still many things that are good about our "farm." There are things that are done well, and many things grow well. But stewardship is a high bar, and it is difficult to achieve. It is a lifelong journey, and I am only six years in.

Mud volcanoes in the Salton Sea, California

RED CLAY

SHAWN PORTER AND LOU ANNE LEWIS

Red clay, what I'm made of
runs deep in my veins
holds fast to my footprint
marks well where I've been
pulls out the poison
and carries the pain
back to the soil
from which we all came

Stiffens the break
like plaster applied
shaped by the hands
hard as stone when I'm dried
Mixed with straw 'twixt
stones and sticks
pressed and daubed
a house or a chimney
the breath of me draws

My sister is yellow
my brother is grey
my mother magenta
in layers we display
the flesh of the earth
compressed over ages
red clay runs deep in these
letters on pages.

SOME PASTURES

A Cowboy's Almanac

SAM RYERSON

January, Upper Summer Pasture—Roscoe, Montana

On New Year's Day ten years ago we rode into Upper Summer Pasture. It was clear and cold in the snow. We had shipped all the cattle in October. Laura Jean and Randy and I were the only ones left of the summer crew. We rode up along the bench west of the creek, on the open ground where the snow had blown clear. I got off and opened a gate out of the horse pasture.

From here, the rest of the ranch was open. In the fall after the cattle had gone, we'd opened all the gates of the pastures and left them open for elk to pass through.

We rode closer into the aspens. The wind picked up and some low clouds started down from the mountains. That summer we'd had about 2,500 yearling heifers and their bulls in this pasture for a few weeks, during their steady drift across the ranch. It was a big pasture for this ranch, maybe two thousand acres, west from Stokke Flat to Shorey Hill, and against the boundary of the Forest Service Permit above. It was a good pasture, and high, and the grass grew later than the pastures below. Through the hot days, the cattle watered in beaver ponds along the creek and shaded up under the trees.

A pasture needs these things to be: grass, water, fences, and cows. In Upper Summer we had them all. We rode every day—checking the water, riding the fences and patching them, watching the cattle for pinkeye and foot-rot and doctoring them, watching the grass grow and be grazed, and looking for wolf tracks on the trails near their old den. After their turn here, when it was time to move, we gathered the heifers and took them west over the ridge where the wolves once had their den, down to the big valley below. We were following a grazing plan we had made, a rotation across the ranch.

A ranch feels empty without its cows. To sustain itself, the grass of a rangeland needs cattle or bison or wildebeest or other large migratory herd-bound herbivores. There in the Upper Summer, in the dormant winter season, the range was resting and it was open again—without fences, cows, or cowboys. We carried memories of the summer over the grass and snow, along the water, waiting for the next growing season. We knew Randy would leave soon. We dropped down into a little coulee, sheltered from the wind in deeper snow. Randy opened his flask of whiskey and we each had a little drink to a New Year. It was warm from inside his coat but it caught in my throat anyway. Then we turned and rode home down along the creek.

The Pasture in Summer

QUEER LAND JUSTICE

VANESSA RADITZ

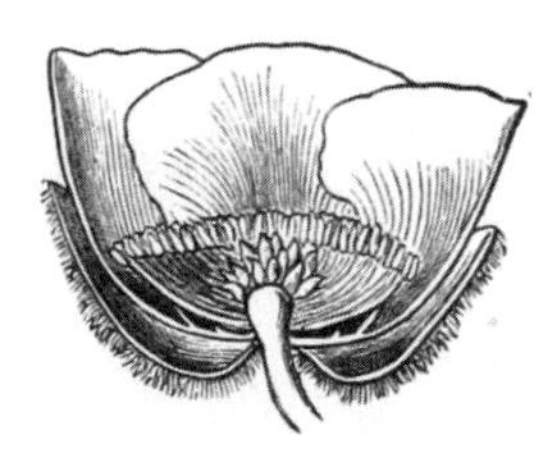

I. BODIES & LAND

Our bodies are not separate from the land. Violences to the land—clear-cuts of the Pacific Northwest, mountaintop removal of Appalachia, fracking of the prairies—are violences that manifest themselves in our bodies: as asthma, cancer, and hopelessness. So, too, do the violences to our bodies reveal themselves through the violences to our ecosystems, to our homes.

As a young queer person born during the height of the AIDS crisis, I came to understand myself and my body in its wake. Then, it was still common to hear that queer people are "unnatural" or "crimes against nature," a belief waged against us for centuries. How many queer people have internalized this? The eco-consciousness of the mainstream LGBTQ+ movement today can be summed up by sweatshop rainbow flags and glitter (a microbead that finds its way into the watershed). The mainstream frequently revolves around the framework that LGBTQ+ people deserve state protection because they are a valuable consumer group in the global economy. Walking the streets after the last New York Pride parade I'll ever go to, I stopped to throw away plastic trinkets and rainbow-colored coupons to a multinational software company littering the ground.

This eco-apathy within the movement leaves impoverished queer and trans people in a precarious situation amid the gathering storm of climate change. Today, almost half of all homeless youth identify as LGBTQ+. This figure does not include all those who are currently "housed" within juvenile detention centers, most often for doing what they needed to do to survive, and it does not include all those "housed" in the devalued foster care system, who stand a high chance of homelessness when they age out or escape. Stripped of many of the resources for resilience, many queer and trans people are exceedingly vulnerable to the crescendo of fires and floods we are likely to experience in this lifetime.

Particularly absent from mainstream LGBTQ+ dialogues are the identities and livelihoods of rural queer people, and those who still tend the land. While venting about this to my partner, a non-binary queer farmer who grew up in rural Michigan, they remind me that amid this reality of dispossession and vulnerability, "there is also all around us a burgeoning movement of queers returning to the land to find that we are loved and revered by a natural world that prizes diversity and variation over sameness and monoculture."

Understanding our queer bodies gives insight into how queer this land is as well. What, then, has brought us to this heightened state of disconnection from our ecologies, our homes?

Indigenous Two-Spirit storyteller, poet, and academic Qwo-Li Driskill writes about the way that the violence of colonization stole people not only from their lands, but from their bodies. It was through violence to the body that the theft of land was able to to transpire: "Homophobia, transphobia, and sexism… these forms of oppression are the result of colonization and genocide that cannot accept women as leaders, or people with extra-ordinary genders and sexualities. As Native people, our erotic lives and identities have been colonized along with our homeland," Driskill writes in "Stolen From Our Bodies: First Nations Two-Spirits/Queers and the Journey to a Sovereign Erotic." Sexual abuse of femmes and children, public harassment and mutilation of Two-Spirit people, the wholesale slaughter and removal of all those who resisted, as well as those who didn't—these are the violences that paved the way for the clearcuts, mountaintop removal, and fracking that we see today.

To reverse these trends, we need healing, and access to the resources for resilience. We need land.

Queer land justice starts with the assertion that queer people have always existed, and hold deep lineage as stewards and tenders of the land. Queer land justice starts with the recognition that poor and working-class queer and trans people are marginalized under the same system that globally drives peasants and indigenous people off their land and into city streets, slums, tenements, mental institutions, detention facilities, and prisons. Queer land justice is a node in a web of interrelated struggles for a future where all beings can exist.

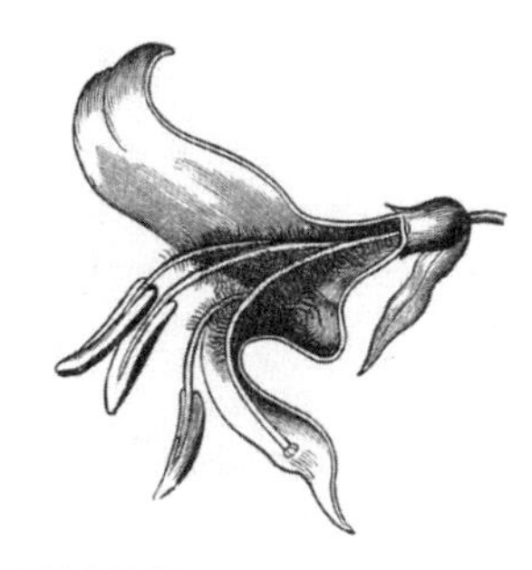

II. DISPOSSESSION

"The body as home, but only if it is understood that bodies can be stolen, fed lies and poison, torn away from us. They rise up around me—bodies stolen by hunger, war, breast cancer, AIDS, rape; the daily grind of factory, sweatshop, cannery, sawmill; the lynching rope; the freezing streets; the nursing home and prison."

—ELI CLARE, *EXILE AND PRIDE*

I had a conversation once with a homeless youth who I used to work with, a Black transwoman from central Georgia. We walked together through a wind tunnel between tall buildings in San Francisco's Tenderloin district. The strong winds that day gave us reprieve from the stench of urine and feces that fills the streets there. She looked up and told me how much she missed the sky, the stars at night, and the quiet. I side-eyed the cop car on the corner. How many would have stayed in rural areas if they felt that they belonged? How many would have stayed if they could have been safe? How many would have stayed if they had access to the resources to survive?

LGBTQ+ people have historically been, and often continue to be, barred from gainful, safe, fulfilling work, which makes housing security—let alone land ownership—nearly impossible. Transgender people today are over four times more likely than cisgender people to live in extreme poverty. This is particularly true in rural areas: queer and trans folks in rural areas have poverty rates that are twice as high as those who live in large metropolitan areas.

Getting to the heart of this economic injustice requires an analysis of the economic system that entangles queer and trans people. The Movement Generation Justice and Ecology Project in Oakland, California traces the word "eco" back to the Greek *oikos*, meaning home. By this definition, the word "ecology," eco + logy, becomes home + knowledge: knowing, reading, and understanding home and the relationships of home. And in the same tracing of history's linguistic threads, the word "economy," eco + nomy, can be deciphered as home + management: the set of decisions that govern those relationships of home. Many of us grew up being taught that there's only one economy, the extractive capitalist economy that has driven us to this point of ecological disruption. But in reality, there are thousands of different ways to manage home, as exemplified by indigenous cultures who managed their relationships with the land, water, minerals, animals, their own bodies, and other beings that make up their ecologies—their homes—in different ways.

The economy that we live in today is a globalized mismanagement of our home. The resources on which we survive are stripped, cut, mined, over-fished, drilled, and pumped; the waste burned and dumped, buried; and the wealth extracted, hoarded, and enclosed in the hands of fewer and fewer people. And the most precious resource that is extracted from the land and driving this engine of dispossession is our own labor. Human beings do this extraction. Queer and trans human beings do this extraction.

When a people are denied the means of survival; when the basics of life—food, water, shelter, safety—are privatized, our cooperation in this extractive economy becomes mandatory. I remember reading *Stone Butch Blues* for the first time, and how my body shook as burning tears of anger and sadness blurred the pages. Leslie Feinberg's story about Jess, a working-class kid growing up in the 1940s and 1950s who escapes home to graft their own butch "he-she" neitherness in the factories and gay bars of Buffalo, New York, is an anthem to the way queer and trans people have been caught in this extractive system. I think of the trans runaways I've known, who fled from rural areas to the city so that they could self-determine their gender, and I've watched them struggle with the way that the only work offered to them perpetuates the violent economic cycles that drove them from their homes.

All wealth originates from this earth, this land, this home. It is our labor that has been exploited for the extraction of that wealth from this land, and it's the control over our bodies that has led to its enclosure in the hands of fewer and fewer incredibly wealthy people. Queer land justice is an affirmation of our right to the food, water, shelter, and medicine that comes from the land. Queer land justice is a reclamation of our labor from the grips of the extractive economy and an assertion of our right to place our labor into a regenerative, caring, and collective management of home.

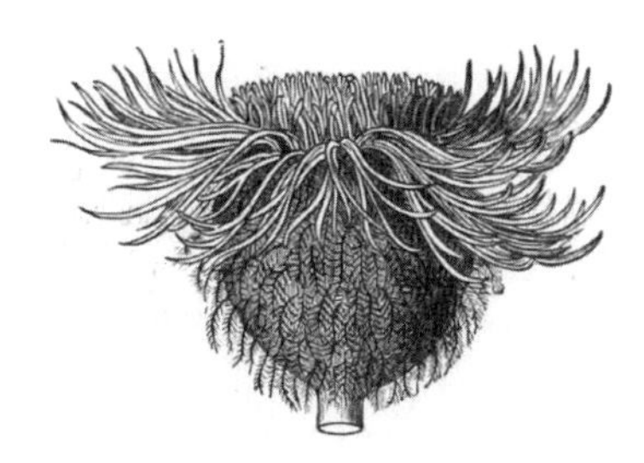

III. OWNING LAND, OWNING BODIES

"My right to be me is tied with a thousand threads to your right to be you."
—LESLIE FEINBERG, *TRANS LIBERATION: BEYOND PINK OR BLUE*

I'm staring out the window at the crocuses growing through cracks in the pavement, listening to the scratches of sparrows building their nests in this three-hundred-year-old barn, and seeking hope in these small acts of resistance. The farm where I work has been owned by the same family for hundreds of years, passed down from patriarch to patriarch. There are streets through our town bearing their last name. In our environmental education curriculum, we build wigwams to study the Nanticoke Lenni-Lenape people who "used" to live here, but nowhere in our curriculum does it ask us to pause to reflect on the violence that the removal of these human beings actually entailed, or the role of a farm such as ours in the ongoing process of colonization.

As a white settler on these lands, descended on three-quarters of my family tree from people who arrived on this continent in the 1600s, I know that my ancestry is bound up in this history of violence. Yet as a queer person assigned female at birth, I also know that my access to land is shaped by this same poison brought to our shores. The roots of colonization have born fruit to a tightly woven social structure of white supremacy and cisheteropatriarchy, which the blog *Decolonize ALL The Things* defines succinctly as "a system of power based on the supremacy & dominance of cisheterosexual men through the exploitation & oppression of women and LGBTQIA* people." These systems perpetuate the extractive economy by shaping who has control over the choices to manage their homes.

My partner, Lindsey, is a white queer and non-binary farmer who grew up in rural Michigan. They are descended from a long line of farmers who've lived on those Pottawatomie, Ottawa, and Chippewa lands since the 1800s, when the family moved there from New Jersey. Many of their ancestors, Christian-reform Dutch escaping the Church of England, likely first came to this continent through the system of indentured servitude. Lindsey's father and uncle still own family land, which they rent out to the neighbor dairy farmer, who grows corn, soy, and alfalfa in rotation as feed. Lindsey's mission as a farmer is in developing a small-scale regenerative system to grow food, soil, and resilience for local communities, part of a movement of farming that many Greenhorns can relate to. But Lindsey will likely never have access to that ancestral land, because their father does not support their life choices and vision. Because the patriarch with the title to the land decides what will be done on it.

The gender binary that was brought to this

continent by colonization was not just a binary of "male" and "female." What was brought to these shores and every mountain, valley, and plain was toxic masculinity and devalued caring labor. Colonization required an aggressive, competitive, and self-absorbed masculinity in order to wage violence against bodies and land, and it justified forced caring labor—sexual labor, child-rearing, cleaning, cooking, growing food—by denying femmes and enslaved people full humanity.

Within the colonial mindset, some bodies belong to other bodies. Within the colonial mindset, black bodies could be bought and stolen from their land, forced to labor by white bodies in their quest for land and wealth, and have no worth beyond accumulating capital. Within the colonial mindset, the desirous, sexual, caring, and creative impulses of the body were meant to be subdued, reigned-in, and trained only for extraction. Puritan codes from the mid-1600s showcase how heavily the pleasure of sexuality was feared and stigmatized, and the legal frames through which heterosexuality for reproduction was made compulsory. Nearly two centuries later, after Congress banned importation of slaves from Africa in 1807, the desire to continue extracting wealth from the bodies of those made to be slaves exacerbated the violent exploitation of the reproductive capacity of black women, and created yet another economic incentive for the strict policing of compulsory heterosexuality. Meanwhile, Indigenous communities continued to be pushed further from their homelands, corralled onto reservations, with their children sent to boarding schools to control their bodies and minds. Within the colonial mindset, the land, and all regenerative, living, and interconnected beings that call it home, can be owned and used at the will of the owner. And necessarily, within the colonial mindset, land cannot be owned by owned bodies.

This colonial mindset is not a historic relic, but the fabric that holds together our mainstream culture and consciousness. The mentality that grants this "right to rule" and make decisions for other living beings also leads to the ultimate means of dispossession: the theft of the body itself. Today, transgender women face four times the likelihood of being killed than do cisgender women. Of the hundred and two cases of lethal violence against trans people tracked by the Human Rights Campaign, eighty-seven of them have been people of color, and the vast majority were black transwomen. Over half of all these murders were in the South. Meanwhile, only three percent of LGBTQ+ annual foundation funding went to organizations and services based in southern states.

I can understand why so many queer people leave the land. The interconnected histories of settler-colonial, white supremacist, cisheteropatriarchy brought to this land continuously displaced queer and trans people from rural livelihoods, as they continuously displace and dispossess all indigenous people from their lands.

Queer land justice is a reclamation of our right to a relationship with our bodies, and a reclamation of the right to the resources to heal the body. It is the struggle to not only fight the bad, but to create in its midst, liberated spaces where this healing can take place. To do this, queer land justice, at its core, requires a commitment to solidarity with all liberatory movements, and all struggles for land and body sovereignty.

I turn again to the audacious purple of the crocuses coming up through the pavement, and strain to hear the chirping of new swallows in the mud and straw nests, a hopeful reclamation of this old barn that was never built for them. Knowing my ancestry, I often wonder to myself if I am the hopeful swallow, reclaiming space, or if I am the colonial barn, repenting my history by providing sanctuary. Knowing that I fit no binaries, I decide that I'm both. As Lindsey puts it, "My ancestors have been part of the history of segregation and slavery, and my ancestors have also been part of the history of desegregation and liberation. I walk in those footsteps."

IV. RECLAIMING THE RESOURCES FOR RESILIENCE

"I have AIDS and need my body to be as strong and resilient as it can be... this land itself nourishes, nurtures, and teaches me. Everyday, I drink fresh spring water from deep within the earth, and feast upon wild plants, homegrown organic vegetables and fruits, and concoctions prepared in our communal kitchen with tender loving care."
—SANDOR KATZ, *WILD FERMENTATION,* WRITTEN AT SHORT MOUNTAIN SANCTUARY IN TENNESSEE

Alongside a long history of dispossession, there is also a long history of queer and trans people finding ways to carve out space and place, to explore and practice a reclaimed relationship with their bodies, to define new rituals of care and compassion for each other, and to reconnect to land and draw from it the power we need to work for its defense, and for our own defense. "This dream belongs to all of us who have been driven from the land of our ancestors and who are seeking ways to reunite with the sun, water, and soil that sustain us all," Lindsey says. "We are building the queer ecological future right now, at this time of climatic disruption, and we are committing everyday to a just and regenerative economy."

As queer and trans people, we live within a legacy of land reclamation. For forty years, Radical Faerie communities and lesbian separatists have created sanctuaries in rural areas across the country, providing safer access to land-based living. During the AIDS crisis, these sanctuaries became critical sites for hospice or healing, a critical reprieve from the contamination of the cities, which disproportionately impacts people with weakened immune systems. Many of these sites are owned collectively through land trusts so that they can never again be bought or sold, highlighting our ancestors' and elders' desires for a different way of relating to the land. As some of the founding documents of Oregon Women's Land Trust, the first women's land trust, proclaimed: "We want to be stewards of the land, treating her not as a commodity but as a full partner and guide in this exploration of who we are."

However, there is much work yet to be done. Most of this collective queer land is still managed by white people, though there is now some work happening to change this under the leadership of queer and trans people of color, such as Building Permanence, a collective shifting faerie lands to QTPOC (Queer and Trans People of Color) leadership at Idyll Dandy Arts in Tennessee. Because these sanctuaries often need to keep a low profile within the largely conservative rural communities where they exist, young queer and trans people growing up in these communities might never know that there are thriving hubs of queer people in their own backyards. As gentrification and displacement creep outwards from the city, some of these faerie lands are in the paths of development, and rural queer livelihoods threatened. These are all pieces of the work we have today to protect and advance our queer land legacies.

We can learn from this history and bring it to our work: we need to get as much land as possible off of the speculative market, as quickly as possible. We need queer land in rural areas, and we need queer land in urban areas. And we need to collectively steward this land under the guidance and leadership of those who have been most impacted by these histories of dispossession. It is not enough for queer people to "own" land—we need to find the ways to steward it queerly.

"To be very clear," Lindsey explains, "my father's choice is not just about my gender and sexuality, but it is still deeply about my queerness and queer politics; it's about what I wanted to do with the land. Me and my cousins, all the people who are slated as the next generation to steward that family land, we want it to have a social purpose. We want to give it back. We want it to be a place where black and brown people—people who have been stolen from the land and had their land stolen from them—can be safe. This is what my dad dislikes the most." But Lindsey's father and uncle have been raised under settler-colonial, white supremacist, cisheteropatriarchy. They hold on aggressively to their belief that they have the right to make decisions for other people. They believe—and defend the belief—that they own the land. Queerness is not merely a description of a sexual orientation, it is a political positionality that questions the constructs that have been made to seem normal. It is a rejection of cisheteropatriarchal logic, and as such attempts to queer land noticeably confront these systems.

As the tides of climate change rise, we will be increasingly confronted with this dilemma. Our communities have been stripped of earth-based resources for resilience, and are among the frontlines of climate disaster. Yet our queer and trans ancestors have also passed onto us clues and tools to building a more caring, cooperative, and healing "management of home" and management of the first home: the body. I have been lucky enough to know queer land. I have walked across it, bathed in its waters, eaten its food, and drunk tea steeped from its medicine. I know what queer land tastes like. It is in me, and it gives me the courage to dream of queer ecological futures. We must cast off the internalization that we are "unnatural" and reclaim our place as forces of nature within a diverse ecological coalition.

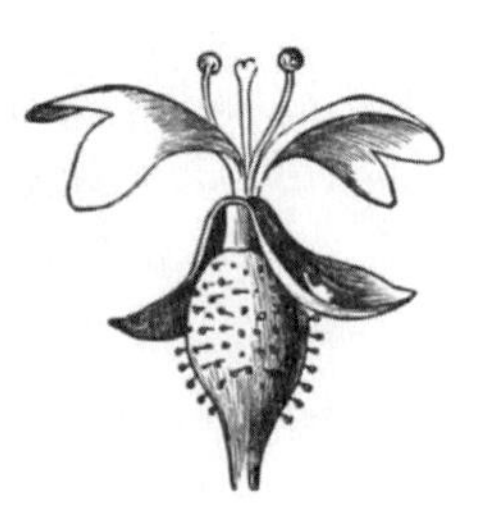

V. DREAMS OF QUEER ANCESTRAL FUTURES

Queer land justice is already here, spreading underground like the mycelium on which we depend. It can be hard to recognize as such. Most of the queer and trans environmental organizers, farmers, and healers I know are not in this movement only for the benefit and survival of the queer and trans community. Traditional ecological knowledge reminds us that everything living off this land belongs to it, and how inextricably our relationships are bound with all the other living and non-living beings that make up our homes. No one is free until we all are free.

Queer land justice is in solidarity with the greater fight against all fruits of colonization: the extractive economy, gender binaries, compulsive heterosexuality, white supremacy, and all other violences to the desires and expressions of the body which underpin and uphold the extractive economic system. There can be no land justice without justice for the body.

For me, to leave behind my genderqueer identity in environmental spaces is to leave behind the homeless trans youth, the HIV positive elders, the incarcerated sex workers, and the impoverished rural queer people, none of whom are seen or supported in mainstream disaster response. To leave behind my identity is to leave the important tools passed down to me by my queer and trans ancestors, stealing myself of critical assets in this moment of transition. Queer liberation as a movement has struggled for the very survival of our people through so many iterations of oppression—from the violence of police beatings, AIDS apathy, poverty and homelessness—and we have developed a culture of care, mourning, and healing to bring ourselves to where we are today.

It is our duty to bring our whole selves to this moment of great transitions.

I have known queer land, and queer land is in me. I can dream and see a future where all beings exist, all beings have the capacity and the resources to heal, where all land is stewarded communally and all bodies are treated reverently as the natural extension of the biological exuberance of this queer planet.

These dreams of our elders and our ancestors live on; they follow me, propel me. In this moment of peril and opportunity, they demand we answer these questions:

How will we reclaim our labor to reclaim the land? When will we, as queer people, join in solidarity with global indigenous and landless peasant movements, and learn from these movements? What will we manifest amid the coming fires and floods of ecological disaster? How will we protect each other, and leverage these moments of change towards the changes we seek? What gifts will the land provide us as we ally with it in our collective liberation?

Queer land justice is here, and the mycelia are spreading. We are waiting for the rain.

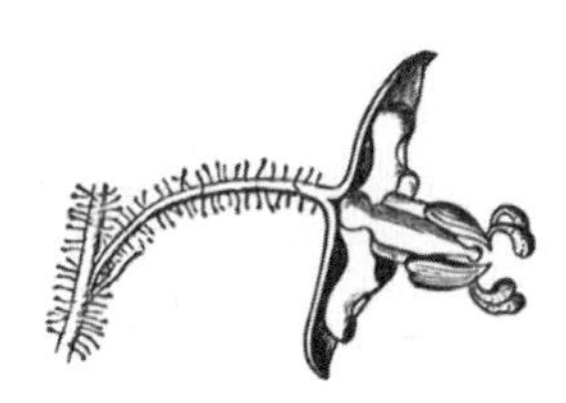

In memory of Amanda Arkansassy Harris. Inspiration for many of the ideas in this piece come from Movement Generation, Black Land and Liberation Initiative, Queer the Land, 40 to None, Queers for Economic Justice, and the book *Land Justice* by Food First. If you're interested in the ideas in this article, want to offer feedback or critique, or want to get connected to more readings and resources to queer the land, please reach out to Queer Ecojustice Project at queerecoproject@gmail.com.

OSTYAK BOATS.

FEBRUARY

Age of Sail

STORY OF AN ELUSIVE PARADISE

The Deep History of Florida

MALORY FOSTER

The place of understanding is not necessarily kind. Heaven and earth meet there, and they will crush you if you do not hold them apart... Nature is fire and ice and gravity, falling rocks and rushing streams, but it moves well through myth and poetry: voices, sounding through the smoke.
—C. L. RAWLINS

The deep history of Florida unfolds like its opalescent beaches: in waves. Only capricious ocean currents and epochal sea level fluctuations could create crushed coral underbellies, quartz sand shorelines, and cavernous grottos. Here, the tides play architect. Over the millennia, waters rose and fell, followed by sea creatures and land dwellers, migrating and layering and leaving their bodies in humble fossils. A cryptic note from the past, forewarning the sirenic and inconstant character of this place.

Florida was, curiously, once a part of northern Africa. Two ancient supercontinents collided to form the well-known grandmother land mass, Pangea. Geologists have gained evidence for this theory by measuring magnetic characteristics in rock from Florida and matching them to that of northern Africa. They have found more harmony there than between rocks from Florida and surrounding North America.[1]

Years passed; continents drifted apart. Sometime between the Miocene epoch—twenty-three through five million years ago—and the Pleistocene epoch —about 1.8 million through ten thousand years ago—Florida reached relative tectonic stability.[2] Rising up the spine of the peninsula is the Lake Wales Ridge, summiting at Sugarloaf Mountain, 308 feet above sea level.[3] The Ocala Platform, a limestone formation, hugs the region to the north, and the Peace River Formation traverses the center of southwest Florida.[4] The Peace River Formation begins at the southern end of the Ocala Platform and stretches south to the Okeechobee Basin. Interbedded sands, clays, and carbonates constitute the Peace River Formation, which is rich in phosphate, a boon to the modern fertilizer industry in Florida, and where eighty percent of the phosphate used in the United States is mined.[5]

Next to leave its mark on this canvas was temperature. Several ice ages came and went throughout the Pleistocene epoch. Due to the command planetary temperature has on glacial activity, sea levels surrounding lowland shoreline varied dramatically throughout these times. Across the Florida peninsula, sea levels varied from one hundred fifty feet above their current level to four hundred feet below contemporary shorelines (see Figure 1). During the times when high sea levels swallowed up earth, the Florida shelf was a shallow temperate zone, ideal for tropical corals to grow and develop extensive reefs (see Figure 2). When planetary waters were frozen and the ocean retreated, the coral reefs on higher lands, then exposed, died off, compacted, and became limestone. This repetitious cycle formed the soft limestone bedrock of Florida.[6]

The lower water table also contributed to the formation of some of Florida's most interesting geological features: caves, springs, and freshwater aquifers. As rain collected carbon dioxide in the air and filtered into the ground, it created a mild carbonic acid, which over centuries dissolved tunnels through the soft limestone bedrock, forming characteristic karst topography and caverns. Streams caused erosion over time, and these "sinking streams" eventually connected with the caverns, opening them up to sunlight to form the enchanting surface springs we see today.[7]

FIGURE 1

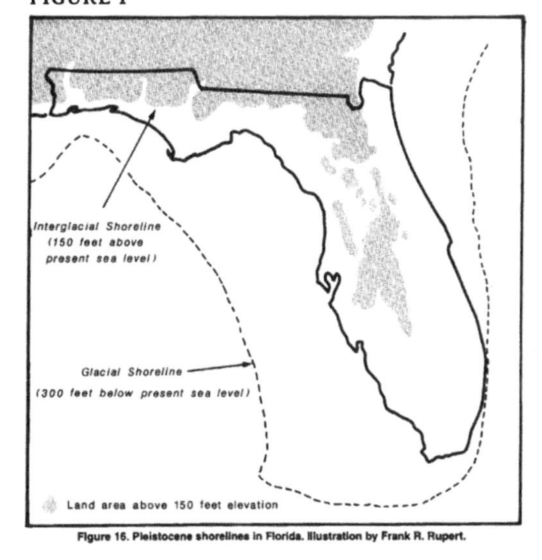

FIGURE 2

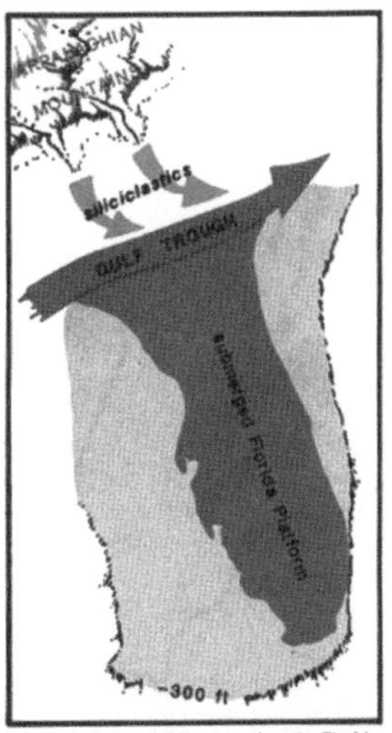

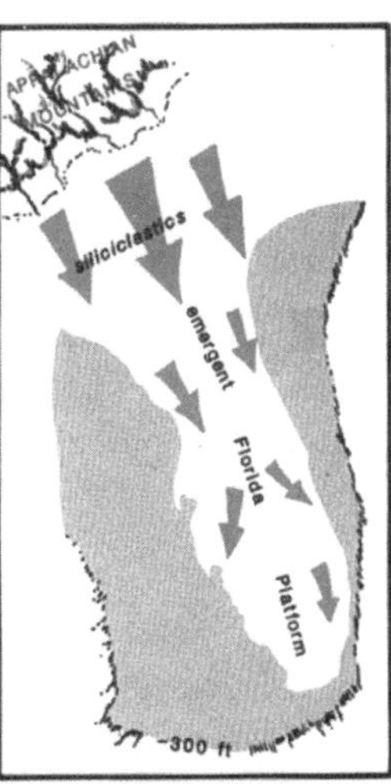

Figure 1. Pleistocene shorelines in Florida. Illustration by Frank R. Rupert.[8]
Figure 2. Florida bobs above and below the surface of the ocean as waters freeze into and out of glaciers through the ice ages of the Pleistocene epoch.[9]

The native Florida bioregion is a wild tale of antilogies: wetlands alongside drylands, muck juxtaposing sand, fire complimenting water, thunderstorms trampling drought. Extremes coexist in the forms of cypress dome swamps and dry pine flatwoods ecosystems, which sometimes neighbor in a tacit solidarity. These habitats showcase the perfection of their species to adapt to Florida's unique demands. Florida's sponge-like foundation and ongoing play with waters ought to give pause to we who work the soil. Myakka fine sand is the official state soil, and extreme porosity is a reality when working with sand and limestone. It is a challenging environment to produce annual vegetables, and soil building is an ongoing effort in this pursuit. Alternatively, some farms treat their fields like hydroponic operations, using plastic mulch beds and an exactitude to fertilizer applications usually only seen in greenhouses. Gardeners south of Orlando may opt for tropical perennials that stay rooted and enjoy sand. We have much to learn about how to apply soil-sensitive practices in such a delicate and responsive climate.

In the grand history of this place, we are but a tiny speck, ants creating a temporary ant hill. As the shallow, submerged Florida shelf lured corals to grow on it millennia ago, it has now drawn humankind to build concrete reefs on its balmy shorelines. Will our reefs someday become fossilized memories of a time long past? As we have learned from those before us, this is but a fleeting spell, an elusive paradise.

NOTES

1. Ed Lane, "Special Publication Number 35: Florida's Geological History and Geological Resources" (Tallahassee, FL: Florida Geological Survey, 1994), 11–14.
2. David W. Scholl and Minze Stuiver, "Recent Submergence of Southern Florida: A Comparison with Adjacent Coasts and Other Eustatic Data," *Geological Society of America Bulletin* 78 (April 1967), 444.
3. United States Geological Survey, "Feature Detail Report for: Sugarloaf Mountain," October 19, 1979, water.usgs.gov/ogw/karst/kigconference/abt_karstfeatures.htm.
4. Ann B. Tihansky and Lari A. Knochenmus, "Karst Features and Hydrogeology in West-central Florida—A Field Perspective," in Eve L. Kuniansky, editor, *U.S. Geological Survey Karst Interest Group Proceedings* (St. Petersburg, FL: USGS, 2001), water.usgs.gov/ogw/karst/kigconference/abt_karstfeatures.htm.
5. J. D. Gulley, J. B. Martin, P. J. Moore, and J. Murphy, "Formation of Phreatic Caves in an Eogenetic Karst Aquifer by CO2 Enrichment at Lower Water *Tables and Subsequent Flooding by Sea Level Rise." Earth Surface Processes and Landforms* 38 (2013), 1210–24.
6. Lane, "Florida's Geological History."
7. Gulley et al., "Formation of Phreatic Caves," 1211.
8. Reproduced from Lane, "Florida's Geological History," 22.
9. Reproduced from Lane, "Florida's Geological History," 19.

HERITAGE LIVESTOCK BREEDS WORK

Historic Sites Help Preserve the Old and Useful Breeds

CHRISTINE HEINRICHS

Colonists who settled in Virginia came to the mainstream of public events when they came to Williamsburg. They may have lived outside of town, but Williamsburg drew them into political and other public events. Williamsburg was the colony's political center and economic heartbeat from 1699 to 1780.

Most colonists were farmers, raising their own food and offering surplus as trade. Those with skills offered their services.

A visit to the apothecary would be in order. Apothecaries advised on medical problems of all kinds, and some were midwives. Remedies included chalk for heartburn, calamine for skin irritations, and cinchona bark for fevers. (Cinchona bark contains quinine and quinidine, now used, respectively, to treat malaria and cardiac conditions.) Visitors might take home some vinegar of roses, a remedy made from steeping rose petals in vinegar, to dab on the head for headaches. They could also buy cooking spices, candles, salad oil, anchovies, toothbrushes, and tobacco.

COLONIAL LIVESTOCK

Farmers might have ridden into town in a cart or carriage drawn by American cream draft horses. They're big, with mares standing fifteen to sixteen hands tall and weighing 1,500 to 1,600 pounds, males even bigger at sixteen hands or more and weighing 1,800 pounds. The Canadian horses, developed from horses sent from France to Quebec in the 1600s, were good strong farm horses too, if slightly smaller. Getting together in Williamsburg gave farmers a chance to compare their livestock.

Wool from Leicester longwool sheep made a valuable product to barter, perhaps for some finished cloth. Spinning yarn was time-consuming. It was easier to trade the raw material for cloth that came in from Britain. Delicious Leicester longwool meat was also in demand.

Sailing ships arrived from ports in India, China, and the East Indies (now Indonesia), laden with exotic spices and other goods. Chickens moved around the world that way. Easy to maintain aboard ship, they provided a steady source of meat and eggs. New chickens from foreign ports became exciting additions to local flocks. Brahmas, Langshans, and Cochins arrived that way, adding size and color to smaller farm chickens. The buff color, in the 19th century, created a craze among poultry breeders, who bred the color into other breeds.

Eggs and live chickens were always in demand. English Game fowl were the familiar poultry that came with the British colonists. Reliable, sturdy, good mothers and protective roosters made them a favorite. They could be relied upon for entertainment as well, with the farmers and townspeople settling in for an afternoon of cockfighting. That aggressive rooster causing trouble in the flock might find his value in the ring. Winning wagers on him might bring in some welcome cash.

Dominique chickens, also called Dominikers, were gaining popularity among Virginia farmers. Their feathers are barred, silver-white, and dove gray, in the color pattern known in other breeds as cuckoo. The colors blur as one watches them, making them look blue. That pattern provides protective camouflage for them when they forage in the barnyard. Their small rose comb has an upturned spike at the back, distinctive and stylish. Dominiques are good foragers. They went on to be considered the first American breed, shown at the first American poultry show in Boston in 1849. The males have longer sickle feathers than females. Their bright yellow legs stand out. Getting the rose comb perfect is a challenge to breeders. It may lack the required spike, or the spike may be misshapen. Tail angle in both males and females can be difficult to perfect. Dominique tails should stand at a jaunty forty-five-degree angle. Dominiques are a general-purpose barnyard breed. They make good roasters or fryers. They are steady, reliable layers of brown eggs. The hens will settle in and brood eggs and raise the chicks they hatch.

The little Nankins made such good mothers, they were put to work hatching eggs for pheasant and quail for the big estates. They are small but have value as broody hens. Not every hen is willing to set those twenty-one days to hatch out the replacement flock.

POLITICAL LIFE

Politicians appreciated having a nice meal over which to write the documents that shaped the future of this new political state.

Thomas Jefferson was a student at the College of William and Mary, founded in 1693 under the sponsorship of the English monarchs William III and Mary II, for whom the college was named. William Small was Jefferson's teacher in several subjects, from rhetoric to mathematics. Jefferson fell in love (but was rejected), studied law under George Wythe, Virginia's first signer of the Declaration of Independence, and got radicalized, even defending a slave in 1770. He continued owning slaves, although he spoke against it: under the law of nature, he said, "we are all born free."

Jefferson, Wythe, and Small often gathered for dinner at the Governor's Palace. Governor Francis Fauquier was an economist and philosopher who regularly hosted. Butter and cheese from American milking Devons, descended from the cows esteemed back in Devonshire, England, were probably on the menu.

HERITAGE BREEDS TODAY

Colonial Williamsburg has a Rare Breeds Program working to preserve these historic breeds. The living animals also help visitors understand the significance of livestock in history, and its importance today. These heritage breeds carry the genes that have been bred out of industrial-production poultry. They retain the sturdy, useful traits that made them valuable in the past and are needed for the future.

Other living history farms, such as Garfield Farm Museum in Illinois, are also committed to keeping these breeds alive and working. Garfield Farm specializes in Javas, named for the Southeast Asian island where they are native. They arrived on those sailing ships trading in spices and became an American foundation breed.

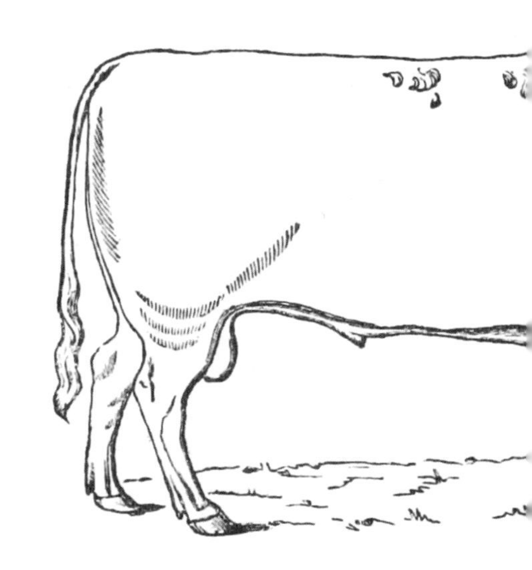

The Association for Living History, Farms and Agricultural Museums is an organization for museum professionals and history buffs. Local historic sites may be eager to collaborate with American Poultry Association members to establish historically accurate flocks. By educating the public about the economic role poultry has played in history, historic sites engage the public in saving historic breeds.

Contemporary farmers can include heritage livestock on their farms. The Livestock Conservancy tracks heritage livestock of all kinds, and they are eager to advise on suitable breeds.

Áine Mullan | Heavy Metal Oysters

Set up by Kirsha Kaechele, curator at Tasmania's Museum of Old and New Art (MONA), Heavy Metal is a joint project between artists and Monash University scientists that is focused on heavy metal pollution in the Derwent River. Approximately four hundred oysters, which naturally filter heavy metals from the water, were introduced around the museum via pontoon. Their heartbeats are monitored to record stress levels in relation to water quality. Once removed, the oysters are kiln-dried, encased in a glass block, and displayed in a mausoleum-like installation inside the museum.

A Beautiful MOOSE.

The curious in Natural History are invited to Major KING's Tavern, where is to be seen a fine young MOOSE of sixteen hands in height, and well proportioned. The properties of this fleet and tractable Animal are such as will give pleasure and satisfaction to every beholder.
Price of admittance, Nine Pence.

FROM THE MASSACHUSETTS MERCURY, DECEMBER 9, 1800.

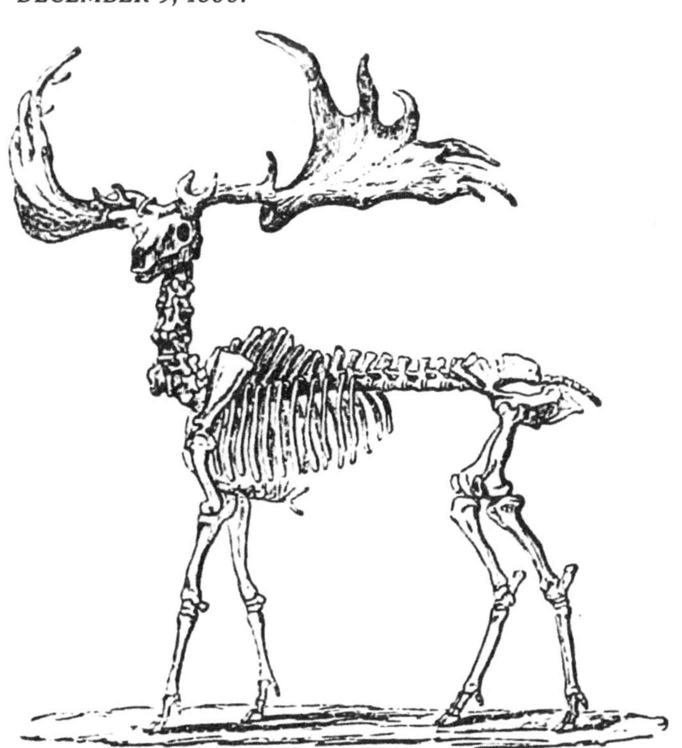

A NEW AGE OF SAIL

FORREST JACKSON

As the cost of a barrel of oil begins climbing again, so do the volume of calls for greater efficiency. While Kansas, in the United States, spent the last decade of low oil prices investing in wind turbines, and Saudi Arabia building solar power plants, one of the world's largest, most globalized industries—shipping—has seemingly done little to prepare for the inevitable change in the economic winds that will come.

If anything, the global shipping industry appears to have doubled down. Ships are bigger, the volumes being shipped are higher, and shipping costs have fallen so low that complex mergers and non-compete agreements are the only things preventing more titans of industry from declaring bankruptcy, as Hanjin Shipping Co., Ltd.—one of the largest shipping companies in the world—did in 2017.

Why is this happening when transnational shipping companies have more accountants weighing the costs of doing business than they have sailors operating their ships? How can they possibly overlook alternatives to a fleet based on oil, a notoriously volatile commodity? How can conservative Kansas embrace the renewable wind energy market while the world's shipping companies shun it?

Better yet, why aren't there modern sailing ships?

HUMANITY'S RELIANCE ON WIND

The basics of shipping have not changed for thousands of years: move a thing from one place to another for profit. The technology required to do this—the displacement hull—predates the celebrated wheel by at least five thousand years. But the displacement hull alone was not enough to send humans around the globe.

When Mesopotamians began putting square sails on their displacement hulls seven thousand years ago, humankind was able to leave the beach behind and escape the constraints of land, and so began the first Age of Sail.

From the grass rafts of precolonial Polynesians through the Egyptian, Greek, and Norse galleys, the wind has dictated where human settlements took root, as well as where and how empires grew. From the galleons that conquered the globe to the tall ships that grew so vast that masts had to be built from steel because there were no trees large enough to meet the power-hungry requirements of commerce and human expansion, the winds of the world determined population distribution, shipping routes, and the viability of goods for trade.

Viewed and measured from the deck of a ship, the size of the earth and her place in the solar system—and our solar system's place in the universe—were observed and measured and have led us to the same technologies now being adapted for further expansion into space: Cartesian coordinates, unified time, calculus, indemnification of loss, unionization of labor, and commodity markets In every way, humanity's success as a species has been wholly dependent on wind-powered shipping.

WHERE DO WE GO FROM HERE?

A century of motor-driven vessels appears to be an interesting anomaly within the course of human expansion, and the demands of commerce, when balanced against economic necessities, guarantee the demise of a petroleum-based global fleet. This isn't speculation—it is demonstrable through the laws of diminishing returns and supply and demand, and shipping companies are investing in technology, preparing to replace the existing fleet. What the next fleet will look like, however, is still unknown.

Architects are currently placing their bets on a model of modern shipping that is an autonomous, LNG (liquid natural gas)–powered fleet. It is a sensible model: efficient and clean, technologically sound and scalable, and unhampered by human-crew limitations; nevertheless, it relies on an extraction-based, finite resource.

LNG also may be the one fuel source more dangerous than hydrogen. It must be cryogenically cooled to -260° to keep it in a liquid state. This condensation of natural gas to a liquid state allows much higher volumes to be shipped, but should that refrigeration process fail, the resulting cloud of gas would rapidly expand and displace oxygen, creating as much as a quarter-mile radius where all life freezes or suffocates.

If ignited, that frozen gas cloud would explode, sending out a violent shockwave with a radius of anywhere from two to seven kilometers. Pools of LNG can burn for sustained periods of time, with flames reaching up to one hundred fifty meters in height. And because the greatest populations in size and density tend to be in port cities, a worst-case scenario involving the failure of an LNG tanker or facility would likely have severe repercussions for affected areas.

If shipping indemnification could withstand the insecurities of dangerous modes of transport, the Hindenburg would have been the introductory bracket to the Age of Airships, not the closing one. If history is any indication, then we are one industrial accident away from the demise of LNG as the de facto fleet of the coming decades.

WHY WE CAN'T SIMPLY JUMP SHIP

The global infrastructure that accommodates ships represents an investment greater than the fleet it actually services. The list of the infrastructure servicing shipping is so endless as to be all-encompassing: deepwater access real estate and rights-of-way; heavy equipment including gantry cranes, floating shipyards, trucks, trains, tugboats, barges, dredges, and other assistance vessels; transportation infrastructures such as vessel traffic lanes and transit authorities, aids to navigation, GPS (and the rockets and launch pads used to get them into orbit), radio and radar installations, coast guards and satellite-based rescue transponders; weather and hydrographic services; and the intangible infrastructures—the intellectual ones—made up of incomprehensible financial instruments, indemnification schemes, and an entirely separate legal structure that supersedes the laws of the land.

These infrastructures were largely built during the post World War II heyday of hydrocarbons and they exclusively service motor vessels. They are also increasingly built up to service the containerization of cargo: a large portion of all cargo is now moved in shipping containers, measured in units known as TEUs (twenty-foot equivalent units).

In the quest for efficiency, the concept of one engine moving more containers at a time makes a lot of sense. Whether it is moving two hundred or two thousand boxes, if an engine burns fifty thousand dollars worth of heavy bunker oil per day, the cost and resulting profit of shipping companies points directly to bigger ships. The largest modern ship is four hundred meters long, almost sixty meters wide and about twelve stories tall. It can carry an astonishing nineteen thousand TEUs, whereas the average ship carries about two thousand. To say that it is the most efficient mode of transportation ever devised by humankind is an indisputable fact—in an economy based on a barrel of oil, that is.

TECHNOLOGIES MAKING WIND COMPETITIVE AGAIN

Size alone hasn't satisfied the demands of investors as a means to achieve greater efficiency and the higher profits that come with it. This insatiable appetite has forced the development of many technologies to retrofit the existing global fleet on a quest for efficiency, and ironically, it is these technologies that create the foundation which will enable the coming transition away from hydrocarbons. The most promising technologies to date are:

- The bulbous bow, a technology already widely adopted in modern times, which increases the efficiency of the hull moving through water, reducing fuel consumption by twelve to fifteen percent.
- Lubrication, a film of tiny bubbles that

decreases the resistance generated by pushing a ship's hull through the water, which can reduce fuel consumption anywhere from five to twenty percent.

- Airfoil hulls, hulls whose topsides not only act to minimize wind resistance but utilize the wind to assist in propulsion by their shape alone, can reportedly reduce fuel consumption by up to sixty percent.
- Photovoltaic sails, capable of harnessing both wind and solar power to assist in propulsion, can reduce fuel consumption by twenty to forty percent.
- High-atmosphere kite sails, which generate up to twenty-five times the thrust of traditional, mast-supported sails, have provided fuel reductions of anywhere from ten to thirty-five percent.

Yet all these technologies that boost fuel-efficiency still have one thing in common: they are all designed and tested to provide greater efficiency to vessels powered by a primary engine burning a hydrocarbon fuel.

A NEW AGE OF SAIL

A sailing ship's masts don't just take away capacity for carrying containers; they also limit the ability of cranes to move shipping containers. The modern container dictates the shape of the hull and the height of a ship's superstructure. Sailing vessels require deeper water and taller bridges over waterways than their motor-driven counterparts. Forward-thinking, modern naval architects are forced to work within these types of constraints when looking to the future.

Lubricated airfoil displacement hulls that use photovoltaic sails and high altitude kites to augment electric auxiliary drives with high-capacity battery storage, pushed by contra-rotating props (propellers), are not far-fetched, just cumbersome to describe. The technology is already tested. A profitable ship of this design could, right now, be produced and put into service carrying modern shipping containers.

The existing infrastructure will have to be upgraded to accommodate this new breed of tall ships. Bridges will need to be heightened. Shallow waterways, dredged. It will take time and lots of money, but it will ultimately be cheaper than the commodity price of a barrel of oil and the externalized costs to bring that barrel to the market.

In spite of the obstacles, a new Age of Sail is coming. It is contingent on the cost of a barrel of oil, not on the design of ships, or desire. There will likely be false starts on the road toward alternative fuel paths. And the strongest resistance will likely come about from governments considering the costs and increased taxes to upgrade the infrastructure to accommodate these new ships—not from the shipping companies building the new fleet.

Sophie Wanzer | Sailing with a Purpose

I spent the spring semester of my junior year in high school at Maine Coast Semester at Chewonki in Wiscasset, Maine. There, I reflected on the beauty and power of sailboats: the sweep of the hull, spray of the water, snap and power of the sails. Sailboats communicate the value of connection, work, and understanding of the natural world. To work a sloop effectively, you must physically connect with the boat and also understand your environment.

When I learned about the Maine Maritime Museum's restoration of their seventy-three-foot schooner Mary E, *I wanted to contribute to their work to open the commons of the water to the wider public. I wanted more people to be able to access the beauty of sailing. I connected with the museum and offered this print, which was part of their 2018 exhibit* Workaday to Holiday: Schooners along the Maine Coast.

Rachel Alexandrou | Fucus

INTERGLACIAL INTERLUDE

BRAD KIK

We—the flickering hominids, breathing the deep air—are bound/wound to terse bodies kindled and hastily extinguished. Far, far beneath, the subterranean rhythmic thrums—the scraping, scribing, kiloannual migrations of ice and rock and ice, basslines below basslines below basslines, the infrasounding abyss of the interglacial inhale. Us, though—our longer generational lines, from the sedimented ancestors to the stainless supersonic myth-promise of the fierce distant unborn, are all but a bright brief stand among these ancient mineral ice spirits.

We pass our lives on the migratory route of glaciers at a ratio of a quadrillion breaths to one.

As we—perched and fluttering—await the certain long-impending geological return, we oscillate in these ultra-hertzed lives and so deposit our artifacts into the stratum: the mound and the arrowhead. The two-man tuttle-tooth-saw, chipped. Tailings and sulfide, black wet Albertan shale among shards of carbon steel. Broken glass and cremation ash and loose change and lint and microplastic beads and the bar-coded wrappings of all the sandwiches and CD-Rs and screwdrivers and salmon steaks and smartphones. The squandered material culture of the mighty. Our flying, failing, and falling.

Abstractly folded, behind and below, lie the lost graces—the phantoms of graylings, thickening in the shades of unsilted streams. The sacred ghost gospel chorus of passenger pigeons, whose muttered murmured prayers rain down as scat to raise the white pines higher. The unplowed bones of mammoths, humming and thrumming with the promise of deep unpeopled rest; a primordial cycling song that hooks "return, return." A breath.

A migration.

My ancestral body stands in the cold, spruce sparking nearby—under the lost canopy of old growth white pine, under the dark glimmer of the new moon, alive and alight and fluttering. My mirror-me stands opposite, below, under the soil, his wet beard fruiting with mycelia, sucking the sweet starch from tongue tangled root-hairs—both of us dreaming of rain and lightning strike and wildfire and rain again. Both of us dreaming of the migrations of water and warbler and carbon and caribou and glacier. Both of our lands, his and mine, above and below, holy.

Both fed—fed!—by a numberless gush of spring—creek—stream—on down the taxonomic fruitful multiplying of the holy nested watersheds, this blessing, this impious temporal stage, this sacred melancholy antique church of dune, swamp, and tall conifer.

He and I, both shouting the same simple fleeting triumph of joy, before the next heaving breath of ice, made mineral dense, scrapes the archival stratum down to bedrock, chucking out Devonian Petoskey into the frozen sweetsea.

Then, beyond us, beyond us all, drawing back, thrumming a miraculous thrum—exhaling—and beginning to scribe again.

Conglomerate core section, taken from a glacier

LAKES AND RIVERS OF IRON COUNTY

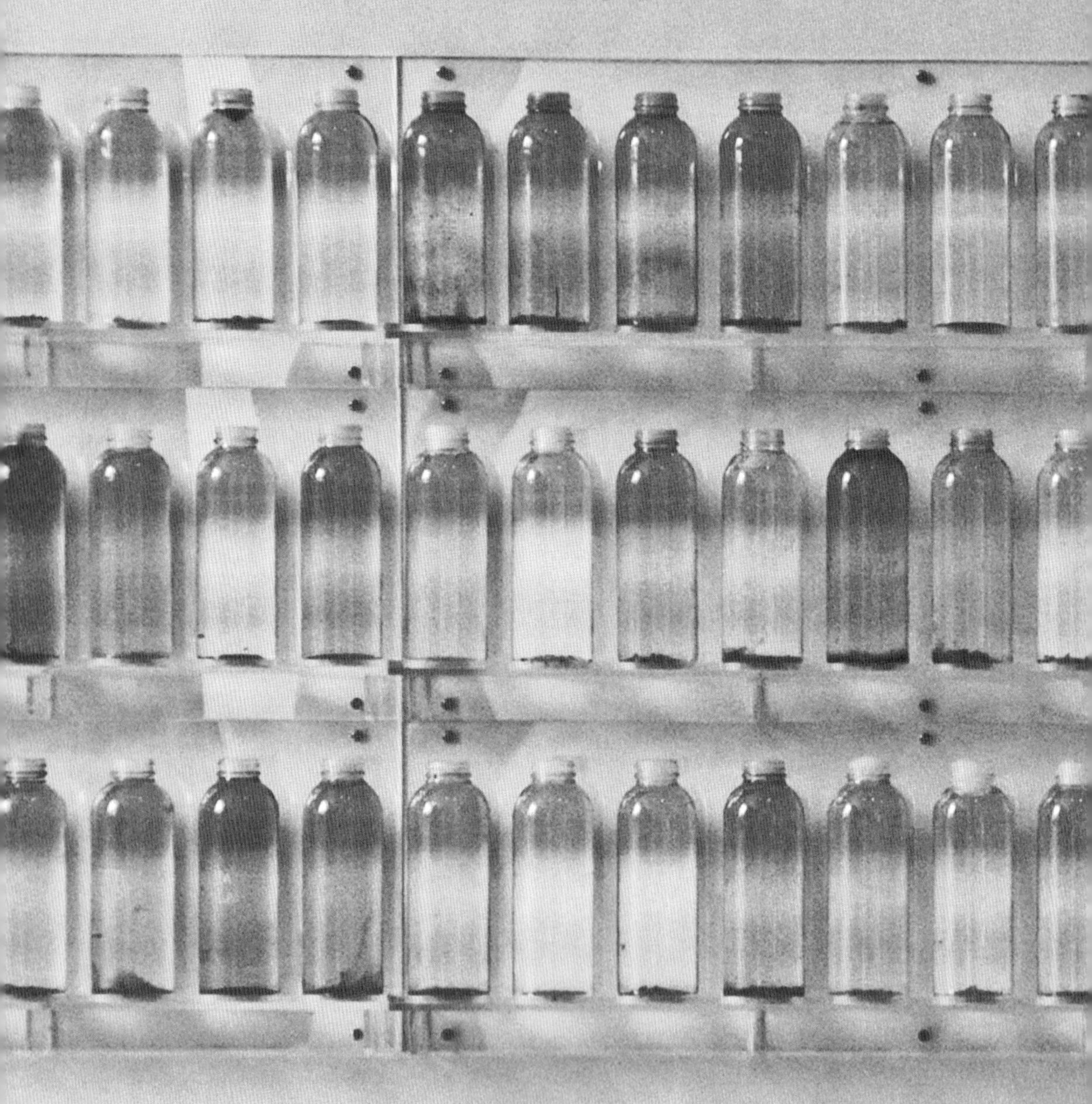

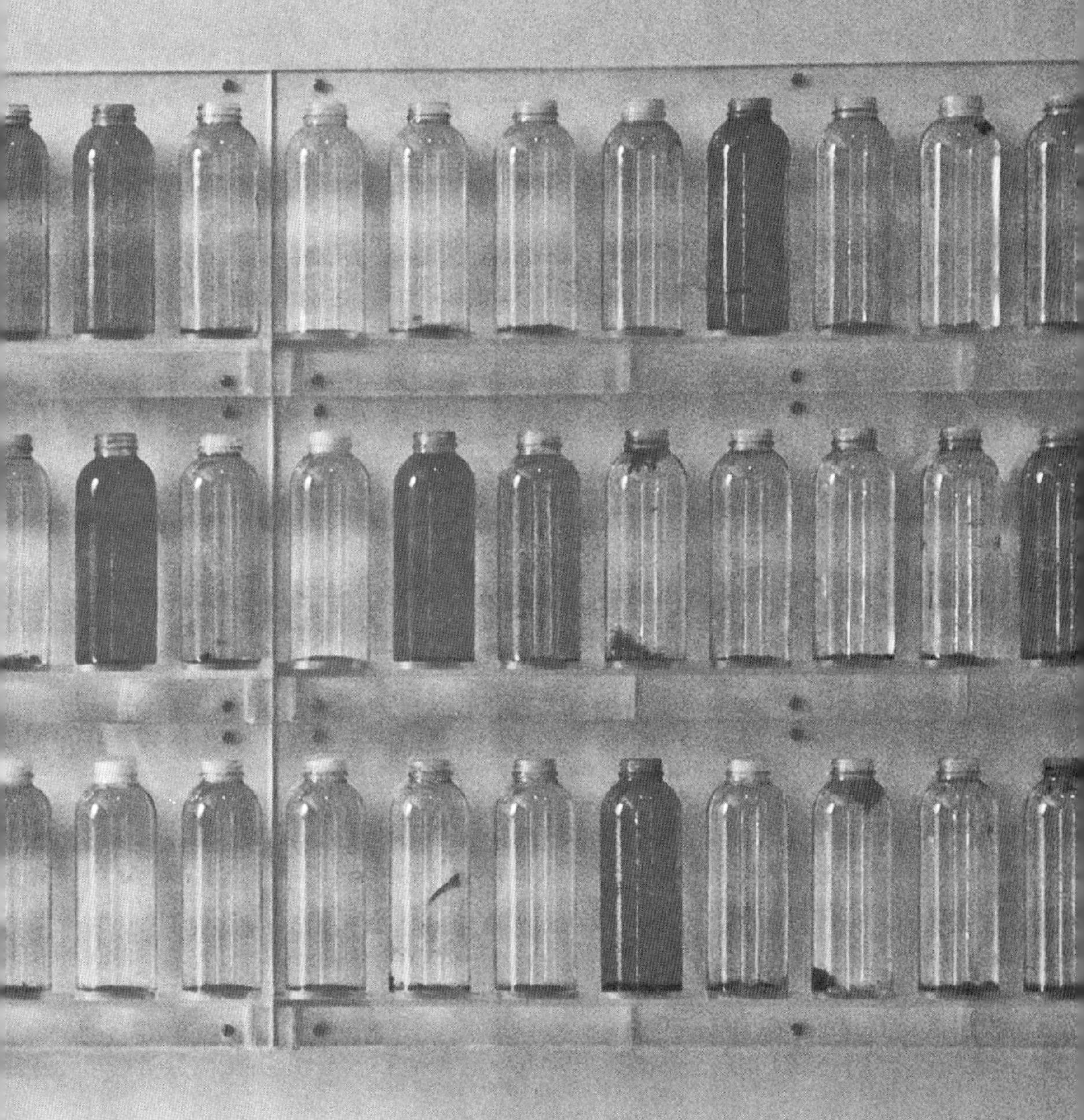

GINNY MCCLURE

Dorothy Lake

Potato River

I wanted to interrogate what I know about the place I thought I knew best, and in doing so examine what we often take for granted—our relationship to home and the sense of belonging, stability, and identity that comes with it. The northern Wisconsin county I grew up in has around 160 lakes and rivers, and over the course of a year, often with family members along to help, I set out to collect a sample of water from each.

I drove on most of the roads, trespassed through people's woods, laid on strangers' docks to collect water, crawled through culverts, cracked through frozen ponds. It was an adventure and a meditation. In the end, this experience brought me to the exact conclusion I anticipated: I still knew very little about Iron County. But that wasn't really the point. When I shared the project by putting the water samples on a wall, the varied hues of amber were attractive, but it felt like the shell of something once living. The meaning was in the time spent exploring this familiar place in a new light and in the way this experience shaped how I consider Iron County now. It served as a reminder of the need for all of us to continually re-examine our relationship to the places we think we know.

Opposite: Installation 2
First Spread: Installation 1
Photographs by Ginny McClure

SPIDER
LAKE
CREEK
CREEK
CREEK
LAKE

A SWEET POTATO RAMBLE

KATHERINE HALE

Beneath the bloody human history of war and dominion, the humble sweet potato reigns supreme, a colorful revolution flaring forth from the underworld and into the light. No matter where I am on earth, I am always home, because the sweet potato is waiting for me, its vines sprawling in the heat, blanketing the ground in a welcome carpet of green, its tasty tubers prepared a thousand different ways, all of them delicious. Wholly cosmopolitan yet miraculously local, sweet potatoes have conquered me along with the world, and I surrender fully.

As cultures collided in the Age of Exploration, and migrations—both willing and forced—began in earnest, sweet potatoes came along for the ride because no one who met them could bear to live without them. Their lifestyle endears them to people on the move: edible at every stage of life, and endlessly adaptable to all but the most extreme conditions. Let a tuber sprout into an endless stream of clonal slips, and you will eat well for the rest of your life. With few pests, little need for weeding, and no special fertility requirements, their only enemy is the frost—a small failing for a plant so perfectly suited to human companionship, and no sin at all if you dwell in the tropics. From its wild beginnings in the jungles of South America, the sweet potato's dominion now spans the globe, from the ultra-modern convenience stores of urban Japan, to the smallholdings of Africa and the highlands of New Guinea, and all the way to my home in the American South.

It's sweet potatoes that dog my farming footsteps every day of the year, marking my place in the great planetary waltz of earth and sun. There's the annual gamble with the weather every April on the best time to put the tender slips out, the nervous faith that marvels will emerge from these small and fragile beginnings. There's the sweet potato roasted whole in its skin, stuffed into my back pocket for lunch in the fields on those summer days when the light and the to-do list are both endless, while the morning-glory flowers blink lavender corollas up from the beds, a reminder to slow down and savor everything as it arises. There's the slow spading among the rambling vines to find every tuber on the eve of frost, the meditative sorting and grading, careful not to bruise the delicate skins, and the miracle of curing, where the starches melt into sugars so the inner sweetness bursts forth when cooked.

Sweet potato patch

There's my kitchen windowsill in winter, laden with sprouting slips to remind me that the sun will return again and that life emerges out of darkness, no matter what the weather says outside, and the cycle begins anew.

If we are what we eat, then I am a sweet potato in all its infinite variety: orange-fleshed, juicy, and dripping; white-fleshed, reminiscent of roasted chestnuts; purple-fleshed, dense roasted cake right out of the ground, no added sugar necessary. I am sautéed leaves dripping milky latex, the only greens left standing in the dog days when the lettuce bolts and collards wilt. We complete each other—the rich flesh sustains my blood and my bones; my sweat and effort nourish the fields.

If we are what we eat, then let the sweet potato be our common ground, the food that unites us if we so choose. In a time wracked with uncertainty, long may the sweet potato reign! Through it, I am connected to all that has come before me; through it, I am unafraid of what the future might hold. When all else fails, the sweet potato will be here—we live and we die together, the sweet potato and I.

THE TALE OF THE OUTLAW HORSES AND THE CREATURE WHO WAS A HANDFUL OF THE WORLD

PETER KIRN

The horses in the forest were declared outlaws because nobody owned them. The town announced they would be removed to die. Here. In the town by the tall shorecliffs where the spirits of the cornered people watch. So, as if they'd been warned somehow, the horses split up. Scattered into the untold reaches of the forest. The spirits convened on the shorecliffs that rose gradually from the trees on one side and dropped into the sea on the other. The cliffs where they chose long ago together to leap and die with their freedom. They fashioned a small boy out of the dirt and blood in the earth and he blinked and came to life. They sent him then to grow up in the forest of the outlaws.

I know this boy. I am a cattle dog here. As the boy grew up and raised animals in a secret clearing, I became his helper. He feeds me scraps and old milk and I listen to the stories of the spirit world the boy mumbles to himself in his sleep. I tuck my ears back, and the boy imagines the drought that plagues his pasture breaking into a steady rain. The boy confides in me. I am happy sometimes, old dog, I am in love. But I have followed him. I know he ventures ever deeper into the countryside at night. As he has tonight—leaving an uneasy design in the grasses.

What I know, and the boy does not know, is the horses are all around us in the shadows beyond the clearing. I can smell them. As sure as I can smell the spirits at the cliffs—they smell like the boy. A creature has materialized here—her skin is made of dirt, her hair a tangle of roots. I cannot smell her, as if she were always here. Always everywhere. As if she were a handful of the world itself. Old dog, says the spellbound boy, this creature says there is a fire burning on the cliffs. A glow up where the boy was made confirms what the creature has told him. I can feel the horses step closer to us. At first I am

afraid. I bristle. But no, they are facing outward. They are a phalanx. They are noticing something neither of us perceive.

The creature hands a small bundle wrapped in old fabric to the boy. They watch each other like a family, their eyes like pearls reflected in a still pool. The glow grows and smoke rises from the cliffs, and the smell of trouble rumbles closer. I do not know what the creature has given the boy, but he holds it and watches the fire. She nudges me toward the boy, turns away, and vanishes.

The pasture glistens. The nightmarish patterns the boy has traced begin to fill with water. The horses stomp and snort. They back closer and are plain to see now. Their muscles press against their skin. The wind blows and the creature does not return. The boy rests his hand on my head and I breathe deeply. We and the horses are standing now in a pool like a shallow dream. Whatever the creature has given the boy, the horses turn each periodically back to him. He is poised and calming with his hand on my upturned coat—even though the elements undoubtedly swell into a menace. For he possesses something mysterious now—and there is peacefulness in the dark calm waters even though there is no peace.

Jaroslav A. Polák | Barbed Wire Study

SOME PASTURES

A Cowboy's Almanac

SAM RYERSON

February, Hay Meadows—Saguache, Colorado

We'd started feeding the Shield-Bar cows on the piled grass hay back in December. We moved them west across the valley toward home, across the highway at Moffat, down the county road to the old Company Ranch. The first pasture was just off the county road where we turned the cows in. There was barbed wire on the north and east boundaries, and a three-strand permanent electric fence across the interior, giving one hundred fifty cows just a little strip of ten acres or so to graze. We moved one strand of a temporary electric fence perpendicular to the boundary fence and a few steps farther south every day, opening a new piece of pasture with new piles of hay—native meadow grasses and rushes and forbs that had been cut and raked late in the summer and left there in the frozen field, waiting for the cows.

Sometimes, making hay, hauling it, and feeding it is a process of moving the nutrients of a pasture away from the pasture itself. George's little hay piles kept the pastures alive by completing a cycle. His cows and fences and hay kept the native grass of the pasture at home and returned its nutrients to its own soil. He was proud of his hay piles and his old tractor he cut them with and the mice that lived in them all fall and the hawks that hunted the mice in the winter. We walked out every morning and moved the fences forward, with the cows following right behind. Then, in the dry springtime, after we moved the heavy cows back to the Home Place to calve, before we turned the irrigation water out of Saguache Creek, we came back with a tractor and drug a harrow across the fields to break up and spread out the old manure: the fertilizing remains of last summer's grass.

I. THE GOAT.

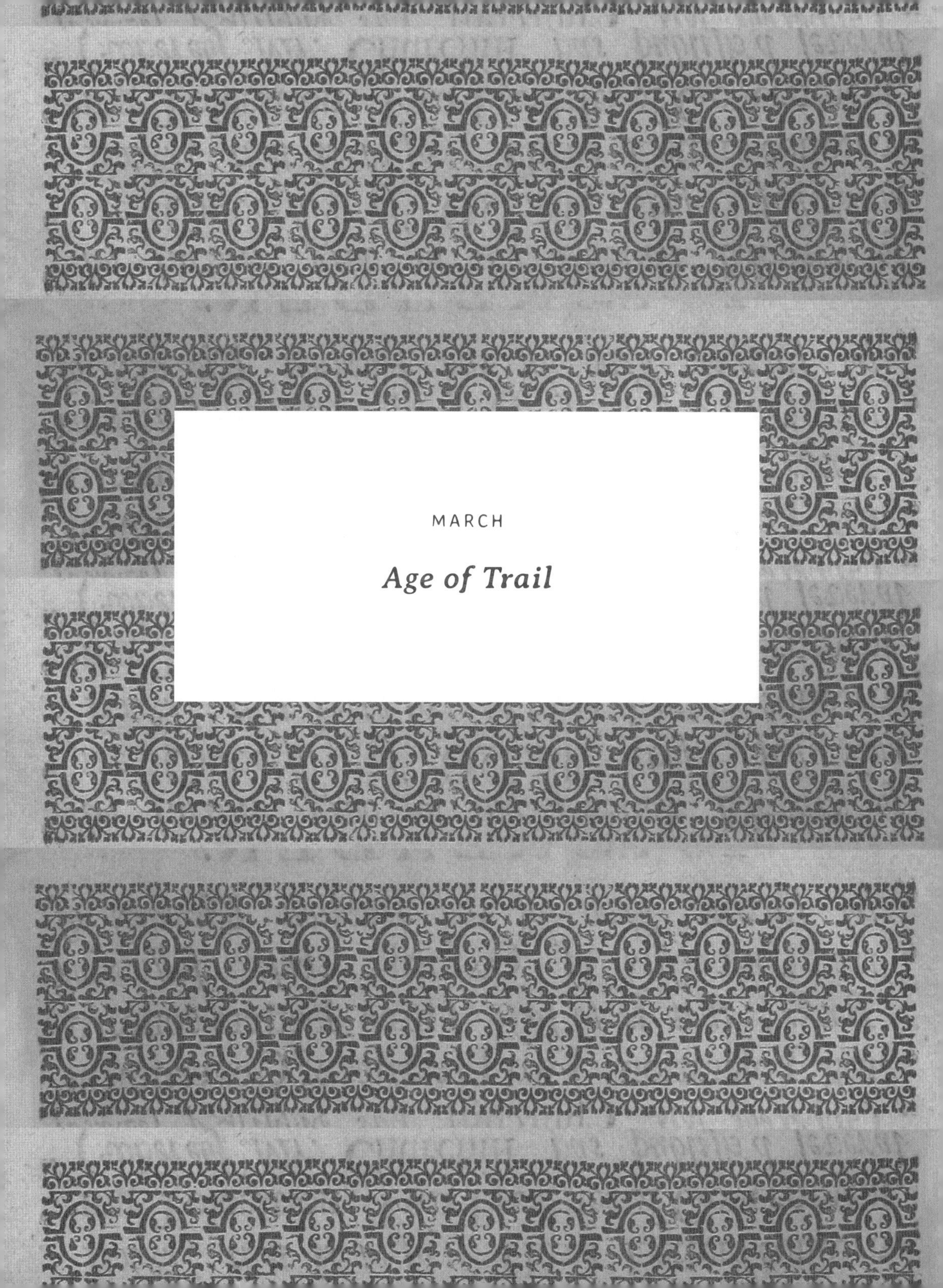

MARCH

Age of Trail

A TOUGH NUT TO CRACK

STEPHANIE MILLS

I sat on the couch in my sun-drenched living room, using a stainless-steel pick to pry the last bits of meat from some hickory nuts. The nuts were a gift from some young habitat restorationists in southeast Michigan, the core of a diverse gathering of anarcho-primitivists, musicians, and radical Christians whose annual winter meet-up I'd been invited to address. Several of them had day jobs in land healing. Many participants were emulating Jesus, shedding possessions and everyday ties, living communally, feeding the hungry, raising their voices, witnessing against a world gone awry, performing songs of love and rage, ferocity and hope. That apocalyptic note, the sense of an ending, resounds in their self-liberation. Most were living at the bohemian margins. A passel of free-range kids was gleefully running riot. The setting was an old farmhouse on some family lands hosting stretches of the region's vanishingly rare but now lovingly tended oak savannas. Along with black oaks and prairie and woodland grasses, shagbark, shellbark, and pignut hickories, with their delicious nuts, are part of that fabric.

There were multiple-band jams late into the night and daylight opportunities for botanizing or learning primitive skills, like constructing a rabbit snare, or hide-fleshing—removing all the muscle and fat from an animal's skin, the process that precedes tanning. That weekend's sunny weather, with temperatures in the mid-thirties to low forties, said the woman teaching fleshing, was ideal, being cold enough to keep odors and insects at bay, but warm enough for one's hands to function.

The dwellers in this land had gathered a goodly quantity of hickory nuts. The afternoon I arrived, I'd heard a backbeat and assumed wood was being split, but it was the sound of someone wailing on a pan of hickory nuts with a hunk of iron. The next day as I was on my way out, Colleen offered me a mug of hot hickory milk. Tawny brown with an unctuous flotsam of nutmeats on the surface, it was rich and delicious, with a wild flavor, tasting of elements deep in woodland soils. Such earth-sprung nourishment and communal generosity has a long American history.

The Quaker botanist William Bartram, in his *Travels in North America*, an account of his late-18th-century explorations in the Southeast, noted that the Creek (Muscogee) peoples esteemed hickory nuts greatly, gathering them by the hundreds of bushels. "They pound them to pieces and then cast them into boiling water, which, after passing through fine strainers, preserves the most oily part of the liquid: this they call by a name which signifies hickory milk; it is as sweet and rich as fresh cream, and is an ingredient in most of their cookery, especially homony and corn cakes."

Back up north in my hickory-less bioregion, after the gathering, I made some hickory milk from the gift of nuts thus: I put them in an old iron skillet, wrapped that with a dishtowel to confine the shell shrapnel, placed another layer of cloth over the nuts, and banged away at them with my dad's old hammer. The shattering took a while, for they are very tough nuts to crack. I put them in a couple of quarts of water to simmer on the stove for an hour or so. Some recipes suggest briefly toasting the busted nuts and simmering them overnight. The shells sink, so there's no need to strain this brew, which I was soon sipping.

The wet residue of shells still contained a fair amount of nutmeat. To discard anything so tasty, so precious, and so imbued with the thought of the young tribe who'd bestowed them would have been wanton. But I had time and plenty of light for winkling out little morsels from the convoluted shells. It was laborious. Long practice, which rural Americans in the habit of nutting had in the old days, might have made the process quicker. Extracting those nutmeats made obvious why we don't find cheap bags of hickory nuts in the supermarket baking aisle.

In *A Natural History of Trees of Eastern and Central North America*, Donald Culross Peattie writes of the shagbark hickory: "The roaring heat of its fires, the tang of its nuts—that wild manna that every autumn it once cast lavishly before the feet—stand for the days of forest abundance." Reading that on an icy cloudless February day, I began to lust for some hickory to burn—they say it's hotter than anthracite coal—rather than the green low-Btu ash and rotten maple that I carelessly purchased for this winter's wood supply; to have the warmth and fat of hickory milk within and the blaze of its stems heating the night, but not to do the work. Bourgeois moi.

I told a friend that I was writing about hickory nuts. She had, some years earlier, gone back to the land in West Virginia and was familiar with hickory. She recalled the nuts being almost more work than they were worth. Talking to another friend, a boatwright, and assuming that the wood must be scarce and as dear as gold leaf, I relayed what I thought must be outrageous gossip about some one-percenter's home hereabouts being lavished with hickory paneling throughout. My wood-savvy friend allowed that while hickory is challenging to work with, it's a common enough tree down South, not some tropical rarity.

The US Forest Service's online database tells us that shagbark hickory is widespread and much loved by many animals. Imagine a pregnant black bear in autumn devouring paw-fulls of those delicious nuts and every other edible calorie within reach to be able to suckle newborn cubs in the winter den. Hickory, noted the Forest Service, naturally recolonized the sites of strip and pit mines in Maryland and West Virginia. Hickory's virtues are many. It's a slow-growing tree that spends much of its youth sending down a deep taproot. It takes about forty years for it to fruit, and can, like many another hardwood under normal circumstances—and where are those to be found anymore on Planet Earth?—live for three hundred years. Durable goods. It could be a perennial crop for the great-great-

great-great grandchildren.

Up here, maple sap is the prime wild tree crop. Our woods don't abound with nut trees, although once chestnuts showered their riches. Still, there are beechnuts and acorns and hazelnuts ripening in the understory. Mast, that periodic spate of nuts, sustains a multitude of beings, from minuscule ants that set up colonies inside the hollowed shells to those bears, deer, squirrels, and jays—no more passenger pigeons, alas—but probably foxes and coyotes, snaffling up the fallen goodies. We too have been forest creatures. Or, more aptly, savanna creatures, living at the edges of woodland and grassland, wanting partial sun and partial shade. One school of thought holds that oak woodlands, not chance gardens of grain, afforded *H. sapiens* the stable, stationary surplus food, fuel, and shelter that eventually led to civilization.

There soon may come a day when the likes of me can't depend on supermarket sustenance. If you acknowledge that we've passed the planetary limits to growth, and that the fossil fuel party's about to be over, it will be adios store-bought, hello grow- or gather-your-own. The foraging life entails constant and varied activity: times of plenty and times of hunger. Yet wild foods nourish the whole being, if but sporadically filling the belly. Before the woods were cleared, mast showered down in wonderful abundance, but nowadays respect and restraint are called for in any wild harvest. Not only can we forage, but we can come to understand, and work to regenerate, the web of relationships that sustain the soils, plants, and animals that are the life of the world.

Jeremy, the habitat restorationist who organized that winter gathering, said the crew working on land stewardship gathered prairie grass seed to sow following their controlled burns of the undergrowth. They decide, by consensus, which of the early successional cherry trees must be cut to allow the native black oaks and white oaks more savanna-style space so they can spread their limbs and bear plenteous acorns to feed their local community of beings.

All across North America, we need to be tending and planting multitudes and multitudes of trees—trees as crops, trees as carbon vaults, trees to harbor biodiversity, trees for beauty and futurity, trees to afford our generosity—a parcel of hickory nuts, a pint of maple syrup, some dry firewood freely shared. Our kind goes a long way back with trees and their providence. We can learn from them still.

Stephanie Mills is the author of Epicurean Simplicity, In Service of the Wild, *and* On Gandhi's Path*, a biography of Robert Swann. She lives in northwest Lower Michigan, foraging in libraries and raising small crops of prose.*

Charlotte X.C. Sullivan and Alayna Rasile Digrindakis | Harvesting milkweed floss at Borderview Research Farm in Alburgh, Vermont, late summer 2017

MAY WEST designs and produces jackets that use milkweed floss as an alternative to feather down. As an insulating material, milkweed is extremely lightweight and naturally water repellent. As a crop, it is low-maintenance to grow in regions where it is native. As a plant, it is essential to the monarch butterfly, which relies on it exclusively to provide food for its caterpillars during its migration from Canada to Mexico.

[SINGALONG]

GINSENG

SHAWN PORTER

Round the mountain on the mossy side,
Where the trees lean 'n the lands a-slide
Underneath the oak and pawpaw tree,
There's a four-prong root just waitin' for me

My name's ginseng, I'm a little shy,
Don't often see the clear blue sky
Favor growin' out in the old growth wood
Little bit of ginseng does a man good

CHORUS:

Ginseng grows on the mountainside,
Ginseng knows just where to hide
Ain't nothing fancy and it don't grow tall,
Hunt for your 'sang in the golden fall.

Yellow root medicine grows nearby,
Cohosh knows where the ginseng lies
Glows like gold in the afternoon,
Dig it in the fall 'neath a wanin' moon

Scatter those seeds all around,
Put a piece of root back in the ground
Come back in another year or three,
Ginseng grows a family

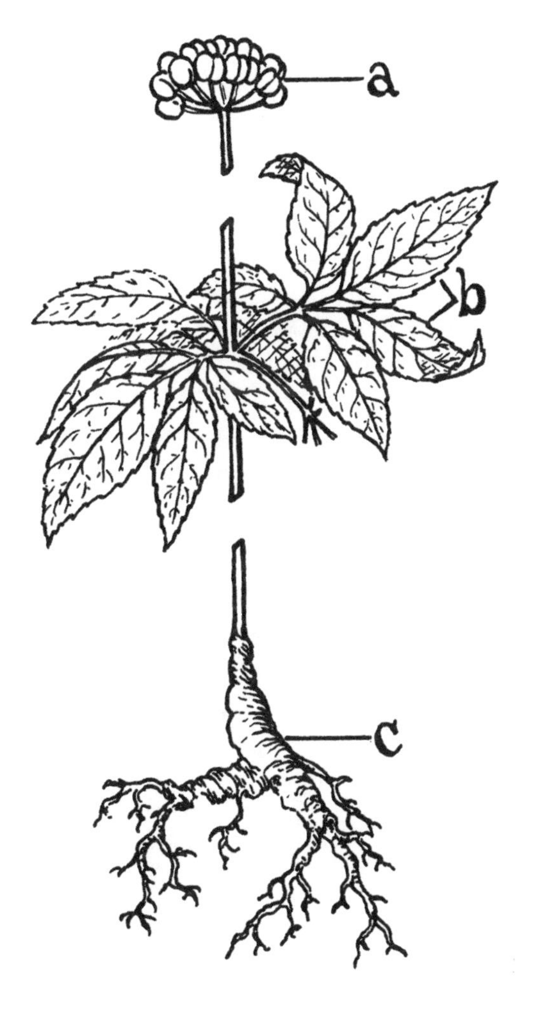

CHORUS:

Ginseng grows on the mountainside,
Ginseng knows just where to hide
Ain't nothing fancy and it don't grow tall,
Hunt for your 'sang in the golden fall.

Down in the holler by the kissin' tree
Ginseng far as the eye can see
Underneath the bluff where shadows fall
I can hear my lover call

Lay her down in a bed of leaves
Trillium trills and the bloodroot bleeds
Hardwood singin' in the mountain breeze
Diggin' that ginseng on my knees

DRYLAND

ZACH WITHERS
SAN ANTONITO, NEW MEXICO

Dryland. A farm, a ranch. An old pickup truck. The crust of malted, brewed barley on a dirty pair of jeans. The Pepto-Bismol pink remnants of blue corn whiskey mash and sorghum straw tickling your belly through a roof-tar stained t-shirt. Six tons of organic matter this week, and it's only Wednesday. Broken shocks and broken wallet and shot springs and dashed expectations. Fighting dragons and internal mental disintegration and sharpening knives and shoveling more mulch. The soft spot behind the big barrow's ear that makes him purr and moan in his half-sleep. The slightly sweet smell of his fermenting bedding, life-giving soil in the making. Lunacy and love and more and more and more organic matter.

It's absolutely insane the things that people throw away.

Pallet city, deep in the valley, where pigs are said to roam in the back of the lot and fourteen dollars will get you one hundred ten gallons of expired milk. Trying to snake a trailer through a side street parking jam under the giant spent grain silo, and the curious bemusement of beer-wielding patio dwellers as cascading flows of protein rich grains rain down, splattering the sidewalk and the back of the borrowed SUV. Trying to figure out what to do with twenty-eight hundred pounds of yellow onions. Tiny blue rubber bands on spring onion bunches and white wire ties on endless bundles of parsley and wax-covered cardboard boxes and plastic wrappers for individual vegetables that like to jump up out of the trash drum and skit away in the wind.

Giant old piñon trees and the sad apple orchard waiting for moisture from the fickle sky. The snake weed and mustard and goatheads crowding out the gramas and kochia and the asters that painted the land yellow and white during the monsoons last year, all standing bone dry on the pastures, abandoned to the drought. For now.

The ephemeral creek that fills the large catch pond my grandma dug, where we caught frogs and sailed on our plank-and-barrel raft from island to island. Invasive olive trees crowd around the culvert and dam, springing back as quickly as you can pull them. Lake Loretta. Dreaming of a school, an education in restoration, resilience, resistance. The outline of ridges and valleys and learning the lay of the land on rambling walks in the dark. Learning to see like water, and following the moisture through the contours of seemingly flat ground until the dips in topography and the mud on your boots can tell you where the oak trees should be planted. Someday, maybe, we will be able to stay home and let the pigs forage for a feast of acorns and piñon nuts, and reward them back in their pens with a dessert of pomace from homegrown cider apples. And in the cold, harsh winter months, when the snow covers the ground, retreat into the warm shed, where the grains from the fall ferment into tasty mash that will keep the critters warm, and the still drips spirits that will keep us warm.

But for now we scavenge. More hunter-gatherer than farmer. Scouring the industrial sectors for food waste, knowing that no source lasts forever. The tortilla dough that the pigs loved so much, gone. Poisoned and turned to toxic waste to protect the corporate bottom line. The local grocery store politics and flimsy plastic drum liners full to the brim of still good produce that usually ends up in the landfill, unless Carlos happens to be working that day. The crisis created by the food bank stopping the supply of rotten vegetables and stale bread to local farmers, instead contracting with someone who, rumor has it, pays for the favor. More rumors that billionaire brewery owners are cutting off the grain silo, prefering to direct the protein stream towards chumming elk on their private hunting estate.

The sinister cloud of urban sprawl spreading its darkness on the landscape, encompassing the small oasis of open space and sanity we've called home for four generations. The looming threat of the local tyrants finally dealing a fatal blow to our small family operation. Chopping the land into two-acre parcels, sucking the aquifer dry. The wells are slowly dying, the rain did not come for five months last fall. Out-of-state license plates on shiny new sports cars and obtuse suburban construction assault the landscape. Water speculators, corrupt politicians, and the mentality of perpetual growth promise a legacy of degradation, desertification, and despair. Realizing our opposition is folly. We do not care. We may lose but we have not lost, not until we surrender, not until the dryland takes us.

Rachel Alexandrou | Compost

TOTAL ECLIPSE

FRANCIS MCGILL

The trail to the summit quieted.
Small camouflage frogs frittered
among loose bark and fern throngs.
We heard the churning creek ahead,
and with time before the eclipse,
chose to scale the crags upstream
where late summer moss outfitted
the shambled-boulder rocks.
Water dropped fast. The terrain rose.
We didn't see the end but had to go
for the moon was coming round
to shield a weary sun.
By your foot on the narrow path
during the last scramble and dash
up the top, a still bird lay, belly up,
its muted, brownish feathers still intact,
a piece of shriveled dead leaf
still locked in its beak. We stopped.
You wrapped her body in birch bark.
I kicked over dirt.

Once in a lifetime the moon buries the sun.

SOME PASTURES

A Cowboy's Almanac

SAM RYERSON

March, La Alfalfa—San Luis, Argentina

One winter I went to Argentina instead of feeding hay in Montana. I worked for my room and board on a few cow outfits around San Luis. It was late summer in the Pampas. The land was mostly flat, with sandy soil, and scattered rolling hills where the native grass and low savanna trees remained. More old pastures were being plowed up every year. On the big farm fields they were harvesting sunflowers. In the pastures the steers were getting fat and ready to slaughter.

Nico, my patrón, was shipping a little bobtail truckload to the slaughterhouse just about every week. Most of the fat steers were raised on one of his ranches. They fattened in mixed alfalfa pastures—twenty- to fifty-acre paddocks surrounded by high-tensile electric fence. We worked early in the mornings until about noon, and then again in the evenings. In the heat of the day we sat in the shade and had maté or took a nap. It was a good pattern. Every day we rode out to move the steers to a fresh strip. Some days we moved a temporary fence or two. Some days one of the gauchos fed corn in a long portable trough. Every day we poured an additive in the water tank to prevent the cattle from bloating. The green alfalfa was dangerous and the steers might have bloated and died if they ate too much too fast. We ran our horses and chased the steers when we moved them to new pastures. The gauchos said it was good for the cattle. They said running helped to break up the bloat.

We loped our horses everywhere we went. We rode their recado saddles of sheepskin and rawhide. When we gathered fat steers to ship, we ran them from the alfalfa to the corrals. The cattle were used to traveling that way. When we sorted them in the pens, Nico came out on his good chestnut mare with his polo saddle and sorted and called the big ones, and the gauchos sent them through the gate at a trot. At dawn we loaded the steers on the truck, up the alley and the ramp, mostly at a run. The men were quiet, just working their horses with their rawhide quirts, and the cattle were quiet, with their heads down, but it was fast work. On the weekends Nico and his friends played polo and I was a groom for his horses. Late at night on the weekends we had asados, of the best beef I ever ate, from fat steers just off these pastures.

A rider hones his skills, 1939

LIVE BURLS

PHOTOGRAPHS BY KIRK CRIPPENS AND GRETCHEN LEMAISTRE

Today less than five percent of the northern hemisphere's old-growth redwood forest survives, most living in the Redwood National and State Parks. Yet even the protected trees of the parks are subject to threat. Burl wood grain is prized on the black market for its exquisite rarity. Poachers sometimes harvest the burls for profit. In 2013, an alarming number of trees in the Redwood National and State Parks were targeted by poachers with chainsaws. One team of thieves was bold enough to fell an entire tree for its burl. Sadly, they jettisoned the five-hundred-pound burl by the side of the road when it proved too heavy for their vehicle. Brett Silver, a supervising park ranger, compared the act to hacking up a church, echoing Theodore Roosevelt, who in 1890 reflected that a grove of giant redwoods is like "a great or beautiful cathedral." Because their burls contain stem cells that enable the redwoods to clone and nourish themselves, wounded trees are more vulnerable to disease. Their ability to reproduce is endangered, and the entire grove ecosystem is compromised. Decades may pass before the full extent of the damage can be assessed.

Ladybird *drew us to the project of documenting the natural beauty and human disfigurement of these trees. Poachers cut around the nine-foot-tall burl's circumference, acting with no regard for the tree's survival. When a media team covered the story, Kirk was struck by it.*

Semper Virens *borders private land. Once harvested for lumber, it was in a regenerative stage when poachers cut out the entire base and facade, leaving behind what resembles the markings of a quarry.*

Swamp Lantern *stands in the center of an old-growth forest, where it has lived for more than a thousand years.*

We returned over many seasons to capture these haunting images. You can find the complete series in Live Burls, *published by Schilt Publishing, 2017 (ISBN 9789053308813).*

Ladybird

Semper Virens

Swamp Lantern

ZOMBIE AGRICULTURE

The Reanimation of Dead Soils and Chemical Drift

KERI BRANDT OFF AND JANINE FITZGERALD

We live in a storied landscape. Keri and Janine's stories are rooted in landscapes which also hold the stories of all who have come before.

Janine lives on four hundred acres, east of Durango, Colorado, in the HD Mountains. As kids, she and her twin sister rode horses, ran from rattlesnakes, and weeded corn with clouds of gnats buzzing around their ears. Her son and daughter remember building a house on that land and have their own stories of gnats and bears. The old timers told Janine's family how the HD cattle company overgrazed the land in 1882, and went broke by 1896, leaving only the mysterious name of the HD brand to the mountains.

Keri grew up in Boulder, Colorado with horses who taught her of the meaningful worlds humans and animals craft together. She now lives with her husband, David, on the Off Family Ranch in the San Luis Valley, where multiple stories of one family reach as far back as one hundred fifty years. While horse is still her best language, she is learning cow and grappling with creating stories with animals who will become her food.

Most of us live on storied lands. Yet much of our food comes from lands without stories. We characterize zombie agriculture as food grown on lands without story. And, as the Aboriginal people of Australia say, a land without stories is a dead land.

Our culture is surrounded by the iconic image of the zombie. The lore of zombies in Haiti is rooted in the real life horror of slavery, and zombies are thought to be the reanimated corpses of those who died a violent, unnatural death. The modern zombie that peppers much of pop culture today is infected by a virus of unknown origin. Haitian zombies work endlessly doing their master's bidding and the modern zombie is mindlessly hungry. Like the zombies of *The Walking Dead*, a similar sort of eternal hunger drives high-yield industrial agriculture. With an emphasis on efficiency and productivity, the high-yield practices of zombie agriculture require a nearly endless cycle of reanimation through chemicals to keep the land in constant production. The outcome of zombie agriculture is ecological degradation of epic proportions that includes poisoned food, polluted water, dead soils, and the brutal treatment of animals.

Under the banner of feeding the world, food in the U.S. is produced with genetically

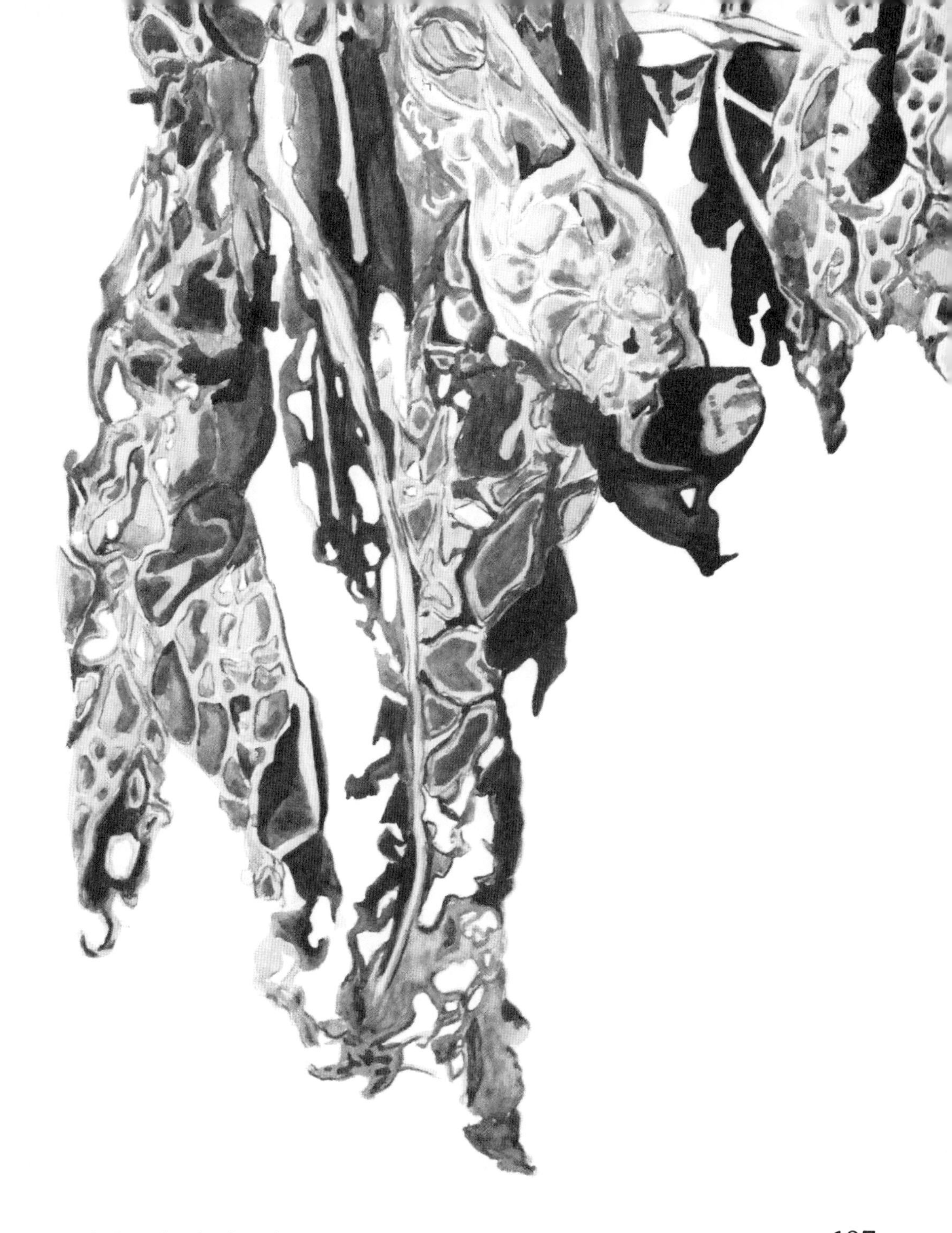

Rachel Alexandrou | Kale in decay

engineered seeds that can withstand a constant barrage of chemicals until harvest. Herbicides like Monsanto's dicamba drift like a virus, infecting neighboring fields and killing plants that have not been genetically modified to withstand their influence. In 2017, dicamba drift is known to have killed or damaged at least 3.6 million acres of cropland, devastating many farmers and their land.

These practices kill our soils and weaken our plants, and because of this, we are in a constant process of reanimation which enlivens that which is already dead. Dead soil, like makeup on a corpse, is reanimated with synthetic inputs, fertilizers, pesticides, and herbicides that construct the appearance of life. Zombie agriculture also creates conditions where animals are housed in torturous, near-death conditions. Most pork products begin with sows in tiny crates, where they will live their entire short lives birthing and nursing multiple litters a year. We reanimate animals by decorating our packages with pictures of happy cows and happy pigs living in an idealistic life in harmony with humans.

The American farmer is also injured under zombie agriculture. Farmers live within a context of crippling debt, public distrust, and increasing rates of death by suicide. In 2016, the Centers for Disease Control and Prevention reported that farmers, along with those in fishing and forestry, had the highest rates of death by suicide among all other occupations in the United States. Yet advertisements for food are filled with pictures of farmers and ranchers as happy and in touch with "nature."

Most consumers today live far removed from the realities and stories of food production. The process of reanimation disappears all the hardship of living. We argue that that it is only by reclaiming the story of agriculture as a cycle of life and death—which includes the trials and tribulations of our lives and the lives of plants and animals—that we can reclaim a radical center, a place where we can stand and see our diverse entanglements.

HUNT LOCALLY

CHRISTIE GREEN

Christie Green

Two men wearing day packs, shorts, plaid shirts, and boots make their way down the earthen footpath, navigating over the lumpy roots of Douglas fir, ponderosa, and aspen trees. I hear their casual chatter though they're too far away to discern actual words. As I spot them a couple of switchbacks up, I take the shot. Their words and feet come to a halt when the arrow makes contact.

"What just happened?" one man asks, as Olivia and I rush up to retrieve the bird.

"Uh, well ... I shot this grouse," I reply, as casually as possible. Olivia and I are wiping tears from our eyes. We gingerly hold the still warm grouse, trying to move up the slope unnoticed. The two men glance at us, quickly taking in the sight of us with the bird. Impassive, their facial expressions reveal nothing, and no further questions or comments are exchanged—just a sort of hurried, polite, "Enjoy the rest of your day," as they continue their descent. But toting a bow and arrow on a hiking trail in what's practically the backyard to thousands in Santa Fe is noticeable, a noticeable threat or danger to some. "Am I going to be OK? I mean, are there other hunters up there? Is it OK to walk my dog now?" the lady in the parking lot asks later when I return to my Prius, bow and arrows visibly in hand. I chose this trail off the Santa Fe Ski Basin because of its accessibility and relative proximity to town—all the wrong qualities for avoiding people; all the right qualities when fitting in a quick hunt between work and picking up my daughter, Olivia, from school.

On this day, my forty-fourth birthday, I've chosen to take Olivia out of school to be outside hunting, hiking, and picnicking. We're both completely surprised when, not ten minutes up the trail, among multiple hikers ascending and descending, Olivia says, "Mom, look!" There she was, a lone grouse nibbling on bearberry leaves near a fallen fir log, twenty-five yards from the next switchback. Once I'm sure the nearest hikers are well out of harm's way, I pull the broad-head arrow from my quiver, place it in the bow, pull, aim, and release. Whoosh. Instant. The bird disappears, falling behind the log. The rarity of this shot, an unlikely fortuitous combination of visible bird, clear line of sight, and people safely far enough away, sinks in. Most of my successful shots occur deep enough in the wilds where no humans may be frightened or hurt.

Racing up the switchback, I'm in disbelief. Did that really happen? I wonder to myself. We crouch down, touching her warm soft body and feathers. In the moments between life and death, the animal often thrashes uncontrollably until death's weight takes over. I have the urge to protect Olivia and the nearby hikers from what could be shocking, but the grouse lies still, motionless. The arrow had penetrated her body easily at the base of the neck. No blood. No guts. A clean, swift kill.

Quickly, quietly, Olivia and I climb the slope, grouse softly in hand. I wrap the gray-brown downy bird in an extra shirt I had brought along and place her on the other belongings in my backpack. Up the slope through the gilded aspens we climb. Olivia stops here and there, perching on curved aspen trunks, gazing through the binoculars to spot more wildlife, all the while chattering as much as the resident squirrels, busily hoarding sustenance in anticipation of winter.

Atop the ridge, off the trail, we choose a large lichen-covered boulder upon which to eat our picnic lunch. There are sandwiches of store-bought lunch meat, chips, chocolate, fancy juices, and snack nuts—all decadent treats in celebration of my birthday, our day together and playing hooky from our weekday obligations. We share more about the grouse and what felt like automatic spotting and shooting. Pure reflex, and without precedent. This was my first kill with a bow, an unlikely one given the context and the small size of the animal. All seems magical, as if the bird and the day are gifts, just for us.

Mid-afternoon, we descend the ridge to head toward our next form of celebration: the luxurious Japanese-style hot baths a few miles toward town. Upon arriving at the spa, it seems disrespectful, casual, even negligent to leave the grouse in the car. I bring her with us in the backpack. Some hunters, like those in Europe, allow the bird to "rest," guts intact, at room temperature for up to ten days, allowing the meat to cure and become its most flavorful. I choose to clean and process the birds the same day they're shot, and then refrigerate them in salted water. But first, I let their bodies cool.

We enter the contrasting world of the highly designed luxury hot baths, custom-fabricated

Japanese woodcraft and stonework, facials, foot rubs, and massages. With the hike and the hunt still at our fingertips, the fancy architecture, warm robes, and exclusive spa room feel grossly superfluous and gluttonous. All I want is to go back up the hill.

We stay at the spa, although neither of us truly enjoy it. Later, we try a well-known Italian restaurant downtown. The multiple adjectives describing local, artisan, handcrafted fare seem over-the-top. "We can make our own food, Mom. It's way better than this. Why does it have to be so fancy? Why do people pay for that? Where does it really come from? Do they know or care?"

Olivia's questions are the same ones I keep asking myself. I'm supposed to feel happy going out to eat. I'm supposed to feel pampered and alive in a spa, relaxing indulgently. I'm supposed to feel the acclaimed chef has connected me to place through his purchase of local food from a local farmer. Yet I wonder about others' access to the luxury of spas or artisanal food, and how those who live closest to the land, whether out of economic necessity or family tradition, feel about how "local" has evolved to a highbrow experience.

I was raised in Alaska and worked on my grandfather's wheat farm and cattle ranch in west Texas during the summers. I know a little bit about being a fledgling local farmer, but have never relied on agriculture to make a living. I grow as many edibles as I can on my two-acre parcel in Santa Fe County. Anyone who knows this area knows the 7,000-foot elevation, combined with highly alkaline soil and extreme weather patterns, is a challenge to growing food on a consistent enough basis to feed a family. It's damn hard work, even if you don't have a full-time job.

To complement my vegetable and fruit cultivation, I decided to try hunting. It's been in my family for generations. All of the men hunted, never the women. I worked a couple of seasons in a hunting lodge in Alaska where I learned to butcher, pack, process, and cook wild game, always for the male hunters who rarely had anything to do with the animals once they were down and dead. I fleshed the hides, scraped brains from skulls, and pulled grizzly bear claws from heavy furred paws. I toiled through the night to get just the right amount of meat, fat, and flesh off the hide in order for the taxidermist to take over, making pretty for the wall what once was wild.

I wanted to hunt to see if I was able, yes, but I also I wanted to be self-sufficient with my food in as many ways as possible, from meat to greens. Why not? New Mexico has plenty of access to public lands and a rich diversity of wildlife species to hunt.

Before my first hunt was over—for cow elk—I knew I was hooked. Over the nine years since then, I have hunted turkey, deer, grouse, pheasant, oryx, and quail. Elk and turkey are by far my favorite animals to hunt. Spring turkey hunting, during the mating season when gobblers and hens separate, coming to each other's calls to mate, is the most exciting and rewarding, though trickier in many ways than during the fall. Spring birds, hopped up on hormones, responding to each other's calls from just before dawn, are hot to mate. The gobblers may gobble back to a hunter's call and may even come within range but, as all turkey hunters know, "It's never a sure thing." The majority of opportunities to shoot are foiled either when the gobbler or hens sense the hunter's presence or when a hen swoops in to lead the gobbler away. Hunting for fall turkeys, when mating calls have ceased, hens are with poults, and gobblers and young jakes group up, is harder due to the lack of calling and hormone frenzy, but one fall I decided to commit to harvesting my own Thanksgiving bird.

After hunting turkeys in six game management units for three and a half hours per day, four days a week, over three months that fall, it took me four minutes to walk into the grocery store one morning and buy an eighteen-pound, tidily wrapped turkey that looked more like an outer space creature in a shiny plastic jumpsuit than anything once living.

To say I was fine offering my Thanksgiving guests this store-bought bird would be a lie. I apologized profusely to Olivia after I returned from the final four hours hunting. Her sweet response, "It's OK, Mom," softened the blow.

Those who plant, cultivate, and harvest their own food, raise their own animals to eat, or hunt for their meat understand the amount of effort—and love—that goes into sustaining ourselves. Speaking for myself, the more I do it, the more I can't not do it. Not because I love to kill or because I'm a masochist sentencing myself to hard work, but because my commitment to cultivating and connecting to life has become an addiction for me. My food tastes better when I know where it came from. I have something of real value to offer family and friends.

I love the habitats in which I hunt—the cougar scat, the elk tracks deep in mud at watering holes, and the squeaky Townsend's solitaire wheel. The turkey hen yelp and excited gobble response,

Olivia Green

the frisky squirrel darting up one tree and down the next, and the musky deep sounds of autumn rut. I fall more in love with the mesas, ponderosas, mixed conifer, and vast desert scrub the more I crawl on my hands and knees, sweat, strain, sit, and glass the horizon and wait—and wait and wait. When I get the "Aha!" gift of a fresh track in the soil, the elegant curve of a six-by-six regal above meadow grass, a distant call, scat, or scratch signs, I'm thrilled by my proximity to the ground and the animals. To sit still, breathing steam into dawn's gray air, and have twelve elk walk so close that I feel their vibration beneath my butt on duff is to appreciate their lives, our food, this place.

If I am fortunate enough to harvest an animal—small or large, hoofed or feathered—its bones, blood, guts, and hide tell me about its life, how it ate, where it lived, and how it will feed me and my family. The intimacy with life and death, with the distinct eco-regions of this state, strengthen my bond with New Mexico as my home.

So the day I came off the hill in my mud-packed pickup, adorned in layers of camo, hair up and shotgun safely sealed in its case, I felt ridiculous waltzing into the grocery store to purchase a turkey. I wondered if the better choice would have been to serve my guests the oryx, grouse, elk, or deer I had successfully harvested myself. Would that have been a more true Thanksgiving?

Once back at the house after our spa and dinner out, Olivia and I have our hands on the now stiffened bird. I feel at home. I push my thumbs into its sternum, gently opening the flesh beneath the feathers. We smell her distinct scent from her open cavity. Grouse smells different from turkey, which is different from quail or pheasant, and especially different from domestic chicken. Sometimes I think I can smell and taste what the animal has eaten that very day, and whether or not it was rainy and wet or dry, whether seeds or leaves have been transformed to make the bird's flesh.

Olivia watches. We cry. I pull the skin back and begin slicing off the fresh pale pink muscle from bone, marveling at its sweet perfection. The skin, feathers, tail, and tiny-scaled feet rest in Olivia's hands. I wonder if Olivia will ever want to hunt, if she'll reflect on our time together hunting as positive or one of those things from her childhood she'll be ashamed of and try to hide as she grows up.

We continue processing the bird, placing the meat into a bowl of water, deliberately arranging the other external parts like an altar around the cutting board. Later that week, we plan what to cook to complement and honor the grouse: sweet roasted carrots from the garden and wild rice from a recent visit to my dad's in Minnesota. Maybe some wilted chard from the garden. And wine. Yes, wine.

A friend joins us for the celebratory meal. We admire the contrasting colors and textures on the plate, the perfectly cooked grouse, whose flesh is now a light whitish pink, juices dripping into the accompaniments. Sitting down at the table, napkins folded on our laps, I start to say a blessing when Olivia remarks, "Look how beautiful, Mom." So many other hunts, successful or not, converge in this moment. So many times I have traversed hills and canyons, calling, watching, trying for the game that will grace the plate. This time, I'm most thankful for Olivia, who was with me on the hill as the arrow flew and who's with me now, savoring the bird and the beauty.

RETURN OF THE BEEKEEPER

PETER KIRN

The young woman returned to the sagging blue house from which she'd been exiled to taste the honey left by her absconded bees. She marched out among the past lives in the snowmelt—the tractor implements and grave markers revealed in the receding sea of ice. Below, the blueberries were a tangle of brown thread in the ash and rotting pine needles. She came to the machine shed. The blue truck was still there and almost invisible. It coughed twice and was revived. She rode out into the night field with a long knife sealed in a thermos of hot water, a large clean bucket, and the void hung around her neck. She threw it in neutral to pass the farm owner's house and rolled down the tractor road flanked by the bogs until she stood there before the hives, the wind collecting in her ears, a timeless algorithm telling of her life as a tree. She was swallowed by her own idea of loneliness—saw nothing but light. The story lingered in the bog. The cranberries were to report it to the bees that came. And come they did. By the end of summer they and the season had carried in a mixture of truth and bitterness, filled it with pectin, and dressed it in a hypnotic shade of red. Everyone noticed. Creatures of all kinds. And within them the shadows of the fruit persisted and told them what they'd seen.

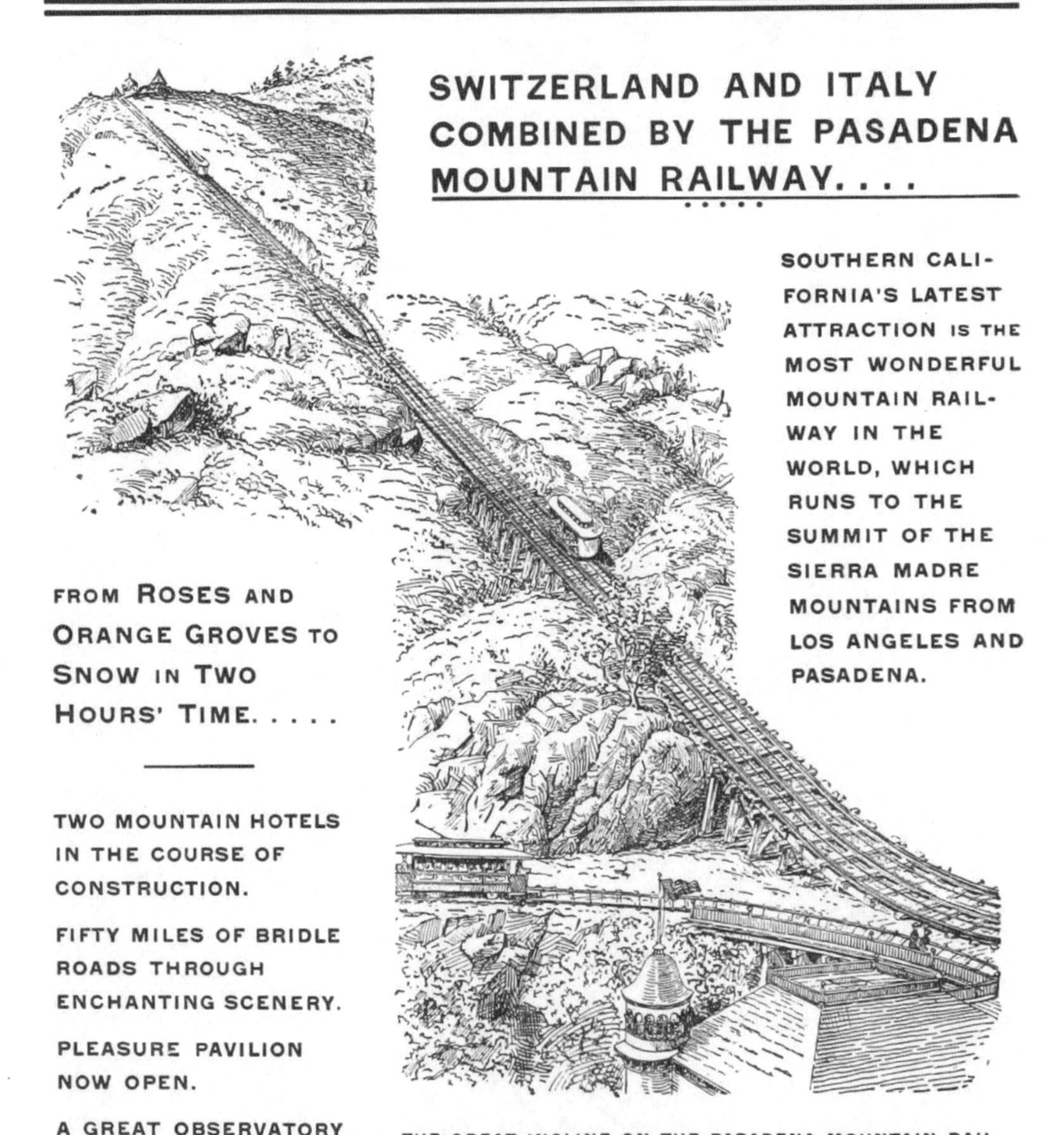

THE GREAT INCLINE ON THE PASADENA MOUNTAIN RAILWAY, WITH GLIMPSES OF PLEASURE PAVILION AT ITS FOOT AND ECHO MOUNTAIN HOUSE AT TOP.

APRIL

Age of Rail

"AND ON MY NAUTILUS SAIL AWAY."

IF ONLY

M.T. SAMUEL

If only all children could run through fireweed like these:
purple petals dazzling their crowns
as they tromp down expired railroad spines
in summertime.

They are explorers, botanists, royalty, sprites.

They hum and buzz in the chilly morning
while we sip black coffee from pottery
on the playground.

These starlings sleep through earthquakes,
yawning with heavy lids as we rouse them
to meet under the tree.

For them there is magic in the soil
and wonder in the waxing moon.

Little midnight sun-dwellers, cottonwood-ramblers,
make your applesauce and your strawberry jam,
gather eggs from your chickens and catch
raindrops on your tongue.

SOME PASTURES

A Cowboy's Almanac

SAM RYERSON

April, Marathon Gas Fields—Cody, Wyoming

That spring after I came home from Argentina, I had a cow camp job lined up to start in May, and Laura Jean was in college for another month, so in April I went to work helping Brandon run two thousand or so wether goats in Wyoming. It was BLM land, all gas wells and sagebrush, cheatgrass, knapweed, and kochia. There hadn't been many cows there in years. Lani and Brandon were working out a contract to reclaim the well pads and roads and restore the range. Lani ran her goats on ranches and in city parks and along roadsides, eating weeds all over the West. They were not a production herd but just weed-eaters, and a lot of them were getting old. When the goats got old, Lani let them die on the range where they lived. She said it was better for an old goat that worked all his life to die under a cedar tree up on a ridge than be hauled to town and slaughtered. She didn't wait for a contract from the BLM either but sent all the goats one day from Lander. We didn't know how many there were—we just jumped them off the trucks—but there must have been close to two thousand.

It was mostly big dry open range with a few barbed-wire fences, but the fences didn't matter. The goats went right through them. We built temporary pastures of a few acres or less, with portable electric net fence and solar-powered energizers to hold the goats overnight or concentrate them where we wanted more animal impact—hoof action to open the hard soil, and manure to fertilize it. We loose-herded the goats with our border collies most of the day, Jesse or I did, while the other moved the fence. The cheatgrass was just greening up, and we tried to force the goats to graze it and prevent its growth. It was mostly cold and windy, with light dry snow sometimes.

I had to herd the goats to water once or twice a day. That was the best part of those days. I could walk in front by the lead and just send a dog to the back and she would bring all the goats right along. A dog gets confident working like that. The creek ran down through a little alkali flat. The goats didn't trust the chalky mud along the edge, but they went down eventually and got a drink. While they drank I would sit on a flat rock above and wait with my dogs. We spent a lot of time waiting. Lani never got her contract, and the BLM told her to remove the goats. I went back to Montana and picked up my horses from their pasture and drove west to Lima.

Gabriella Marks

QUERIDO JACINTO YAÑEZ

JASON BENTON

I once read B. Traven's short novel, The White Rose, *recommended by Ed Abbey in one of his essays. The novel is a fictional account of the oil drilling business Traven witnessed in Mexico during the 1920s. In that story, Jacinto Yañez is the ancestral leader of an indigenous farming hacienda in Mexico named the White Rose. As the story goes, Jacinto was murdered by an American oil company thug after refusing to sign his farming community, a place sustained by his ancestors for hundreds of years, over to oil drilling. The company then forged the dead Jacinto's signature to obtain the deed to his land, even though Jacinto did not read or write. The story was originally published in 1929 and later made into a movie in Spanish,* Rosa Blanca, *in 1961. No English version has been made. It was withheld from American audiences and English readers until 1979. But as it turned out, Americans were no longer listening. I happened to read it exactly three weeks after an Exxon land man knocked on our door in 2013, and, feeling moved at the uncanniness of it all, wrote this poem.*

They started drilling just last week. My house is shaking. My neighbor is wringing her hands with worry and leaves her TV blaring to drown out the traffic and noise. The chickens stopped laying for a week, but picked up again. The goats play in the snow-covered meadow all night long under the bright light of the rig tower. We plan our garden for the next year and do our best to ignore the racket and drone, and not worry about our air, our water, our future.

Our daughter overheard us talking over adult things, as kids are so apt to do, and heard about our thoughts of possibly moving if things got unbearable. It was horrifying to her. "Will we take our animals and flowers with us? What about our trees?"

There is life yet in this place, the place in between today and tomorrow, between the substance of this wondrous earth and sun and moon and that invisible vernal power pulling up the stalks of corn toward the heavens; between this death and tomorrow's life. There is life. So we plant, we live, we dream.

Querido Jacinto Yañez,

I was dead when I signed
School, Mortgage, rights, Debt and Debt,
It has all been planned and insured
Feeding the Monster
To the death—to my death

This land is an illusion,
My body is not me.
Reality lingers in the vernal warm life
Burgeoning from the place in between—
That is where
I   will   be.

Property and possession,
Mere illusions of control: calculated greed
Delusions of power and money
These ancient sexy myths
Of superiority and all is well.
Of all is well.
All is well.

A place—a real place is what I seek
A flowering of the in-between
A flowering of the in-between
That place which flourishes between
The Earth and me.

Does such a place remain on Earth?
Is there a place left to be—just be?
To know and to see,
To live and die and live and die
And die and live freely?

Where the morning sun and dew greet me
The birds call and plants seduce me,
Call and mingle and commune
And create.
Commune and create.

Dear Don Yañez,
Dear Don Yañez,
Where do we go?
Fighting or fleeing is futile—
Yet another revolution?
No, I can see the end there:
All will be spent.
All is spent.

Oh these boxes, these
Cubicles of anguish!
Places without a dream
Life without a wish,
With myopic dreams obscene—
Zombies indeed.

Bacon, Locke, Newton, Einstein,
Useful ideas they lent
But like a child with a gun
A child with a gun
A child with a gun
We destroy ourselves again ...

O, Don Yañez!
Today human and tomorrow more
When will we learn
That what is in store
Lies not on the moon or Mars
But just outside our door?

Dear Don Yañez, Dear Don Yañez,
Where do we go?
To find a place wild and free
To live simply and simply live.
Pray tell, who knows?
Was there ever, and
Will there ever be
Such a place
As the White Rose?

DEATH, DEBT, AND ZOMBIE AG

JANINE FITZGERALD

VAMPIRE BAT.

Farmers and ranchers live close to the process of death. I have written odes to my milk cow, Clemencia, whom I butchered because her large udder was making it difficult for her to walk. I gave the meat to friends, but it still makes me sad to think of her—it was Clemencia who taught me how to milk a cow. She trusted me and I butchered her.

Ironically, the proximity to death means we live close to debt, as I am indebted to Clemencia. This sense of indebtedness is distinct from the morass of financial debt which opened the way for zombie agriculture, an outcome that has led to the near-extinction of the family farm and ecological catastrophe.

Anthropologist David Graeber has shown how it is part of the human experience to understand the life and death cycle as debt. This understanding of debt puts us in debt: to each other, to our ancestors, to our soils, our water, our food, and our gods. The payment of these debts are played out in rituals and spiritual traditions. We never expect to pay off these debts, as our relationship with the world around us is never-ending. The 20th century shifted the focus of our debt from "all our kin" to the "Ag Credit Union," distorting a fundamental part of what it means to be human.

The class and race dynamics of early agriculture in the United States were extremely problematic. The federal government stole land from Native Americans and gave it to poor Europeans fleeing persecution and famine. African Americans were trapped in a system of slavery or sharecropping, and Mexican Americans were dispossessed by the Treaty of Guadalupe Hidalgo. While these actions are unforgivable, the community dynamics of mutual aid should not be discounted. Farmers and rural peoples of all groupings were suspicious of outside debt and relied on an ethic of mutual aid through grange halls, barn raisings, and communal efforts.

The shift from community-oriented agriculture to agriculture built on financial debt came about through three historical events. The first was the fossil fuel revolution of the early 1900s that invented machinery and inorganic fertilizers, allowing farmers to produce more with less labor. The second was the ecological disaster of the Dust Bowl in the 1930s that created a huge outmigration of farmers to the cities and left rural communities abandoned and devastated in the United States. The third historical event leading to zombie agriculture was mass starvation in Europe during and following World War II. US policy through the Hoover Initiative responded to this famine by calling for farmers to follow their Christian duty and help feed the starving people of Europe, a global narrative that is enshrined in Monsanto's campaign to "Feed the World."

The 1970s and early 1980s can be seen as the last stand against zombie agriculture by farmers in the U.S. The growth of global agriculture created a commodity price bubble that burst following the Carter Administration's grain embargo to the Soviet Union in 1977. Farmers and ranchers who had borrowed heavily during the boom were left upside down on their mortgages. They resisted "death by debt" by organizing disruptions of farm foreclosure auctions and a massive tractor march on Washington D.C., where farmers brought their tractors from as far away as Oregon and Texas. But agricultural policy sided with zombie ag and let financial debt destroy the family farm.

Today, only about two percent of the US population is in agriculture. The consequence of these events is that acreage size of US farms has increased, while the number of total US farms decreases. According to Census Data, the U.S. lost eight thousand farms in 2015 alone.

Financial debt continues to be a driving factor in favor of zombie agriculture. Farmers and ranchers find they cannot enter in or stay in agriculture without borrowing heavily, while massive production keeps commodity prices low. Rates of death by suicide continue to be disproportionately high among farmers and ranchers, a fact directly related to financial debt. Even as farmers and ranchers struggle to make these debt payments, the deeper debts to the land, water, animals, and one another are ignored and forgotten. And yet, despite an uphill battle, a renaissance in agriculture is placing our debts to one another and our earth in the center.

TEN FACTS ABOUT COOPERATIVE ENTERPRISE

Benefits and Impacts of Cooperatives

JESSICA GORDON-NEMBHARD

1. Cooperative enterprises address market failure: they provide rural electricity or other utilities in sparsely populated areas and increase access to affordable healthy and organic foods, credit and banking services, affordable housing, quality affordable child or elder care, and markets for culturally sensitive goods and arts.

2. Cooperatives overcome historic barriers to development in the ways they aggregate people, resources, and capital. Of the one hundred sixty-two nonagricultural cooperatives that participated in one study, forty-four percent of respondents said they could not have opened their business had it not been organized as a cooperative.

3. The economic activity of the 30,000 cooperatives in the U.S. contributes an estimated $154 billion to the nation's total income. These co-ops have helped to create over 2.1 million jobs, with an impact on wages and salaries of almost $75 billion.

4. Cooperative businesses have lower failure rates than traditional corporations and small businesses, both after the first year (ten percent failure versus 60–80 percent) and after five years in business (ninety percent still operating versus 3–5 percent of traditional businesses). Evidence also shows that cooperatives both successfully address the effects of crises and survive crises better.

5. Since most cooperatives are owned and controlled by local residents, they are more likely to promote community growth than investor-oriented firms. Since cooperative business objectives are needs-oriented, cooperatives are more likely to stay in the community.

6. Cooperative businesses stabilize communities because they are community-based business anchors; they distribute, recycle, and multiply local expertise and capital within the communities where they are based. They enable their owners to generate

income and jobs, accumulate assets, provide affordable quality goods and services, and develop human and social capital.

7. Co-ops and their members pay taxes, and are good citizens by giving donations to their communities, paying their employees fairly, and using sustainable practices.

8. Cooperative start-up costs can be low because they are eligible to apply for loans and grants from a number of federal and state agencies designed to support co-op development, and they are often provided relatively low-cost loans from nongovernmental financial institutions like cooperative banks that are chartered or established to make such loans.

9. WAGES (now Prospera) in Oakland, California, finds that owning a house cleaning co-op led Latina worker-owners' median income to increase from $24,000 to over $40,000. Ownership in the co-op has put their income higher than the national average of $38,000 for Latinos and Latinas.

10. Food co-ops spend more revenues locally, buy more products locally, buy more organic produce, recycle more plastic, and create more jobs than conventional grocers. For every $1,000 spent at a food co-op, $1,606 goes to the local economy; for every $1 million in sales, 9.3 jobs are created.

An excerpt from "Benefits and Impacts of Cooperatives," prepared by Jessica Gordon-Nembhard (John Jay College), with Charlotte Otabor (Howard University), for the Center on Race and Wealth, Howard University. For references and more information, see raceandwealth.coas.howard.edu/reports&publications/0213-benefits-and-impacts-of-cooperatives.pdf

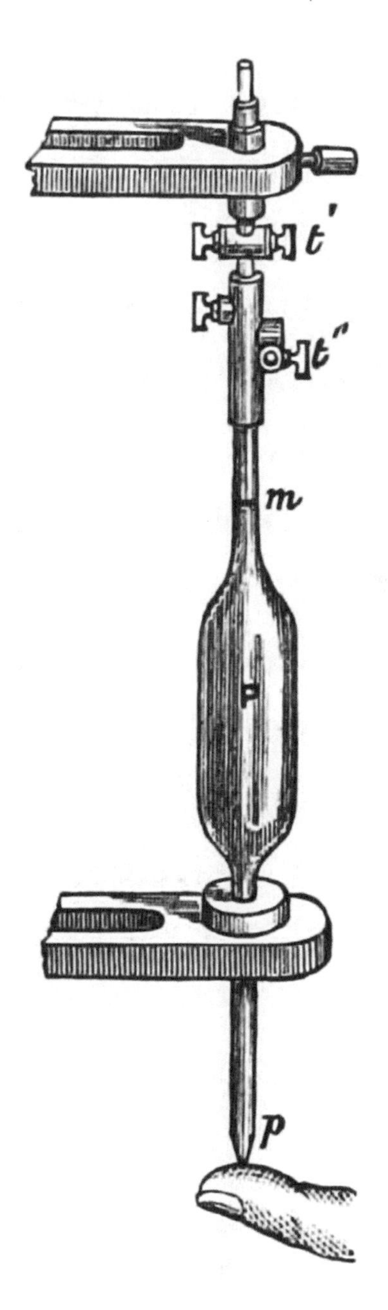

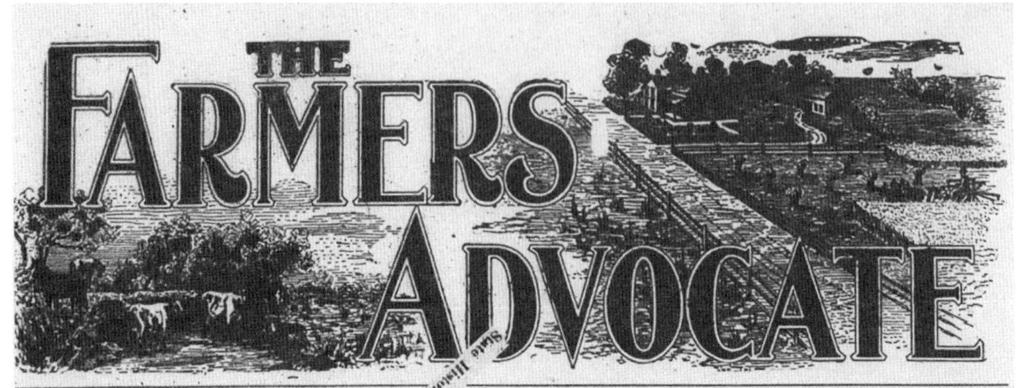

THE FARMERS ADVOCATE

Volume 28. No. 2 TO...KA, KANSAS. July 27, 1905.

Educate, Agitate, Concentrate, Co-operate.

Don't overlook any bets.
Eternal vigilance is the price of liberty.

We have been so busy producing wheat, beef, and pork, wool and other human necessities, that we have overlooked their distribution to society, which has become perfectly frenzied, until the laborer hardly ever, if ever, gets the full product of his toil (but he sees the amount he has produced.)

Now, Comrade, all you want is what you produce; then you must own the machines, and you want every other person to have as much; so let's quit philosophizing, quit chewing the rag.

While we have been doing this, the other fellow has got control of the ownership of the machines of production and distribution, he has concentrated, has co-operated, until there is scarcely anything that you produce or consume but what has to pass through his hands and that he does not put the price upon.

But, don't BLAME HIM. He is simply doing what you should have been doing collectively long ago. So here is your Uncle Johney with the organization in so far as woolens are concerned, who is agitating (and who hopes is educating).

Please drop him a card requesting plans and you will soon be free from the system. THE TIME TO DO THIS IS NOW. Don't wait for all to agree with you how to own the machines. If this is your kind of medicine write for prescription. It's sent free and calls for VOLUNTARY CO-OPERATION ONLY.

RIO GRANDE WOOLEN MILLS CO., (Co-operative)
ALBUQUERQUE, NEW MEXICO.
JOHNEY H. BEARRUP, President.

STILL LIFE

CARL SANDBURG

Cool your heels on the rail of an observation car.
Let the engineer open her up for ninety miles an hour.
Take in the prairie right and left, rolling land and new hay crops,
 swaths of new hay laid in the sun.
A gray village flecks by and the horses hitched in front of the
 post-office never blink an eye.
A barnyard and fifteen Holstein cows, dabs of white on a black
 wall map, never blink an eye.
A signalman in a tower, the outpost of Kansas City, keeps his
 place at a window with the serenity of a bronze statue on a
 dark night when lovers pass whispering.

LOOKING FOR SIGNS OF THE CHILI LINE

WILLY CARLETON

If you found yourself on a steam-powered locomotive on a hot summer day as it crawled through the canyons and mesalands along the eight-hour journey from Española to Alamosa, you'd definitely want to open the window. You'd want to take care, however, that as you looked out the open window at the expansive blue sky for signs of distant rain clouds, the nearby clouds of dust didn't cause you to sneeze hard enough to send your expensive dentures flying out into the endless plains of sagebrush below. That, of course, would be an unqualified disaster. Unless, that is, the train stopped and the entire load of passengers got out to help you look for the fallen false teeth.

Luckily for Mr. Albee, the engineer who lost his teeth through the train window somewhere between Taos Junction and Embudo in the early decades of the 20th century, he was conducting just that sort of unhurried and congenial train. Or so goes the nearly forgotten story, told by an eyewitness to the Santa Fe New Mexican in 1965, several decades after the incident. It was just one of many locally famous episodes in the history of a narrow-gauge branch of the Denver and Rio

Grande Western Railroad, known as the Chili Line, which ran from 1880 to 1941 from Alamosa, Colorado, to Española, New Mexico, a small town that was founded in 1598 and served as New Mexico's first capital. The train hauled the eponymous peppers along with other produce, such as potatoes and onions, livestock, mail, and countless passengers. It was a lifeline for many and connected the predominately Hispanic and Native communities throughout the agricultural valleys of northern New Mexico and the fertile basin of the San Luis Valley in Colorado. Shortly after it closed in 1941—by many accounts a sad day in northern New Mexico—nearly all its track was pulled up and sold as scrap metal to Japan. Just months later, the Arizona sank into the fiery waters of Pearl Harbor.

Today, there is no longer a rail line through the agricultural valleys between Alamosa and Española. Since the train's heyday over a century ago, the northern parts of New Mexico produce a much smaller percentage of the state's agricultural products. The axis of traffic no longer tilts north from Española to agricultural markets surrounding Alamosa, but more south to the labs of Los Alamos and the tourist hub of Santa Fe. In 2006, when New Mexico built the Rail Runner commuter train from Belen to Santa Fe, the tracks stopped about twenty miles short of Española. Every weekday morning a long line of traffic leaves from the former capital of New Mexico as commuters crowd the roads to get to work. There are several small bus routes throughout this neck of the high desert, but most who are able, it seems, choose to drive themselves.

The public land where the Chili Line once ran is now home to a stretch of scenic blacktop that unfurls through the remote western landscape like a backdrop for a car commercial. It is a vision of independence and individual freedom amid boundless acres of solitude. It is one of my favorite roads.

I was on it not long ago, cruising north to buy some seed potatoes from a farmer near Alamosa. I was meditating on the passing piñon trees and the occasional cow, and kept an eye out for any signs of the old tracks. Occasionally, I thought I could make out the soft linear mounds of an old rail grade, rising subtly from the landscape like the body of a large submerged snake pushing the earth up for brief stretches before receding again deep below. Somewhere between Ojo Caliente and Tres Piedras, I spotted such a rise in the ground and parked off to the side of the road to stretch my legs. As my dog sniffed around for the perfect spot to relieve himself, I idly searched beneath the piñons for some sign of the former railroad—just a short piece of track, or a splinter of a rail tie, or a rail spike, or a lost denture. I found nothing but a few recently discarded cans of Bud Light. I got back into my truck, and together with my pup, returned to the solitary path north.

Frank Utpatel | After the Threshing
Used with the permission of Stanton & Lee Publishers, Inc.

OUTSIDE OF WENDOVER

FRANCIS MCGILL

Pure delight: the moon lit your face.
I was sleepless most nights

Thinking about that Utah canyon,
The evening breeze, and satellite.

Surveillance: our naked bodies
Under the Pleiades skies.

The coyotes mating, howling.
Red dawn began disappearing,

Curve of the Earth
Spun round the horizon

And you observed with mirth
As the salt flats toppled

The salt mountains.
We walked barefoot

The day of the races
Breaking the sound barrier

(Silently, a monarch fluttered.
We remember precisely.)

Outside of Wendover.

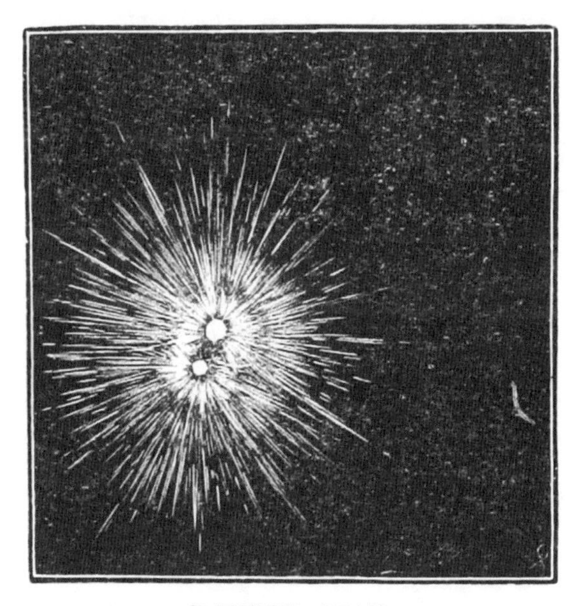

DOUBLE STAR.

A RAINBOW RIVER

SAVANNAH J. VOLKOFF WITH CLAUDIA K. GUNSCH
DEPARTMENT OF CIVIL AND ENVIRONMENTAL ENGINEERING, DUKE UNIVERSITY

On a hot July morning in Virginia, I found myself suiting up in waders and a safety hard hat. This trip to one of our field sites was already different than any other. The summer sun had been blaring down on the riverbank for a few hours, beckoning to the surface a sour concoction of aged creosote.

In the early 1600s, the Elizabeth River—a tributary of the Chesapeake Bay in Virginia—was one of the first sights for many British colonizers, including Captain John Smith. Named in honor of King James I's eldest daughter, the Elizabeth River was an emblem of opportunity. Over the many years of settlement and industrialization, the Elizabeth River became one of the largest shipping channels in the United States, hosting many ports, a naval base, and manufacturers of various shippable commodities.

In the 20th century, three creosote manufacturing companies operated along the banks of the Elizabeth River. Creosote, a product formed through the chemical transformation of various tars and plant-derived materials, was primarily used as an effective wood preservative to coat telephone poles and railroad ties up until the late 1900s. Over time, creosote manufacturing byproducts were released into the Elizabeth River and, because of their persistence, these compounds rest in the sediments to this day. The amount of creosote pollution found at these sites is some of the greatest in the world.

On the day of our visit to collect sediment from the Elizabeth River, the long-ago dumped creosote was making its presence known. I had never seen anything like it. It truly was a rainbow river. There was a huge sheen floating on the surface, hosting pinks, oranges, and purples. When disturbed, this floating cloud morphed into crisp metallic colors. The sediment under the water—my primary research interest—shone similarly. Using my gloved hands, I carefully loaded this smelly gold dirt into my collection tube for later analysis. Some of my colleagues also collected fish from the Elizabeth River.

The fish would be returned to our laboratories at Duke University, where our research team analyzes the effects of chemical exposure on the fish's ability to function effectively. In the sediment, we would be looking for the smallest forms of life: microorganisms, capable of surviving in the presence of heavy pollution. Microorganisms, such as bacteria and fungi, have adapted incredible metabolic systems and thrive in some of the harshest environments on earth, including hot springs, deep ocean trenches, and the tundra. In the context of my project, I wanted to find organisms that could consume creosote and break it down into nontoxic by-products.

Using microorganisms to remove chemical pollution from the environment is a treatment approach known as bioremediation. Bioremediation serves as a promising noninvasive alternative to typical sediment remediation strategies, which rely on dredging the river bottom and sending the sediment to a landfill or incinerator. In the case of the Elizabeth River, polluted sediment is relocated to an unsettled island—saving the problem for another day, in another place. However, if efficient microbes can be identified, a bioremediation strategy could be implemented and all the sediment could stay just where it is, preventing further damage to the local ecosystem.

The Elizabeth River is already home to a diverse community of microorganisms—bacteria and fungi that are working hard to maintain a functional ecosystem. As we better understand the dynamic nature of these microbial communities, we can learn how to help them consume creosote with greater efficiency and at a faster pace.

Thankfully, the Elizabeth doesn't always host such diverse colors and pungent smells. In the dead of winter, she seems just like any other river—cold, calm, and quiet. My hope is that my research will help rehabilitate the Elizabeth River and prevent it from ever being a rainbow river in the future.

Savannah Volkoff | Aged creosote in sediment and surface water at former creosoting facility in Elizabeth River, Virginia

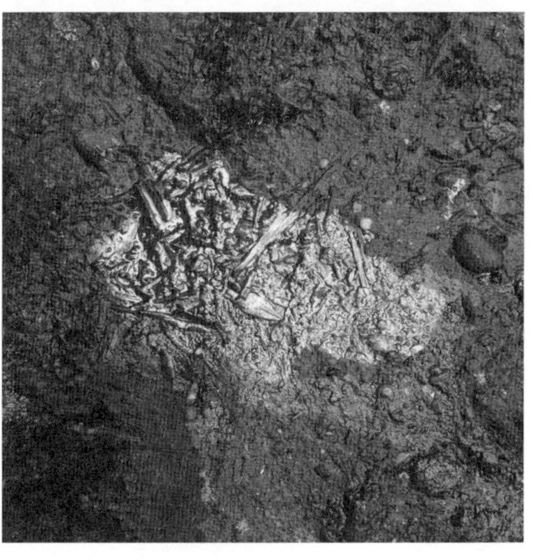

Spore print

MYCELIUM RUNNING

SHAWN PORTER

From the spore
mycelium
grows with every
dew and strain
burrowing
entangled forms
nets of silver
gilt and chain

Dining on
both waste
and ferment
sipping on
some last decay
tracing shadow
paths it follows
fuses into
living skein

When the web
can hold no more
the life in death
at last proclaims
fruit of darkness
revelation
mushrooms
in the pouring rain

THE BALLOON NEVER HAD A CHANCE.

MAY

*Fail*Failure*Failing*

ETHICAL EATING
and the
DENIAL OF DEATH

KERI BRANDT OFF

Nearly a decade ago, armed with my right/wrong thinking about how humans ought to be in relationship with animals, I went on a date with a rancher named David. At the time, I ate mostly vegetarian and was conflicted over the question of eating animals. As I drove to meet him for the first time, I wondered, how will this work?

David and I fell in love, and we both went through a challenging process of learning to trust each other and loosening the grip of our tightly held convictions. Like many American consumers, I had developed strong beliefs around "good food" and "bad food" in response to the horrors produced by zombie agriculture. I believed that if I stuck with eating Tofurky, I could get around the necessity of death in order to eat, and distance myself from a food system that hurts and denigrates humans and animals alike. My "Tofurky" worldview, however, could not integrate the realities of food production that I found on the ranch. I had to move towards a deeper understanding of death and food.

Only two percent of the population in the United States is in agriculture today. Most of us do not live close to food production, nor will we witness the death of animals whose bodies become our food. While death is hidden, the harsh realities of conventional agriculture haunt us: sows in gestational crates, chickens whose beaks are severed, cattle knee-deep in manure and mud, and genetically modified corn that kills monarch butterflies. All of these images drive us toward a desperate search to eat food that is not so wrong, or so evil. We therefore frame eating around rigid moral choices, but even these choices are rooted in visions of right and wrong that deny death. Marketing campaigns capitalize on our angst by constructing narratives about eating and food that offer idealized imagery and romantic taglines of agriculture. Pictures of serene farmers hang above the meat aisles and happy cows and pigs decorate meat packages. This narrative, whether intentional or not, constructs a veil between consumers and producers that weakens trust between us and distorts the realities of zombie agriculture.

Third-party certifiers have filled the space of broken trust, offering consumers affirmation that the meat they're buying is from animals that really were grassfed, treated humanely, raised free range, local, non-GMO, and so on. Ranchers and farmers—on top of planting, calving, vaccinating, harrowing, irrigating, hoeing, harvesting, butchering, and selling—have to keep detailed records and pay outside certifiers to verify that they are honest in making such claims. One rancher told me he has to jump through endless hoops just to prove that he is kind to his animals and is not poisoning people. This certification and marketing process brings the ethics of right/wrong eating to the forefront while denying the basic fact of life that humans kill in order to eat. The decentering of death makes it difficult for any of us to have the meaningful conversations necessary to support and heal problems in agriculture.

The process of eating is paradoxical and an emotional struggle that none of us can escape. The desire to eat "right" is entangled with deeply conflicted feelings over life, death, and eating. After a decade on the ranch, I have an up-close, personal relationship with food that I could have never fathomed. I know, intimately, that animals must die in order to become food. I live close to life and death. I witness birth and first breaths, and I witness death and the decomposition of bodies. I both eat and care for the animals who will become my food. I do not make it through the year without being heartbroken by the slaughter of a favorite cow, and I revel in the joy of checking at first light to find newborn calves. Holding this tension of opposites together reminds us that life and death are central experiences to all living things. The dismantling of zombie agriculture will take more than ethical debates—it will require us all to dialogue about death, the impulse to live, and our deep entanglements with the whole of the world.

The character for "heart-mind" as an enso

 Briana Olson | Little Yosemite Burn

NOTES FROM THE FIELD

Living in Fire Country

MARIE HOFF

We live in fire country. The Californian ecosystem has evolved over millions of years with regular wildfire. According to *Oaks of California*, we should expect major fires every thirty to fifty years in our grasslands (most of our communities and ranching lands are on grassland and chaparral).[1] And yet, this fire season—the season of 2017—was particularly intense, with very quick-moving and very destructive fires across many counties.[2] Was there something different about this year? Or is this what we should expect as Californians? Is this going to happen once every fifty years, or more frequently?

On October 8, 2017, I woke at midnight, just as I was drifting off to sleep. Suddenly, the hillside right across the valley was ablaze. It was a dramatic sight. We didn't pause but rather chose to load up our sheep and get them out of potential danger. We weren't sure if we should take them to my parents, an hour and a half's drive south to Santa Rosa, or to my future in-laws in Red Bluff, three hours' drive northeast. We finally chose Red Bluff, as my in-laws have a bigger backyard. After about two hours on the road, I got a call from my mother: they were being evacuated for fire. It came as a surprise to both of us at the time that we were both escaping a fire, she in Sonoma County, me in Mendocino, she from a suburban area, me from a rural one.

We were both very lucky. We did not lose our homes, nor did I lose any livestock. We were surrounded by people who did, though, and I felt in turns both helpless and empowered to support my friends and neighbors. Over the following days, stretching into weeks, I diverted much of my time to checking in with my fellow ranchers. Many livestock producers opened their homes and pastures to other producers under evacuation. Many people with trailers offered help with moving animals out of danger. Any shift in the winds threw everyone into uncertainty. People began developing backup plans for their backup plans, and some people evacuated their evacuation sites. The communities rallied, aided, waited, watched, prayed, and hoped.

We Californians have developed a sort of geographical amnesia in our modern times. With eighty percent of the population living in cities, few of us are as intimate and familiar with our landscape as with the vantage point of a strip mall. We live predominantly in a grassland ecology, a diverse ecosystem that has adapted with fire for millions of years. Grazing has been a part of that system for as long as that—a version of the llama was indigenous to the US plains going back forty-five million years. Yet I wonder how many Californians are even aware that we live in historic grazing terrain, much less of the relationship between fires and grasslands.

In a grassland, every year there are cycles. In Northern California, we are mainly in a Mediterranean mesothermal climate, meaning we have wet winters and dry summers, with relatively mild temperatures overall year-round. Thus, plants begin to grow during the winter, with rapid growth in the late spring, and taper off during the summer. By late summer and early fall, a lot of plants have grown, dried out, and lay brittle, awaiting some wind to ignite a fire. Unless, that is, they are grazed.

Grazing management is thus of ultimate and basic importance to our state, to our environment, to our homes and businesses, to our lives and families. One of the ways we can reconnect with and support good grazing management in California is to return our material needs to those that come from our local landscape. Locally raised wool is of vital importance to this effect. Not only does local wool support our flocks that steward our lands, but it also supports our local land managers—ranchers and shepherds—in their work to care for the land. Grazing animals are the link that keep us in harmony with our homes and the ecosystem we live in. I am honored to call this place my home.

NOTES

1. Bruce M. Pavlik, Pamela C. Muick, Sharon G. Johnson, and Marjorie Popper, *Oaks of California* (Sacramento: Cachuma Press, 2002).
2. For Mendocino County, the fire season of 2018 would be even more intense: we again had to evacuate, this time for the Mendocino Complex Fire, which burned more than 450,000 acres in northern California.

PRACTICAL NOTES FROM FIRE

CONNOR JONES

Some clear observations from the Thomas Fire:

- Homes and structures made with earth, cob, and adobe don't burn. Only the exposed timbers or flammable roof materials do. Changes could be made to make them entirely fireproof.
- Fire moves slowly when going downhill, even in chaparral. Wind can accelerate this, but for the most part valleys are quite safe places to be.
- Ridges are most dangerous for settlement. For every ten-degree increase in slope, the fire speed can increase by double while ascending a hill. The updraft and radiation front prepares the vegetation above for combustion before the flames even reach it.
- Oaks and most trees (even eucalyptus) with reasonable thirty- to one-hundred-foot spacing don't seem to fully burn as long as ladder fuel is removed.
- Wetting structures and landscaping immediately before the fire approaches can be successful. Timing is important: too soon and it will dry out.
- Irrigated orchards of avocado, olive, and citrus proved to be blatantly effective fire breaks. Certainly other species would qualify as well.
- Mowed dry grass moves somewhat fast but at very low intensity, often only burning the surface and leaving unburned material underneath.
- Water storages such as ponds, dams, pools, and even tanks can be very helpful for defending home and landscape if paired with a pump capable of producing a high flow rate, such as a fifty-foot stream of water out of a fire hose.
- Having clear and wide access around structures for fire-engine access and as a landscape fire break is tremendously helpful.
- Off-grid power potential to operate water pumps, light sources, and emergency communications is crucial. The fire crew might not know your house is there if the lights are out.
- High quality respirators and home air filters are a must.
- Extra storage of gasoline or diesel for vehicles and tools such as chainsaws are crucial. Store them away from inhabited structures. Also, remove propane tanks from around your house if possible.
- Have a week's supply of food and water on hand.
- Wear long-sleeved thick cotton or wool clothing if you end up close to flames. Get it wet.

I'm sure there's more, but that's what I've gathered so far.

NOTES

The Thomas Fire burned 281,293 acres in Santa Barbara and Ventura Counties from early December 2017 to January 12, 2018, and was at that time the largest fire in California's history—a distinction now held by the 2018 Mendocino Complex Fire.

DIRECTIONS *for preventing* CALAMITIES *by* FIRE

Boston, Massachusetts, 1799

The following was inserted in the Old Farmer's Almanack at the request of the Massachusetts Charitable Fire Society.

1. *Keep your chimneys and stove-pipes clean, by sweeping them at least once a month.*

2. *Never remove hot ashes in a wooden vessel of any kind, and look well to your ash-hole.*

3. *After sweeping a hearth, see that the brush does not retain any particles of fire, before you hang it up in its usual place.*

4. *Oblige all your servants to go to bed before you, every night, and inspect all your fire-places, before you retire to rest.—For fear of accidents, let a bucket of water be left in your kitchen every night. The writer of these directions once saved his house from being consumed by fire by this precaution.*

5. *Do not permit a servant to carry a candle to his bed-room, if he sleeps in an unplastered garret.*

6. *Cover up your fire carefully every night in ashes. Let the unburnt parts of the billets of chunks of wood, be placed next to the hearth, but not set upright in the corners, by which means no sparks will be emitted from the wood. Pour a little water upon the burning ends of the wood which are not completely covered by the ashes. Place before the fire a fender made of sheet iron. This contrivance was well known in England many years ago, by the name of Coverfeu. It has lately received (from a top being added to it) the name of Hood.*

7. *Remove papers and linen from near the fire to a remote part of the room.*

8. *Shut the doors of all the rooms in which you leave fire at night. By thus excluding the supply of fresh air, you will prevent a flame from being kindled, should a coal or spark fall upon the floor, or upon any of the combustible matter in the room. The smoke which issues from this smothered fire, will find its way into every part of the house, and by waking the family, may save it from destruction.*

9. *If sickness or any other cause should oblige you to leave a candle burning all night, place it in such a situation as to be out of the way of rats. A house was once destroyed by a rat running away with a lighted candle for the sake of the tallow, and conveying it into a hole filled with rags and inflammable matter.*

10. *Never read in bed by candle light, especially if your bed be surrounded by curtains.*

11. *Strictly forbid the use of segars in your family at all times, but especially after night. May not the greater frequency of fire in the United States than in former years, be ascribed in part to the more general use of segars by careless servants and children?—There is a good reason to believe a house was lately set on fire by a half consumed segar, which a woman suddenly threw away to prevent being detected in the unhealthy and offensive practice of smoaking.*

In case of fire attend to the following directions, to prevent or restrain its terrible consequences

12. *Do not open the room or closet door where you suspect the fire to be, until you have secured your family, and your most valuable effects, nor until you have collected a quantity of water to throw on the fire, the moment a fresh supply of air excites it to a flame. Where water cannot conveniently be had, try to smother the fire by throwing two or three blankets over it. A British sea captain once save a king's ship by throwing himself with a spread blanket in his arms, upon a fire which had broke our near the powder room. He was pensioned for life, for this wise and meritorious act.*

13. *In case it be impossible to escape by a staircase from a house on fire, shut the door of your bed chamber, and wait until help can be brought to secure your escape from a window.*

14. *If safety does not appear probable in this way, wrap yourselves up in a blanket, hold your breath, and rush through the flames. If water be at hand, first wet the blanket.*

15. *To prevent fire descending from the roof, or ascending from the first story, form by means of blankets or carpets, a kind of dam on each of the intermediate stories, near their staircare, that shall confine the water that is thrown upon the roof, or into the windows. It will effectually check the progress of the fire downwards or upwards in brick or stone houses.*

16. *To prevent fire spreading to adjoining houses, cover them with wet blankets or carpets, or old sails.*

17. *To extinguish fire in a chimney, shut the door and windows of the room. Throwing a quart or more of common salt into the fire. Hold, or nail a wet blanket before the fire place. If these means fail, throw a wet blanket down the chimney from the roof of the house.*

Reprinted from The Old Farmer and His Almanack (1904).

Fire-Engine at Work

(From a Broadside of 1760)

SOME PASTURES

A Cowboy's Almanac

SAM RYERSON

May, Sand Pasture—Moffat, Colorado

We put the Shield-Bar cows back over on the old Baca Ranch one spring and ran them through the growing season there. It was a federal wildlife refuge, but George and I put together a permit to run a prescribed grazing program on the upland pastures there. Before that the refuge had only run cattle during the dormant season, cleaning up around the edges of the hay meadows. To offset the grazing fee for using the refuge forage, the cows would perform ecological services by grazing down overgrown grasses and forbs, opening new habitat for ground-nesting birds.

We used the existing barbed-wire fences where we could and tied on a single strand of temporary electric wire and strung it along the rabbitbrush or greasewood to make smaller paddocks within. We tended the cows the best we could, and tried to follow a cowboy's rhythm of the season. We branded the calves with some neighbors and friends—they brought horses and held the cows and calves outside the old corrals, then roped the calves and drug them to the fire. We brought the bulls and turned them in with the cows for the breeding season.

But we really pushed the cows there, like we had pushed the goats in Wyoming. The water from the artesian wells stank of sulfur, the old grass was too weak to nourish the lactating cows and their calves, the better Sand Pasture where the grass was good and stout was full of locoweed, and some days near the big ditches on the wet meadows the mosquitoes made us all miserable. Eventually it wasn't worth pushing the cows around like mowing machines, and we let them back into the better sand country. We gave them bigger pastures with Indian ricegrass, and plenty to eat without grazing the loco, and space to roam away from the mosquitoes, and good water from Crestone Creek. Then they were happy. Cows can't be productive and raise calves and gain weight and breed back if their few basic needs aren't met; if the pastures we provide them don't furnish these, they might not be complete pastures, nor have happy cowboys to tend them.

Project Work.

"Building fence for pasture land."

FARMING WITH QUOTES

SAM HEDGES

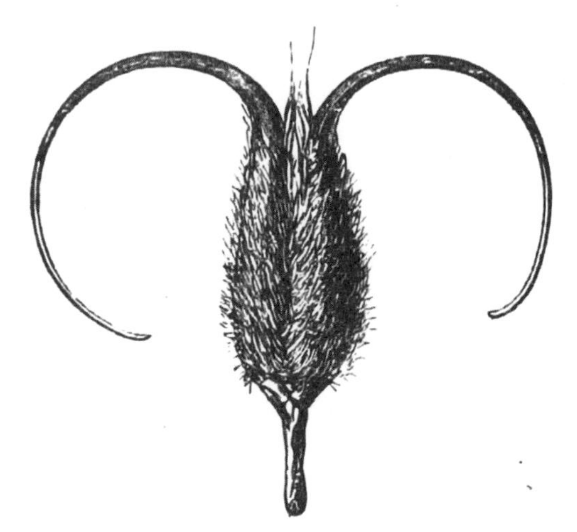

Devil's Horns.

I follow a mix of farmers on social media. With time, a two-tone distinction has become obvious. The inspirational farmers' fields always look perfect—densely packed, soulful Edens overflowing with food. These people smile through their seventy-hour work weeks; they are engines for elevating manual labor into a higher form. I know I'll never be like them, but they remind me of why I started doing this in the first place.

Bright-eyed and innocent, I encountered my first gritty farmer just as I started farming, in a *New York Times* article titled "Don't Let Your Children Grow Up to Be Farmers." I knew I was in trouble before I clicked on the link. The farmers' message was all about the hard financial realities of farming. It was but the tip of an iceberg of honest voices out there, trying to inject some reality into the popular image of farming as Zen urban escapism. Occasionally, they overdo it and themselves forget the why of what they do. In a movement so heavily laced with moral positivity, it's easy to assume everything's fine. I was an intern making six hundred dollars a month without a thought as to why farmers couldn't afford to pay their staff members a living wage.

Five seasons later, I'm still lost somewhere between the two polar caps. Since I started, I've worn my healthy back like a badge of pride, with the occasional free YouTube yoga video for maintenance. Then, on an unassuming February Tuesday, just a month after turning thirty-one, I felt a rubber band stretch ever tighter across my back until, fearing the sound of a pop, I lowered myself onto the ground and lay there. For four days, I hovered in self-imposed horizontal arrest and contemplated a famous line of poetry that had lain balled up in some corner of my brain for years like a stranger's shirt that might someday fit. "This is the way the world ends / Not with a bang but a whimper." Suddenly, T. S. Eliot was relevant to a farmer. The terror in a back going out doesn't lie wholly in the discomfort but in the prospect of a new beginning. The owner of the farm I manage developed his chronic back issues my first year, and they return every season with increasing regularity. I talked to my parents, my in-laws, everyone I knew with recurring back problems, and each could pinpoint that first event that triggered it. I painted a self-portrait of lifelong disability. My prospects as a farmer, potentially ruined in a long, low whimper.

Farmers, I believe, learn melodrama quickly. It's something about working so closely to life's essentials; we are perpetually coasting between rhapsody and trauma. The better farmers I know take it in stride, but the highly sensitive child within me can't create a good buffer. My back is improving in small leaps between long stillnesses. It'd be easy to use a bad back as my excuse to get out of commercial farming, but I feel in some egotistical way that my failure would condemn the whole enterprise. As long as I keep on, I can't say that small-scale organic farming is doomed, but my giving up farming would be a partial admission. Since moving to Albuquerque, I've watched a host of farmers quit shortly after hitting their ten-year mark, while waves of newer, younger farmers step in, wondering where all the older farmers have gone.

I got into farming because of another article I read, titled "On the Phenomenon of Bullshit Jobs," by David Graeber. "Jobs have had to be created that are, effectively, pointless.... The moral and spiritual damage that comes from this situation is profound." At twenty-six, I was already tired of every indoor job I'd ever had and only one frivolous email away from denouncing the modern workplace completely, and farming delivered as expected in this case. No farmer I know ever spins her or his wheels; in regard to questions of value measurement, our minds are spotless. Yet, as the years amass, a sick reversal has occurred, and now the desk is my fantasy. Oh, what would it be, to labor in a workplace designed for people? The office equivalent of a farm field is coming into work every day to discover that a different business moved in overnight. Maybe in the olden days it was easier to stay the course, when folks had no choice but to survive, but as a modern farmer, I'm searching for a strong enough motivator to get me through the looming decades. A nice island, maybe, between the ice caps; my own little unproductive, dysfunctional Eden. Like an addict, even as I contemplate quitting, I know I don't really want to, because, for all its spiritual damage, farming is one of the few professions left that gives as much as it takes and leaves a real imprint on this planet. Perhaps the true challenge of farming is to remember that on a 102-degree day. It reminds me of another quote I heard once in a trailer for a Woody Allen film, a piece of Platonic wisdom with a Hollywood addendum: "The unexamined life is not worth living. But the examined one is no bargain."

RESILIENCY IN THE FACE OF FIRE

RUTHIE KING

Summer term at the School of Adaptive Agriculture was winding down. With under a month to go, students were finding their stride, choosing their areas of focus, taking on responsibility at the farm, and reaching out to potential employers. The weekend was coming to an end, and students who had gone away were filtering back in Sunday night, with plans for a two-day on-farm welding class scheduled to start on Monday.

I spent the night twenty minutes away in Potter Valley, and went to sleep to the sound of wind roaring over the yurt's roof. CJ, a practicum student out of Boston, drove back from San Francisco after midnight and watched the sky curiously brighten with a distinct glow, until she saw the visible flames a few short miles from the Ridgewood Ranch, home to the school campus.

One of the wettest winters on record resulted in thick brush and dense grass. Our characteristic five months without precipitation dried this fuel to a crisp. The winds that came barreling through on October 8, 2017 carried sparks at alarming rates through valleys and hills dotted with homes, farms, and lives.

CJ alerted the nine other students at the school as the rest of the hundred or so residents of the ranch began evacuating north. They grabbed belongings and tried to snatch the cats. Students had very little in the way of stuff, given that they each inhabited one hundred twenty square feet while at the school. The phones and laptops that they grabbed would soon become obsolete, as the power, cell towers, and internet went out. They ran into me on their way out, and I transferred my dog into their escape vehicle, then headed in to report to our volunteer fire department. Eliot, a capstone student and member of our volunteer fire department, remembers the night: "It was alarming seeing everyone freaking out, a wall of fire, ash floating everywhere. It felt surreal."

Over the next week, hundreds of homes were lost and hundreds more lives were displaced by the Mendocino Lake Complex Fire, classified as a firestorm. It consumed over thirty thousand acres. The practicum and capstone students had come from across the country to attend the School of Adaptive Agriculture, and they witnessed a community that was utterly shaken by natural disaster. Their experience and response proves how important the local movement and community resiliency is.

The first evacuation center was the White Deer Lodge, an old hotel on the ranch that no longer operates. Many congregated there in the early hours of Monday morning, until the glow of the fire came closer and the group made the decision to evacuate again. This time the group sought refuge at Green Uprising Farm in Willits. A third evacuation warning sent them even farther north until they settled in for a night at the Little Lake Grange. All the while, communication lines were cut and students had to think smart and communicate clearly on very little sleep, with heightened levels of cortisol coursing through their systems.

By Wednesday, most had come back to the ranch to take part in a week of food preparation and donation distribution for the displaced people who were without water or power at the ranch and beyond. We were fortunate to have a gas-powered generator running the community dining hall, with a walk-in freezer and cooler to keep our hard-earned stores of meat, vegetables, and fruits saved through the season. That generator, plus our intrepid internet provider, who kept the tower up using solar power, allowed us to maintain contact with the outside world with a Wi-Fi signal on the ranch while cell service and power was out all around us. Our supplies of gasoline and propane to run the generator and hot water heaters were nearly exhausted by the end of a week. Gas could only be purchased in town with cash, as the card readers would not operate. The thirty kilowatts of solar panels, installed months before, kept the lights on and gave life to the gathering place.

Our community harvested from the garden, shuffled hundreds of pounds of fresh and frozen meat around, and cooked up many delicious meals for hundreds of people over six days. We met people who have lived on the ranch as neighbors for years but never had the chance to connect, we bumped into each other as scores of volunteers clambered into the kitchen to do dishes, and we recognized how many utilities and resources we take for granted.

Every day for the next week, I worked alongside four other volunteer firefighters of the Ridgewood Ranch fire crew and the rest of the first responders in Division Sierra-Tango. We spent countless hours on high alert, putting thousands of gallons of water down on the front line and jumping from spot to spot trying to stay ahead of the movement, guessing where the highest risk of spilling over was. We came back to the ranch each day for meals that were a true representation of the importance of a close-knit and resilient community.

Students may have missed a week of classes and learning, but we all walked away with a deeper, more intrinsic sense of connection to the whole. We were fortunate, but we were also prepared with backups, preserved foods, stored energy, and equipment.

The sense of mourning after the smoke cleared came in large part from the horrific losses all around us. Loss of lives, homes, and property are all very real and tangible. The harder-to-define loss that came crashing down in the weeks after the fire was the loss of purpose and clear objectives. Maybe it was just the adrenaline crash, but I felt a crushing sadness when this moment was over and we went back to regular life, with objectives that are years long and threats that are invisible. The threat of climate change or of aging farmers is real, but without radiant heat and walls of flame moving toward us, it is hard to mobilize. But the community that we fought for is still here, and every day we still fight for resiliency, life, and growth. The sense of urgency and community connection are feelings I don't want to let go of, and I intend to hold onto that memory as inspiration for moving forward.

ONE FARMER'S JOURNEY WITH LYME DISEASE

GLORIA DECATER

I live in prime Lyme country, with many in our county—and our state—suffering from Lyme, but Lyme disease took me completely by surprise. You can live with Lyme for a long time without knowing it, and I had no idea that I had it until I had massive dental work followed by a full hip replacement surgery.

My husband and I have been farming over forty years on fifty acres in northern Mendocino county in California. In 1988, we started one of the first CSAs in California and the country, serving Mendocino and San Francisco counties. We have been farming with horses since 1980, have raised three strong entrepreneurial sons, hosted between seven and ten school class visits a season, trained between four and six apprentices each season, and fed two hundred households a year with our biodynamic produce and meat shares. I guess you could say we have been rather busy. And, for the most part, I was a healthy, active, fully functional human being throughout all these years.

I was recovering from the hip surgery and gaining more movement in my legs and body when I started to experience sharp, jabbing strikes of pain in my back, numbness and tingling in my legs, pain pretty much all over my body. It was immobilizing but erratic. I kept thinking it had something to do with my hip surgery. I was also very anxious and fearful. The physical therapist I was working with said the fear was preventing me from getting well. I ended up in bed off and on for months, disconnected from friends, stopped doing the work I had done for years, spent days and months doing nothing, and lost over sixty pounds. I stopped milking the cow, feeding the chickens, connecting with our CSA members, collecting payments, writing newsletters, and purchasing all the food for the farm community. Who was I?

It was at least six months before I was finally diagnosed with Lyme disease, given four different antibiotics, prescription painkillers, and anxiety medication, as well as antidepressants, supplements, and more. They took every blood test imaginable, along with CT scans and MRIs of my entire body. Nothing was left undone. All the testing said I was healthy, but in reality I was a mess. And after, my condition only got worse, with more erratic pain and swelling, neuropathy in my legs and feet, sweats, fear, anxiety, screaming fits, and depression. I finally ended up in the psych ward at Stanford Hospital and was told I did not have Lyme, that I had no symptoms and the blood tests were negative. I

Rachel Alexandrou | Ticks Falling from the Sky onto Knotweed

spent a month there, and went home believing I did not have Lyme.

Yet, back at home I could hardly function. I had frequent anxiety attacks and was unable to think clearly, often feeling like my brain was out in the universe somewhere and I could not access it. Again my symptoms increased and diversified. The neuropathy continued to get worse. I had swelling in my legs and was losing thoughts as fast as I thought them. I went to the emergency room at the hospital three times, imagining I had every malady in the books. I saw several local doctors who only wanted to give me more drugs to deal with the symptoms, not treat the causes. I had appointments with a psychiatrist, who told me I would need to stay on the antidepressant for a long time. What I experienced with differing opinions from a variety of doctors is not uncommon: chronic Lyme disease is not acknowledged by a large portion of the medical profession. And yet so many are seriously sick with chronic Lyme.

My husband Stephen supported me all the way, as difficult and challenging as it was. He dropped out of the farm operations to be with and care for me, take me to appointments, feed me. The crew that was here did the best they could, but the farm was missing the owners. Last year, we were fortunate that our eldest son and his partner, with their knowledge and experience, were able and willing to take on running all the operations of the farm. What a godsend. Watching them but not participating, not being a part of all the work and joy, made me feel so strange after forty years of running the farm. At the same time, I was finally getting rest, which I sorely needed.

Many people were praying for me, many people were sending their love, and with that caring I was able to gain enough clarity and initiative to find a nutritionist who started me on a supplement plan. Six months after I was told that I did not have Lyme disease, I connected with a friend who apologized for welcoming me to the "Lyme Club," as she was certain I had Lyme. And the community of individuals with Lyme helped me find a doctor who could direct me on the right path.

I weaned myself off the antidepressants since the psychiatrist would not help me, I found a Lyme-literate doctor, and through "Lymies" (people who have Lyme disease), I found several tools to use at home without having to travel two to three hours to see a doctor. As often as I can, I sit in an infrared sauna, and I work with sound frequencies. I also take various homeopathic remedies, nutritional supplements, and antimicrobials. I educate myself with online webinars, read books on Lyme, and talk to others dealing with Lyme. Everyone has their own way of coping with this insidious disease.

There is so much to look at while treating Lyme: do you also have parasites, heavy metals, EMF sensitivity, or mold toxicity? All these things affect your immune system's ability to protect and heal you. And there are multiple co-infections, each with their own symptoms, though many of them overlap. I do not remember being bit by an infectious tick, but I was scratched by a feral cat and ended up with serious cat scratch fever, which carries Bartonella, one of the co-infections of Lyme. I took antibiotics at the time to get rid of the infection, but apparently it did not get rid of the disease.

I still have healing to do, like being able to remember words. I know I will continue to get better. And I will share with others what I have learned, and recognize that I struggle a lot with why I stayed in bed so much, why I could not get past the pain, the fear, the depression, why I did not participate in life during that year, and feared seeing friends and relatives.

The greatest healing is spending time with my sons, and now my beautiful, happy grandson, getting hugs from friends and relatives, and doing things I have not done in more than a year. In the last month or so I have been milking my cow, and wow, does that feel good. A few months ago, I made an oak leaf compost pile all by myself. This past week my husband and I spoke in front of a farming audience of over one hundred people and people understood what I was saying—and actually liked it.

I made sense! Now I am getting ready to host third grade school classes here on the farm. We have seven of them scheduled for this spring. And I can chase and pick up my grandson! I am so happy! I am alive!

To learn more about Lyme disease, read New Paradigms in Lyme Disease Treatment *by Connie Strasheim, or anything Stephen Buhner has written on the subject. If you have Lyme disease and need some direction, email me at Livepower@livepower.org. It gets so confusing out there.*

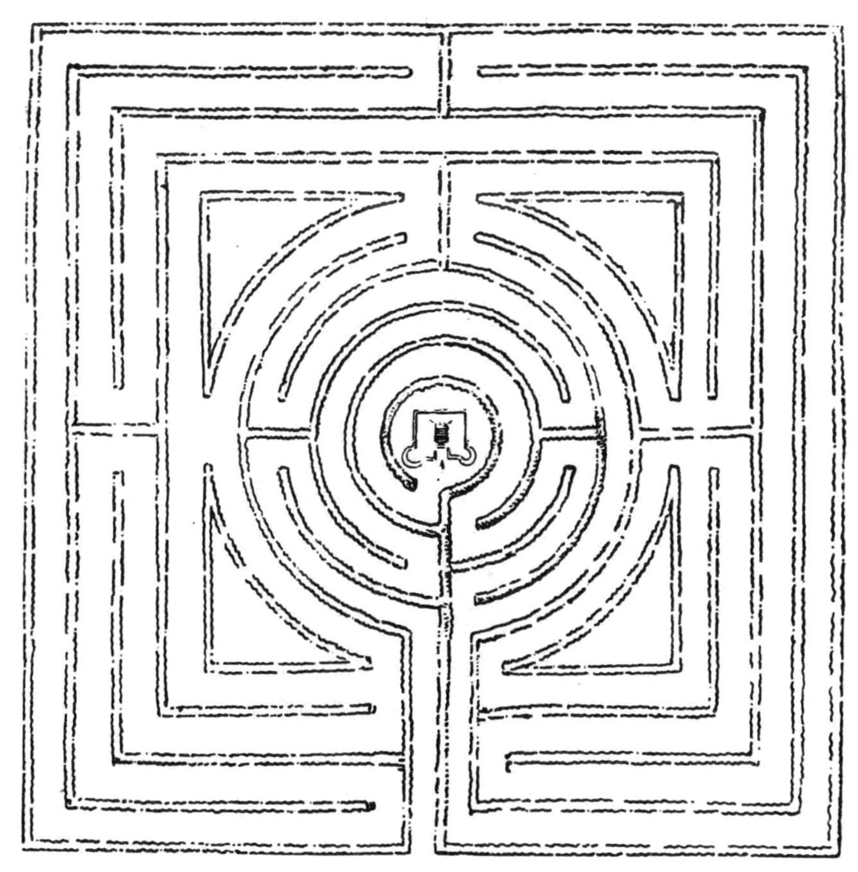

FAIR ROSAMOND'S MAZE.

WAR FIELD

BINYAMIN KLEMPNER

Three p.m. arrived at Tel Saki.

A little butte off of Highway 98—maybe five meters high on the western face and fifty on the eastern slope. Slightly steep, then a steady gentle roll into Syria, 1,000 meters distant.

Gathered onlookers, about twenty of us, sitting atop the bluff. Some with binoculars. All with cameras. Awaiting the spectacle of Syrian Civil War.

Thunderous boom. Pillar of black smoke rising over the skyline. Followed by thunderous boom. Pillar of black smoke rising over the skyline. Some pillars are small puffs. Some pillars are enormous splashes of black dimming out the sun. Some booms come within a succession of instants. Some booms come within a succession of minutes.

Amidst the shelling are green field crops growing peacefully. Perhaps the crops are oblivious. Perhaps they are tearful. I wonder who harvests? Who tends to the growth? Next to the chain-link border fence is an orchard. Looks like almond trees. How pleasant, perhaps even now, to sit under their branches. More booms. More pillars of black smoke. Everything, even war, is peaceful, from a distance.

I think about the buildings being hit. Apartments and homes. Fathers, my age, holding children, my children's age. The children cry. The fathers do their best to protect their precious children. I try to hear the children's terrified, confused cries. I try to touch the wetness of their tears. I try to hold them close.

Kilometers near. Worlds away. It's five p.m. My younger children are in the safety of Tiberias. My older son is with me on our field. Irrigating. Water flowing. All is quiet. All is serene.

FRUIT OF THE BREAD-FRUIT

JUNE

Faithlands

Beaters working back a burn

SOME PASTURES

A Cowboy's Almanac

SAM RYERSON

June, SA Pasture—Springerville, Arizona

Sometimes a pasture needs to burn.

Springtime is fire season in the Southwest. When the mountains don't burn regularly, when the ponderosa forests aren't thinned or the forage grazed and managed—sooner or later they will burn too hot. One year we were branding calves at the Boswell on the first of June when we saw a cloud blow up behind Escudilla Mountain. It looked like a big thunderhead, but it was just gray smoke, looming and building all day in the wind. By afternoon we could smell it. That turned into the biggest fire in Arizona history, and it stopped just short of our fence up near Jenkins Creek. We spent weeks moving cows down out of the timber, and then, after the fire burned itself out, weeks taking them back up into the high country.

The next spring, the biggest fire ever in New Mexico burned up from the Gila Wilderness right onto the yearling place, right into the Bearwallow pasture where we had just turned out a thousand steers, up to the meadow at the ranch headquarters, and around on to the north. We lived alone in a smoky haze for weeks. The horses coughed constantly when we rode. I fired two of my cowboys and the other one quit. We gathered as many steers as we could ahead of the fire, but we missed more than a hundred in the timber. The fire moved slowly through the big trees, and I rode along its edge, along the creeks of the cool green canyons in an eerie dark calm, with a glowing roar on the ridges above where it burned hot and fast uphill. We left the gates open when it was too late to gather more, so the steers could drift out of the pasture if they needed, and we went back to get them when the fire was out.

Gary came to help, and we found them all; not one steer died in the fire. They were grazing in the cienagas or sitting in the ashes chewing their cud next to smoldering stumps, where the smoke kept the flies away. The western fences of the pastures were ruined—the barbed wire turned brittle in the heat and crumbled. After the rains came in July, all the elk and wild cattle in the country found the lush new grass and stayed in the burned pastures through the fall. Old springs that hadn't run water in years came back to life. The creeks ran black at first but soon ran clear, and stronger than before.

LOVING THE WORLD TO LIFE

GRACE G. HACKNEY

"I take literally the statement in the Gospel of John that God loves the world. I believe that the world was created and approved by love, that it subsists, coheres, and endures by love, and that, insofar as it is redeemable, it can be redeemed only by love. I believe that divine love, incarnate and indwelling in the world, summons the world always toward wholeness, which ultimately is reconciliation and atonement with God."
—WENDELL BERRY

In March 2003, as I waited to hear where the Bishop of the United Methodist Church of North Carolina would send me to serve as a soon-to-be graduated, seminary-trained clergyperson, George W. Bush declared war on Iraq. Troops were deployed to Iraq on March 20, 2003. Just two weeks later, I learned that I would be deployed to Cedar Grove United Methodist Church, a small church in the "rural buffer" of Orange County, North Carolina. The paradox was apparent to me right away. Having grown up a pacifist, I wondered how I would be faithful to who I understood God to be while leading a politically diverse congregation. How would God lead me to seek peace in the place I was being sent, even as the United States went to war?

From Bush's declaration of war until the end of the school year, I had gathered with faculty, staff, and students at Duke Divinity School to pray about that very question: how do any of us find peace in a world that is not at peace? We prayed through the holy scriptures, knowing that none of us could sufficiently answer this question, but trusting that God would show us each the best way to live as ambassadors of peace in our particular contexts. I was not given a clear answer right away. It would take opening myself to the complex lives and complicated narratives of a place and people before God would show me that both peace and lack of peace are deeply rooted in the land and the people and the multitude of other creatures who call that piece of land home. I would come to see that my vocation was one of leading people to choose peace. I would learn that choosing peace requires sacrifice, and a deep change in the ways people who follow Jesus see the world. I would learn that such peace-seeking is not easy for human beings—we are so set in our ways. I would also learn that no one could be forced to choose peace; there would be those who would say yes, and others who would adamantly refuse.

It would take me a number of years to realize my true calling: to lead others to love the world to life. In the process, I would learn (and am still learning) that reconciliation involves not only the restoration of a right relationship with God and our human neighbors, but also with the land.

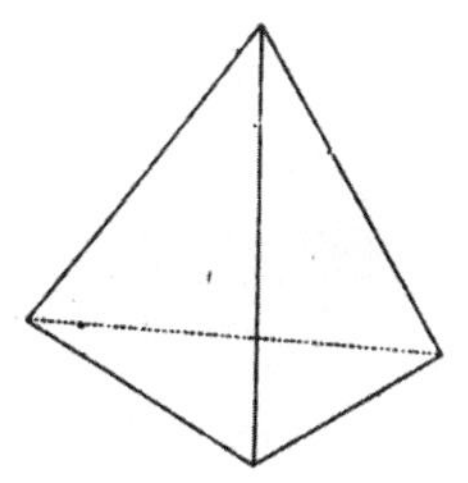

George W. Bush declared war on Iraq because of weapons of mass destruction at the same time I was sent to serve a church in a land that suffered from the use of weapons of mass destruction in the form of chemicals that produced tobacco so that farmers could support their families, send their children to college, and ensure a "better" life for them.

The war in Iraq was unfounded; no weapons of mass destruction were found. But weapons of mass destruction had depleted the soil in the township of Cedar Grove. While I have no scientific data, I wondered if tobacco production in this community had contributed to the high number of pastoral visits to, and funerals of, parishioners who had suffered from cancers of various types. And I was not the only one who believed that the poisoned land would take generations to heal. Farmers who had previously grown tobacco using methods imposed on them by a commercial industry may have sent their children to college and built fine houses from the proceeds, but they were also very much aware of the cost of such "success." One farmer told me, "You don't know how impossible it is for me to farm something else on this land; it will take generations for it to heal."

I recognized early in my ministry that the war I found myself deployed to was a war against the principalities and powers. These are the powers that have tricked us into believing that the world is here for us to use up as needed; the powers that name "success" in economic terms; the powers that have created societies of producers and consumers. I realized that the church itself was complicit in this way of thinking. The words of the prayer of confession we prayed most Sundays struck me to the core: "We have failed to be an obedient church."

As a pacifist Christian I knew that the war I was called to fight was not a war of spears, but a war of plowshares; not a war of hate, but a war of love. As a leader in a Christian congregation, I was called to "fasten the belt of truth around my waist, put on the breastplate of righteousness, wear shoes that would allow me to proclaim peace, and carry the shield of faith." Only then could I offer the people I had been entrusted with a path away from dis-ease and the status quo, and toward life and peace and wholeness.

It wasn't easy. The love that we are called to embody is a sacrificial love. We give up one life so that we can live another. Not everyone will understand; some will protest or resist violently, because their worldview will be shaken to the core. This is not the kind of love you wake up with one morning and live the rest of your life. It is a lifelong endeavor, and sometimes we get it right. It is the love Wendell Berry speaks of. It is the kind of love with which God has loved the world, the love I strive for every day.

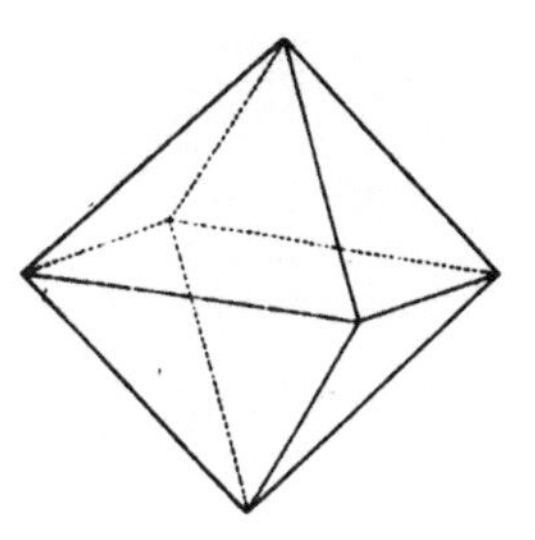

My deployment to rural Cedar Grove was a complicated one. The stone church built in 1939 had burned to the ground in a mysterious fire two years previous, and the charred skeleton stood next to the cemetery as a constant reminder of what had once been. The congregation and community argued over how and where the new church would be built at the same time they grieved the physical loss of the iconic building, which had been constructed by local builders with the same quarried stone that was used for the Duke Chapel on the campus of Duke University in nearby Durham.

Couple this with the fact that barely a year after my arrival, two tobacco buyout bills were approved by the US House and Senate, adding up to more than twenty-one billion dollars, with North Carolina farmers being the largest recipients.

The small Northern Orange County *XTRA*, a free paper delivered to every household in this part of the county, listed names of the farmers who would receive the most dollars and the amount they were due to receive as a result of the buyout. Many of those names were people I knew well as members of the church I served. The paper also reported that small farmers, who account for up to eighty percent of all farmers, would see virtually no benefit from the buyout. I knew these farmers as well; many were black.

Some saw their dreams come true, with a big check in the mail. But far more were angry and even devastated. Farmers were paid not to do what they knew to do. A farmer would tell me, "My soul was ripped out. Pastor, you may have been called to be a pastor, but I was called to be a farmer."

Land lay fallow. Briers, thorns, and sweet gum trees grew up on some. At least two farms turned to growing turf. Still others put land into conservancy. Some land was sold, but not much—this was land that had been in families for five, six, seven generations, and to let it go was a tragic kind of death. It was a hard time in northern Orange County.

And then there was a murder.

Two miles down the road from the church, Bill King was murdered in his little store in broad daylight. Bill was white; his common-law wife, Emma, was black. Some thought the murder might have been racially motivated. Others were certain the murder was drug-related. Bill had cleaned up this corner of the community, disallowing drug trade and instead offering food on credit and ice cream for kids who rode their bikes to the store. The peace that Bill had built on the corner of Carr Mill and Mill Creek Roads had been shattered. And in the back of my mind was the same question I had come to Cedar Grove with: how do we seek peace in the place we have been sent?

Valee Taylor knocked on the door of the parish house just a week after Bill's murder, asking if our church could do anything to help raise money as a reward for information about the mysterious murderer. "People are afraid," he said. The local sheriff's department had already shut the case, claiming there wasn't enough evidence to move forward. Together, we talked about the most just response the Christian community could make. As Christians, as people who have put on truth, righteousness, peace, and faith, the answer was not to sequester ourselves out of fear. The answer was not to condemn or to judge or to protest. Our call was to love, to stand up against fear and hate. Valee, a black man, and I, a white woman, agreed to stand up against fear, and to call the community together. He would contact the black churches, and I would contact the white. One week later, we would stand up against fear on the very site of the murder.

And that is what we did. We gathered the community together in front of Bill's cinder-block store. We told stories, prayed, cried, and sang. Together, labels of rich and poor, black, brown, and white, employed, underemployed, and unemployed fell away as we united to stand against fear and for community. Together, we stood on the site of an unsolved murder and firmly and peacefully bore witness to love. Some said that such a gathering had never happened in Cedar Grove.

But what happened next was even more unexpected.

Scnobia Taylor, Valee's mother, was in the crowd of about one hundred community members that evening. She had a vision that resulted in her giving five acres of land to the church I served as a means of bringing the community together. Did I mention that I am white, and that Cedar Grove United Methodist church had been referred to by some as the "rich white church?" The daughter of a black sharecropper, Scnobia gave five acres of land to a white congregation in the hopes that peace could be a reality in our community. As Scnobia would say, "People are hungry. The land is growing up in thorns and sweet gum trees. What if we grow food together?"

We had been preaching from the prophet Jeremiah that summer. The new church building was beginning to rise up from the ashes at the same time we read about the prophet Jeremiah's call to "build and to plant." What if God was calling us not only to build a new church structure, but also to plant a garden?

Even before Bill King's murder and Scnobia's offer of land, a small group had been meeting to talk about what it meant to live in a place with the gift of abundant land. We learned that the new look of poverty was obesity; the fact that residents lacked access to healthy food in a place that had land seemed to be an anathema. If there is land, and if our food comes from the land, and yet, we live in what has been called a "food desert," then something is not right. As a people of faith, we named this sin. We are not in right relationship with God when land is not allowed to be what it was created to be. We are not in right relationship with God when people are not granted the resources to be their best selves.

How can we seek the peace of the place we have been sent? Maybe, some of us dreamed, we have the opportunity to bypass the industrial food system in this community, to bear witness to a different way of life in which both land and neighbor are honored. Orange County had been in conversation about how to end homelessness. But what if that was the wrong question? What if the question was not "How do we end homelessness?" or "How do we end hunger or poverty?" Perhaps the question we needed to ask was "How do we build a community in which homelessness (or hunger, or poverty, or lack of good food) is not an option? How do we build communities of love?"

Scnobia Taylor's gift of land was the perfect response to the murder of Bill King. Instead of putting our energies into raising money to find a killer, we would put our energy into creating a community in which no one would be hungry, in which no one would be told they are less than anyone else, in which everyone would be loved

toward life—including the land. Land that had been farmed by sharecroppers was given to the descendants of slave-holders, so that together we could find some peace in Cedar Grove.

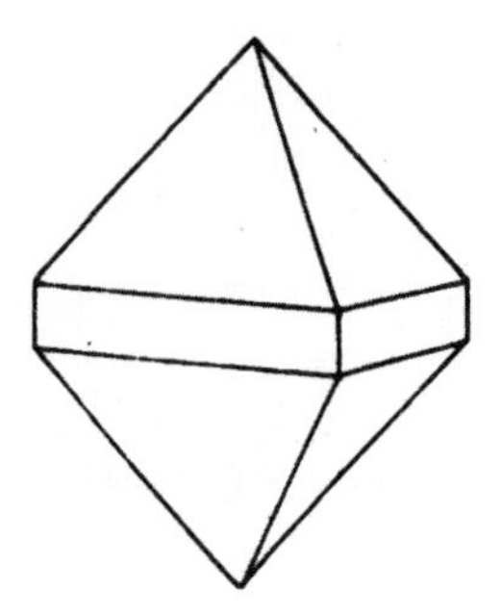

The congregation I served was immersed in teachings of the Old Testament, learning from the history of Israel. God's people were exiles in Babylon, and yet, through the prophet Jeremiah, were called to "build houses and live in them; plant gardens and eat what they produce … seek the welfare of the city where I have sent you into exile, and pray to the Lord on its behalf, for in its welfare you will find your welfare."

The garden was named Anathoth, the name of the field God directed the prophet Jeremiah to purchase during a time of war. This was the way God would lead me to seek peace in the place I had been sent. This was the garden that would say no to the ways of the empire, and would show that there is another way, a way of peace. The mission of this new Anathoth would not only grow friendships with the land, each other, and God, but would grow hospitality, gratitude, and a way of life based on love.

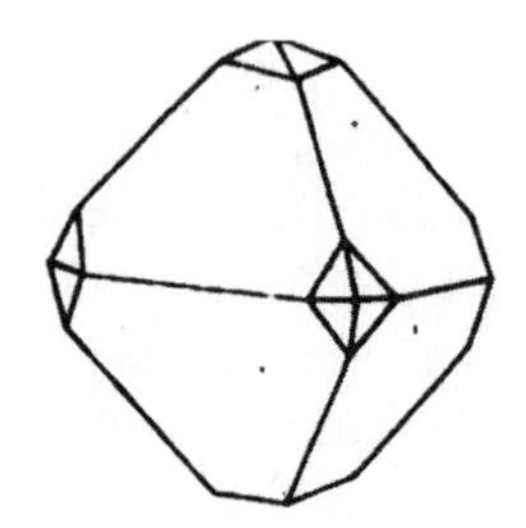

Almost thirteen years since the first sweet gum trees were mowed down on the corner lot of Lonesome Road and Millcreek Road, and garlic and blueberries were planted, Anathoth Community Garden continues to flourish. It was funded by the Duke Endowment, the same Dukes who grew tobacco, the same Dukes who helped build the church. Two of Anathoth's many interns found their vocations by working at the garden; Julia Sendor and Chas Edens are now the codirectors and farmers. Chas and his young family bought land in Cedar Grove, and use part of it to grow food for the two hundred recipients of HarvestShare, a community-supported agriculture program. HarvestShare offers a great diversity of people who would be without fresh food to receive weekly boxes, often delivered by one or more of the garden interns who work, live, and learn together in Cedar Grove during the summer. A summer youth internship program thrives, and friendships are made across racial and economic lines.

Every Saturday you can find a group of people—black, white, and brown, rich and poor—eating together after working in the garden all morning. Labels fall away as the hands that worked together suckering tomatoes or harvesting okra are joined before eating, and the ritual of naming takes place: Around the circle, people speak their names to one another, and needs or thanksgivings are voiced. Glowing with sweat, friendships are made and nurtured as the community shares a meal—and more than a meal—together.

Is there peace in Cedar Grove? Yes, there is. But peace is a tender and fragile thing, and there is always work to be done. Eight years ago, the new store owner and his colleague were shot and killed in the same store where Bill King was murdered. The property owners razed the building, and the bloodshed there is overgrown with weeds. Reconciliation is an ongoing and complicated endeavor. It is not finished, but there are glimpses of the kingdom that is reconciled and continues to be sought. Slowly the real weapons of mass destruction—fear, greed, racism, and unsustainable agricultural methods—are being dismantled.

Scnobia Taylor's granddaughter, also named Scnobia, has joined the Anathoth Board, and the women of the predominantly black Lee's Chapel Missionary Baptist Church are seeking ways to be more involved. The old tobacco farmers are dying, and with them, a way of life that built a beautiful, yet unsustainable, community in Cedar Grove. A new community is being built there. New regenerative farms are thriving. Blueberries, beef, fresh turkeys for Thanksgiving, responsibly farmed pork, and chicken from Cedar Grove farmers are regulars next to vegetables at area farmers markets. Yet, even while these farms thrive, black farmers are losing their land and are unable to sustain a way of life. Until we address social injustice, and right the inequality of land ownership, peace will remain out of reach.

Thinking about all that has happened, I find myself drawn to Gregory of Nyssa, a fourth-century theologian and author of *The Life of Moses*. Gregory of Nyssa writes about the ways Moses's life mirrored a life of virtue. Called to lead the Israelites through the wilderness and into the Promised Land, a land flowing with milk and honey, the reluctant Moses would be called a friend of God's. Through the story of Moses's life, Gregory reminds us that the only thing worthy of our desire is friendship with God. In fact, Gregory wrote that the worst thing imaginable was to fall out of friendship with God.

Reconciliation is hard. In fact, I confess that every day I am learning anew what the word even means. This I know: to be reconciled is more than a legal agreement, or a handshake, or even a coming together to agree to disagree. I am beginning to think about reconciliation as "becoming friends again." I am coming to understand that to be friends with God is to also be friends with the land, and with the diverse humans who farm the land, and who are its stewards. If the worst thing imaginable, as Gregory writes, is to fall out of friendship with God, then there is also tragedy inherent in falling out of friendship with the land. When we as humankind disregard the land, we do so at the peril of all creation. Yet when we love the land, befriend the land, care for it, and nurture it, we find ourselves bound up in the mysterious life of the Divine—however or whether we name the divine or not.

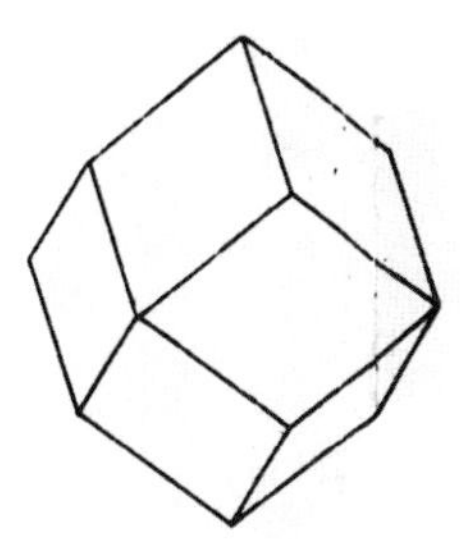

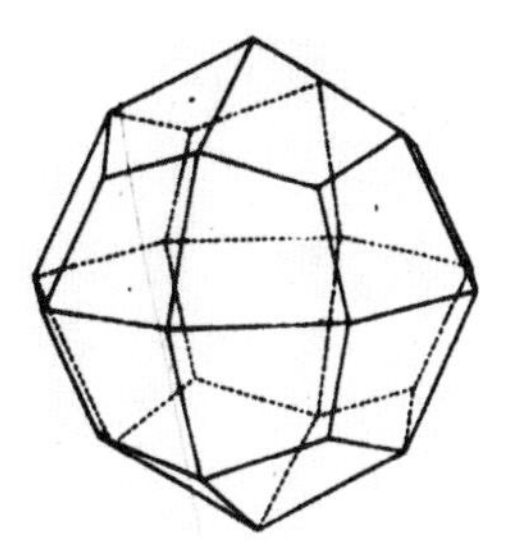

PRECIOUS FORM

DOUGLASS DECANDIA

The stars have made
not one body,
one branch
or spore
the same as any other.

For every seed
there is a universe
unfolding
into Spring.

I am
not
man,
but human.

Earth,
water
and light
taking precious
form.

Ursula MacFarlane | On Reflection

CREATING AN ECOSYSTEM OF FAITHS FOR THE FUTURE OF THE EARTH

ROBERT KARP

I recently attended Faithlands, a remarkable gathering of faith leaders and sustainable agriculture leaders convened by Greenhorns and the Agrarian Trust, and hosted at the beautiful Paicines Ranch in California. The purpose of the gathering was to explore the potential for lands held by faith communities to be placed in the service of the sustainable agriculture movement and beginning farmers in particular. Many of us left this gathering, however, feeling that something much deeper and more far-reaching had occurred during our three days together.

To understand the deeper experiences that many of us felt, it may be helpful to call to mind the concept of the farm ecosystem or organism as it is understood in the biodynamic farming movement. This concept suggests that all the diverse elements of a farm—the woodlands, wetlands, cropland, pastures, orchards, farm animals, wildlife, and so on—can be woven together biologically into a higher synthesis through the wise management of the farmer. Each of these elements must be managed as a whole unto itself, but each can also contribute to a whole that is greater than the sum of the parts. This greater whole is the farm ecosystem.

When a farm begins to achieve this quality of wholeness, remarkable expressions of health and vitality begin to emerge on the farm and in the food produced on the farm. The farm, you could say, becomes a vessel for something working at a higher level than biology. In biodynamics, we would go so far as to say that the biological organism of the farm can become a vessel for a spiritual being to work in a new way on the earth, a being that can bless the whole surrounding landscape. To use a sacramental image, we could say that the farmer raises up the elements of the farm like a priest raises up bread and wine, allowing them to be infused with new life, bringing nourishment and healing to the whole community.

This same potent image can be applied to interfaith and ecumenical dialogue. Imagine for a moment our gathering there in California: a room of Muslims, Jews, and Christians, of Episcopalians, Lutherans, Baptists, Presbyterians, and Quakers, gathering not so much to compare theological perspectives, but rather to share our unique ways of loving and caring for the land, for farmers, and for our own communities.

Imagine a Muslim farmer sharing how his love for the earth called him away from an academic career and inspired him to partner with the New Jersey Muslim community to create an organic farm and educational center inspired by the environmental and social values of Islam.

Imagine a Dominican sister sharing how she feels her traditional religious vows have been transformed into a new healing chrism for the earth, inspiring her community and many others to place their lands into the service of small farmers and local land trusts on behalf of a bioregional vision of the future.

Imagine a leader in the Jewish community farming movement sharing how work on a farm invested in community supported agriculture helps reconnect young people to the roots of their faith and to the deeper social and environmental dimension of lived Judaism.

Imagine a leader of the Lutheran church in Milwaukee sharing how the community garden she founded, located on land historically connected to the Underground Railroad, has become a sanctuary for the African American community, where the arts of growing, preparing, and sharing food are restored to their cultural and sacred dignity.

Imagine a United Church of Christ pastor sharing his calling to create a new church rooted in the land and culture of rural communities as well in the radical agrarian values of the Bible—a church founded on wonder and reverence for the working of God in all of creation.

Imagine these and many other offerings and conversations over three days creating a social vessel filled with a mood of devout open-mindedness, of reverent curiosity, of sacramental solidarity with one another and with the earth. Imagine, through these conversations, an ecosystem of faiths beginning to form among us, woven from our goodwill and inspiring us to place our diverse talents and gifts—and lands—in service to the future of the earth.

What emerged clearly for me in this gathering, and what I think many others felt in their own way as well, is that there is an ecological and agrarian spirituality manifesting in our time that both embraces and transcends the beliefs and practices of individual faith communities, a spirituality that can serve as a new source of unity among all people who love the earth, whether religiously affiliated or not. This spirituality shines toward us from soil, plants, and animals; air, light, and rain; clouds, stars, and seasons; and work, food, and fellowship. It inspires us to unite spirit with matter in practical work on the land, work that blesses and is blessed by the holy, unifying embrace of the web of life.

You might say that we felt our seeming differences dissolved like so many materials in a compost pile. Yet I think the image of the farm organism is ultimately the better analogy. For it was not so much that our differences were dissolved, but rather that they were brought into connection with a higher source of unity that gave to each person and each religion and denomination represented a new sense of purpose and responsibility within a larger ecosystem, the full nature and texture of which none of us had felt before.

And yet there was an essential ingredient to this gathering that must not be overlooked: those who attended not as representatives of faith communities, but rather as representatives, you could say, of the earth herself. These were the farmers and conservation biologists, the agroforesters and environmental lawyers, the staff of land trusts and farm organizations.

Through their presence, we stayed grounded in the urgent needs of young farmers, in the loss of species, and in the plight of the glyphosate-poisoned earth.

Without the urgency felt by these leaders, this gathering would likely never have happened, and thanks to them, we walked away from this event not only with our souls full of the new wine of an emerging ecosystem of faiths. We also walked away with a host of new tasks and commitments:

- Conducting an inventory of lands held by faith communities that could be suitable for sustainable agriculture and beginning to engage those with responsibility for these lands in conversations about their future;
- Exploring the feasibility of a national land trust for holding and managing these lands, managed together by faith leaders and sustainable agriculture leaders;
- Creating a circle of coaches to support farmers negotiating leases with faith communities;
- Organizing other faith lands gatherings, small and large, across the country, such as at the 2018 Biodynamic Conference in Portland, Oregon;
- Beginning to articulate the shared values and theological principles and insights that inform and inspire this work;
- Identifying the many voices who were not part of this conversation but that need to be for the future;
- Forming a group to steward this ecosystem and the work flowing from it into the future.

We came to this gathering as representatives of diverse communities and interests, but with these commitments in our hands and with these experiences in our hearts and minds, we left feeling as members of a single community, a new ecosystem of people, faiths, and organizations devoted to the future of the earth.

For more information about this gathering and the work following on from it, please contact Jamie Pottern at the Agrarian Trust: jamie@agrariantrust.org.

The following organizations were represented at the 2018 Faithlands Gathering: Greenhorns, Honoré Farm and Mill, The Jesus Center, Religious Lands Conservation Project, Biodynamic Association, Agrarian Trust, The Abundant Table, Mount Grace Land Conservation Trust, Black Church Food Security Network, Quivira Coalition, Savanna Institute, Sustainable Economies Law Center, The Conservation Fund, Hazon, Islamic Society of North America, Green the Church, Vandersall Collective, Life Around the Table, Plainsong Farm and Ministry, Saint James Episcopal Church, Land For Good, Alice's Garden Urban Farm, Interfaith Sustainable Food Collaborative, F.A.R.M.S. Agricultural Law, Urban Vegetable Project, Santa Cruz Farm, New Hope Catholic Worker Farm, The Good Tree Inc., The Keep & Till, Stony Point Center, Presbyterian Hunger Program

Vincent Sheridan | Murmuration II

FIELD LITURGY

JAMIE HUNYOR

In the high tunnel pulling weeds,
hands sweeping over a yellowing earth,
I wipe sweat from my forehead, dust
from my glasses & glimpse the goats
in the pasture, Alfalfa at rest
surrounded by grazing does.

The goats don't stop to contemplate
why the sycamore rises above settled things
or retains its fruit throughout winter but
instead climb atop one that had been felled,
ignoring routine spores & the worms cloistered
inside its darkness.

Each living thing is restless in its own way;
goats ruminating surprisingly ripe late
January grasses, & me, ruminating the
difference between pasture & sidewalk,
pulling roots deep from the dirt of plants
whose names I strain to remember
    (or simply don't know).

Here, the planet seems muffled,
some trick of wind or meandering
consciousness, out of reach
of the men bunched up in their
fortresses, deliberately speaking old
tongues & harvesting violence.

For the time being, I will return
to the sun beating down on dried-out
onion skins, but not until I speak
dutifully to the billy: "Get up, Alfalfa,
there is work to be done."

Originally published in Jettison *(jettisonmag.com/poetry/2043/field-liturgy/). Reprinted here with permission of author and publisher.*

Marisol Grandon, United Kingdom Department of International Development (DFID)

Mother support group leader Gladys Nagilai stands among her tomato crop in a greenhouse in Lodwar, northern Kenya. Support group members actively train other mothers in the community about feeding and nutrition. The greenhouse farming project–which is supported by UNICEF and the International Organization for Migration with funding from the UK–is underway as a means of improving their household food security. There are thirteen mothers in the group who meet once a month to learn from each other about infant and child nutrition.

MOTHERS IN FOOD, FARMING, AND FAITH

A New Generation

JOHANNA M. KEEFE

SLOW LIVING AND THE RETURN OF HEALTH TO OUR NEW GENERATION

I have deep respect for mothers who live their lives deliberately. Mothers whose choices are informed by an ecological sensitivity see that everything is connected. They understand that choosing to live sustainably and delicately, with a reverence toward nature and the earth, allows the earth to nourish them back. I have discovered a whole subculture of mothers who carry this eco-ethic into their daily lives, who make decisions about how to source, prepare, and nourish their families, and in turn, nourish their souls.

In May 2018, I attended the annual Slow Living Summit in Brattleboro, Vermont, where small businesses and individuals with big ideas came together with mentors to discuss how to grow a thriving food system for the future. The summit was hosted by Strolling of the Heifers, a nonprofit that organizes the eponymous annual event and was founded to bring awareness to the local dairy and farming industries in Vermont and to the New England region. Orly Munzing, the spokesperson for the summit, explains that the purpose of Slow Living is "not to do less, or to do things more slowly—it's to do your work mindfully, with the good of the community, the bio-region, and the planet in mind."[1]

This personal eco-ethic is also held by the mothers who'd shared their stories with me as I worked on a doctoral thesis revealing the connection between restorative food, regenerative farming, and maternal and child health. While each had their personal faith, they all moved through their daily lives with a deeply held eco-spiritual worldview. Ecology—the living, breathing, animated natural environment—intersects with both personal and universal spiritual identities.

Satish Kumar, founder of Schumacker College and the charitable Resurgence Trust, shares this holistic worldview. In his recent book, *Soil Soul Society*, Kumar offers a trinity of interconnected paradigms that he says are essential to inform one another as we move about our lives. We must at once, he says, take care of our spiritual lives (soul), the earth (soil), and each other (society).[2] His framework fits what I found while collecting narratives from a small purposeful sample of women who expressed sincere gratitude and appreciation for the health of their families and who have gradually learned to cultivate lives rooted in their eco-centric values and their preference for a life with nature at its center. Each of the mothers' stories was indeed infused with an ecology of soul, soil, and society. Through my research, I hoped to learn more about how and why these women, in an effort to raise healthy vibrant children, were able to access the great personal initiative, self-discipline, and resolve to remain true to their ideals.

LIVING WITH HEART, HEAD, AND HANDS, A NEW LIFESTYLE IS EMERGING

Head, heart, and hands ~ I see this trinity often expressed in today's literature as the new ethos of natural living. One uses their head, or intellect, to stay informed and to avoid being carried into the status quo or blindly adopting any dominant narrative of lifestyle. The heart is at the center of a life well lived and makes work, play, home, and life outside the home feel authentic and joyful. Greater numbers of young adults and families are choosing a lifestyle that incorporates more creativity and work that requires use of the hands, like cooking, gardening, and fiber- and woodwork. There is a willingness the mothers I interviewed possess to move in a direction that gives them more freedom, more personal choice, and agency when it comes to how to nourish and support their families' health. We have moved from the Industrial Age to the Information Age and, as many others have suggested, are in the process of welcoming in the Ecological Age, marked by a focus on making lifestyle choices that support the health of individuals and the planet as a whole.

OUT OF HARM'S WAY (FOR OUR CHILDREN)

Today, the world's children suffer from the industrial practices of food and farming and from the profit-driven directives of our health care systems. Children are developing an array of increasingly severe and chronic illnesses. A sobering list of statistics on these illnesses can be found on the brilliant website of mother, author, researcher, and advocate Beth Lambert, who founded a nonprofit called Epidemic Answers and has conducted research that focuses on fourteen children recovering from chronic illnesses.[3] Those of us in our sixties and seventies remember when chronic illnesses in children were not so pervasive—ask any seasoned teacher or school nurse if they recognize rising rates of autism, depression, nervous system dysfunction, severe allergies, attention disorders, asthma, and food intolerances, and they will most likely shake their heads "yes." And yet, there are mothers and young women of childbearing age who are discovering alternative paths to health: deliberately choosing a lifestyle of slower, eco-informed, intentional living, offering their children a chance to thrive naturally, from conception onward.

A new subculture of moms who want their children to have a naturally resilient immune system and a flexible down-regulated nervous system is emerging. Mothers, pregnant women, and young women of childbearing age are deciding that it is wise to reacquaint themselves with a nutrient-dense, ancestrally informed, well-sourced diet, and to clean up their own life choices to rid themselves of accumulated environmental toxic loads. They are smart, informed, and committed to ongoing education and research, reading such works as Dr. Catherine Shanahan's *Deep Nutrition: Why Your Genes Need Traditional*

Food. They are awakening to the dangers of genetically modified foods contaminated with the herbicide glyphosate (first sold under the brand name Roundup), which contributes to the destruction of a healthy gut lining and, according to Dr. Stephanie Seneff from M.I.T., to the rise in autism spectrum disorders.

As they find more and more self-determination and personal agency in crafting a healthy lifestyle for their families, many mothers are finding they are naturally inclined to homeschool. These mothers are curious and seek out research before accepting claims of a "new normal" of chronic illness in their children. In some cases, they decide to homeschool to protect their children from medical protocols that they believe threaten their children's health, like the national vaccine schedule mandated in the United States. I have met with mothers who think critically about this issue and believe that the natural immunity of the body can thrive with a real-food, nutrient-dense diet and a clean lifestyle. They believe the human body has an innate ability to fight off infection. These mothers are acutely tuned into their children and their values, and they do their own research in making informed choices.

They look for and find community and connection embedded in their everyday lives. Some find their "tribe" in the local food and farming movement; others design support groups of religious and spiritual study to further fold their personal faith into their deliberate, consciously chosen lifestyles. They pay attention to other mothers who approach them with curiosity and find ways to share their experiences as they serve as examples to other women inside and outside their own social networks. An observation echoed among the many women I interviewed was that the process is a gradual one, and "one small step at a time" is a way of life. All expressed extreme gratitude for the health and healing of their families as they followed a real-food diet.

STORIES OF RECONNECTION AND COMMUNITY: CLOSE TO THE GROUND

The stories that emerged from my interviews with mothers found through the Weston A. Price Foundation (WAPF) chapters and conferences revealed that each mother held a deep sense of maternal autonomy and agency as well as a deeply held personal spirituality. These qualities emerged through a process of discovery as each of these women gradually returned to a lifestyle of slow food: cooking from scratch, with only real food, sustainably and organically sourced. This approach to nourishment informed all other aspects of their daily living. Connection to family and local community became paramount.

In her book *The Ecology of Care*, Didi Pershouse states, "Healing often starts with a desire to reenter the wholes we are born to be part of: landscapes, family and community, and the 'is-ness' of life itself."[4] Many projects, including this *New Farmer's Almanac*, are cropping up to inform, inspire, and celebrate an eco-ethic that offers a door into the wholes of connection that we desire. These projects demonstrate lifestyles and activities that can drive and shape an ecological civilization—an antidote to the age of industry. This new paradigm increasingly makes sense as the path to flourishing for a whole new generation of adults and emerging adults.

Taproot, now rooted in Portland, Maine, publishes a bi-monthly magazine that focuses on the choices of individuals to create with an eco-ethic and an eco-aesthetic, interweaving their cottage industry with a sense of place, of community. Articles are on food, farming, family, and craft. Each issue has a theme; the recent spring issue, "Bloom" (No. 27), is full of stories, as editor Amanda Blake Soule writes, of "socially and environmentally responsible businesses planting themselves in communities to help them thrive."[5] What I love, having experienced so much joy with my own fiber art, is a return to working with one's hands: farming, food, craft.

Another bi-monthly magazine, *Resurgence and Ecologist*, is published by the Resurgence Trust and offers considered and intelligent articles on the environment, activism, social justice, arts, and ethical living, with a focus on the ecologically informed professional and personal experience within the European Union and beyond. The May–June 2018 issue, titled "Working Together & Working With Nature," covers topics such as "slow poetry, co-ownership, and the power of play." The stories emerge from individuals making a difference, such as environmental lawyer and author James Thornton.[6] The magazine's contributors are as interesting as their interview subjects: for example, Colin Tudge reviews *Food Sovereignty, Agroecology and Biocultural Diversity*, a critique of modern food systems and agriculture,[7] and is himself a co-founder of the College for Real Farming and Food Culture, based in Britain. So many threads of intentional work and living weave together to demonstrate a global movement, and a sense of urgency, a conviction that healing needs to take place on many levels at once—in the home, in the community, and on the global stage.

This is what I find to be so encouraging: there is creation, beauty, and joy in living with a slow, deliberate, and ecology-minded approach to daily life, no matter what professional or personal enterprise you find yourself in. Personally, I look for opportunities to pick up my knitting, to regard all of my daily endeavors with a cadence of slow. Even as I sit in my porch in the early spring evening to write these words, I recognize that this is a creative act. Everything is connected. This is what creates the feeling of leading an "authentic" life. Like the mothers I interviewed, and the contributors to projects and publications such as those noted above, I strive to live in the world in a way that allows for a constancy of value: my inner child is allowed to thrive and flourish by weaving in daily activities imbued with ecological wisdom from the heart, head, and hands. Life is not fragmented but filled with meaningful activity.

And now, having connected with the mothers who told me their stories, I sense a quiet revolution taking place, a return to simple values of balance: where the children flourish, often learn at home and co-participating in the family activities of daily living. Their lives are full of embracing wholes in their spiritual lives, communities, farms, and home gardens.

My own wish can be seen in the form of a prayer: for the whole health emerging in the lives of these children to become the new normal for all families to come. May our next generation of parents take notice of the real choices before them and endeavor to join the new ecologically informed lifestyle embraced by many, crafting a life of beauty, health, joy, and simplicity. May those of us fortunate enough to be living in such a world offer a hand, our hearts, and our mentorship. This is my prayer. May it be so.

Johanna Keefe is currently envisioning a book and a film on the lives of mothers who follow a real-food, ancestrally-informed diet. Contact jmkeefe@endicott.edu.

NOTES

1. Orly Munzing, "Welcome Statement," Slow Living Summit, Brattleboro, VT, May 2018.
2. Satish Kumar, *Soil Soul Society: A New Trinity for Our Time* (Brighton, UK: Leaping Hare Press, 2017).
3. Epidemic Answers, epidemicanswers.org.
4. Didi Pershouse, *The Ecology of Care: Medicine, Agriculture, Money, and the Quiet Power of Human and Microbial Communities* (Thetford Center, VT: Mycelium Books, 2017), 11.
5. Amanda Blake Soule, "Bloom," *Taproot* 27 (May 2018).
6. James Thornton and Martin J. Goodman, *Client Earth* (Brunswick, Australia: Scribe, 2017), clientearth.org.
7. Colin Tudge, *Six Steps Back to the Land: Why We Need Small Mixed Farms and Millions More Farmers* (Cambridge, UK: Green Books, 2016); Michel P. Pimbert, ed., *Food Sovereignty, Agroecology and Biocultural Diversity: Constructing and Contesting Knowledge* (New York: Routledge, 2018).

food

1- buy it with thought
2- cook it with care
3- use less wheat & meat
4- buy local foods
5- serve just enough
6- use what is left

don't waste it

U. S. FOOD ADMINISTRATION

RECOMMENDED READING

1. Johanna M. Keefe, "Mothers' Stories of Their Lived Experience on a Real Food Diet for the Health of Their Children: A Feminist Narrative Inquiry of the Restorative Food And Farming Movement." Doctoral dissertation, California Institute of Integral Studies (Proquest Publications, 2018).
2. Satish Kumar, *Soil Soul Society: A New Trinity for Our Time* (Brighton, UK: Leaping Hare Press, 2017).
3. Beth Lambert and Victoria Kobliner, *A Compromised Generation: The Epidemic of Chronic Illness in America's Children* (Boulder, CO: Sentient Publications, 2010).
4. Didi Pershouse, *The Ecology of Care: Medicine, Agriculture, Money, and the Quiet Power of Human and Microbial Communities* (Thetford Center, VT: Mycelium Books, 2017).
5. Catherine Shanahan, *Deep Nutrition: Why Your Genes Need Traditional Food* (New York, NY: Flatiron Books, 2017).

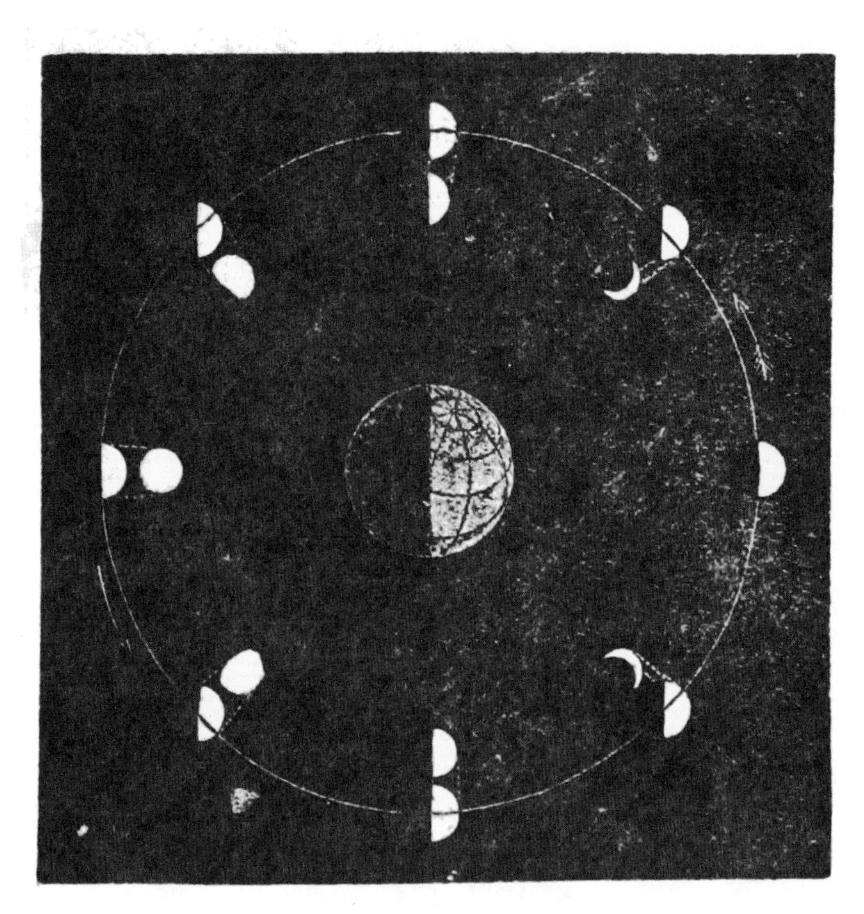

Phases de la Lune.

GASTRONOMY

JULIA SHIPLEY

Take the bladders of red deer and stuff them with yarrow flowers and bury them in the fall and recover them in the spring.

Take cow manure and smear it in a cow horn's hollow; take oak bark and fill the sheep's skull. Bury these in the soil and retrieve them in the spring.

Take crushed quartz and pour it in the other horn and bury it in the spring. Unearth it in the fall.

Now stir the disinterred silica into water for an hour, as fast possible in one direction, then just as quickly in the reverse.

This you'll disperse in the fields—less irrigation, more baptism—to encourage something, to discourage something—I forget; it's archetypal, the residual of a ritual. The inverse of cuisine. A meal you prepare for the soil itself.

DEBT AND MUTUAL AID

Transforming Zombie Agriculture

JANINE FITZGERALD AND KERI BRANDT OFF

Cuneiform tablet detailing assumption of debt by guarantor

The modern zombie has been able to move beyond the struggle of being human. The zombie does not grapple with death, because it is neither alive nor dead. It is not beholden to anyone. It does not have to deal with the difficulties of love, relationships, and loss. It only has to continue on its endless quest to eat more brains. But even as this monster resolves the duality of being human, we know that we do not want to be zombies. It is our contention that the antidote to zombie agriculture is in rebuilding relationships based on trust and reciprocity as a context for a healthy food system. Vital to this process is transforming the nature of debt and reestablishing the primacy of soil.

The book of Leviticus in the Old Testament called for a jubilee every fifty years where all debts would be forgiven and the land would be allowed to lay fallow for seven years. From a spiritual perspective, this practice reminded the Israelites to reflect on their debt to God rather than on financial debt to one another. From a practical perspective, this practice minimized social inequalities and allowed for the restoration of soils. Modern day jubilee movements have called for the forgiveness of all matter of debts, from those held by poor countries and owed to international banks, to the forgiveness of student loans. All jubilee movements understand that financial debt represents a barrier to a sustainable future.

The Young Farmers Success Act, originally introduced in 2015, would allow farmers with student debt to count their work as farmers toward public service and allow for loan forgiveness. Making this program a reality, along with other debt relief programs, is a step in the right direction. If we want to truly transform zombie agriculture, however, we must as a nation make it possible for farmers and ranchers to produce without the constant and crushing worry of financial debt. The primary concerns of food producers must be in remunerating debts to the water, the soil, the animals, the air, our families, and the human community.

Our denigrated and dead soils are both a source of great hope and despair. Soil is the skin of the earth and, like the ocean, healthy soils teem with life. Unlike zombies, soil is a living element in an elegant relationship with the world. As Wendell Berry observes, soil is the only organism that can make life out of death. Soil has a reciprocal relationship with life and death that is the foundation for all living things above the ground. Indeed, soil grows humans and animals as much as it grows trees and plants.

Through mono-cropping, deep plowing, and extensive use of chemical inputs, we have disrupted soil regeneration and borrowed heavily on the generosity of soil. Soil is in a sophisticated dance with organic matter. Chemicals take away the music and replace the animate with inanimate, yet push the soil to keep dancing. But, alas, much of the soil around the globe is tired and has lost its ability to twirl, thus endangering the future of humankind. That's the despair part.

Healthy soils thrive in mutual aid. If we take our debts seriously with a reverence for giving and taking, the complexity and fertility of soil will return. That's the hope part. Former coal miners are planting fields of lavender on top of destroyed mountain tops. Orchard keepers are planting companion trees to restore mycorrhizal fungi in the soil. Ranchers are protecting wetlands and engaging in holistic grazing practices. Young people are starting gardens in the backyards of their college rentals. Farmers and ranchers who practice regenerative agriculture honor the soil as a living being and engage in practices that nourish the soil. When growers implement no-till methods, composting, and cover cropping, compacted, dead soil comes alive in a wriggling black moist mass of dark loam. Consumers at farmers markets and farm-to-table movements are starting to join the dance as well. By rebuilding a relationship of trust between consumers and producers, we can move beyond the dark era of zombie agriculture. We humans are like soil in that we thrive in relationship with one another and all of our kin. This new era of agriculture that is restoring connection and reciprocity will bring us full circle back to what it means to be a producer, consumer, and living being.

LIGHT PASSING THROUGH MANY PRISMS.

JULY

Negotiations

MARVIN'S STEPS

JOSH SLOTNICK

A new young farmer called the other day looking for a pump
5 horse, 3 phase
did I know of one?

Years ago
the turn of the millennium
on a bluebird fall day
I walked this place with Marvin Gallagher
while my young family waited across town

We passed listing corrals, walked through a slotted-with-daylight barn
around a sway-backed milk house
gazed at his soaring new RV garage
with gleaming metal roof
and he told me what I needed to know
about the underground lawn irrigation, the clean-out for the septic
the well for the house

I told him how we would pull those corrals
till up the pasture
build a greenhouse, a wash shed, chicken coop, 8-foot fence
We were going to move our farm here

That spring, to dodge the bindweed seeds from the ditch
we put in an irrigation well
Marvin's 5 horse 3 phase pump
that we never used
has sat perched over the ditch, at the ready
ever since

I might have a pump for you
I said
I'll call you after I pull it

A corrugated metal cover wraps the pump like a quonset roof
arches just right
and slides into grooves
notched
in a platform of treated 2 x 6s

I needed a cheater bar to break the 4" union
Marvin teflon taped every junction
double hose clamped 2" poly over metal barbed fittings, greased the zerks
He did every step
just right

I am fifty years old now
Our big kids have moved on, those things I built when we first came here
have weathered
I am 2/3 to where Marvin was, that shining fall day, at the turn of the millennium
when he stopped short my fast-paced list
of what I would do
to his place
our place, right then

He cut me off
with the gentle, hard hand of a schoolteacher

.......Yeah
He said
I moved here in 1961
in my early 30s, had a couple little kids, a new business
was all full of piss and vinegar

He tapped a fat finger on my chest, sucked in a long
emphysemic breath

Just
like
you

Come pick it up
I said to the young farmer

Wish me luck
He said
Sure, but that's not it
Do what you know you ought to
each step
and it'll work out

All of it
could turn out
just right

THE BUSINESS SIDE OF FARMING

BINYAMIN KLEMPNER

I was on the phone earlier today with a friend who operates a bed and breakfast, and he was complaining about his neighbor, a local sheep farmer. Unfortunately for my friend, the smell (or according to my olfactory senses: aroma) of the same manure that will put me in business is threatening to put my friend out of business. It happens to be that as I build my worm farm, this farmer's sheep poop will be the main entree for my worms' hungry appetite. As my friend and I spoke, he lamented the brazen stubbornness of his farmer neighbor, and took the liberty of extending that quality to all farmers. After we got off the phone I got to thinking: Well, after all, I am stubborn and brazen, and if I weren't, I wouldn't have stuck it out in agriculture after being forced to close the farm I began my agricultural career managing.

Worm farming is straightforward and chilled out. I think that's what attracts me to it. That, and the fact that I love making vermicompost. But I am running a business. Just starting out in this endeavor, I'm in the process of acquiring a five-foot by sixty-four-foot continuous flow worm bin, land, a hoop house, a compost tea brewer, and a small plethora of other worm farming equipment—and of course, worms. I need to get the best prices, which causes inevitable disappointment for the sellers who don't get my business. It makes me feel bad. I don't want to cause disappointment. My father and my wife tell me, "You're not disappointing anybody. That's just the nature of business. Don't like it, don't do business." But I feel conflict and assume many other young, especially greenhorn, farmers do, too. I want to be a nice guy who grows worms for their castings. I want people to like me, to tell me my work makes the world a better place to live. But if I don't get the best prices the market will bear, if I don't get the best contracts, I'm hurting my farm, a farm I will eventually bequeath to my children and grandchildren.

In my early days as a farm owner, there were a couple guys who were in the habit of driving their off-road buggies between my rows of tomatoes and cucumbers. I asked them repeatedly, nicely, to stop. But it was too much fun, and they continued. They were destroying the farm I was tasked with protecting. With patience running out and desperation sinking in, we called the police. There was little they could do. The following week, this group of young men invited a bunch of friends over, each with his or her own off-road vehicle. As one charged at me with his vehicle, I grabbed a pole, stepped aside, and held it out like a clothesline. Nobody was injured, or even touched. By the grace of God, this boy, as well as his friends, realized that they had trespassed one too many times, and were fortunate to be alive and well. While we all want to be nice, a farmer needs to protect the wellbeing of the plants and animals living on the farm. Unfortunately, it wasn't the last time I had to deal with an unfriendly trespasser.

Reflecting on the unpleasant aspects of running a business, I come to realize that protecting the fiscal integrity of the farm through solid, savvy, and, at times, unflinching and unapologetic business transactions is no less a part of protecting our farms than protecting our farms from trespassers set on making donuts in our cabbage fields with their Polaris tires.

I enjoy availing the farm as a resource for volunteers, researchers, worm aficionados, and solace seekers, but if someone needs to be asked to leave, I don't hesitate. And as I get older, already past forty, I know it's time to quit the fiscal flakiness and be solid. As farmers, we need to understand that while we may love what we do, we are running a business—and the tighter the business, the better the outcome. When a business decision is made that supports and promotes the best interest of the farm, there is no place for apologies or regret, only a pat on the back (and a cold beer) for a decision well made.

"THIS, SIR," SAID MR. ANTON'S GUIDE, "IS OUR FARM; AND WE MAKE IT PAY."

TAKING COMMUNITY SUPPORTED AGRICULTURE TO THE NEXT LEVEL

The Case For Investment Crowdfunding

CAMERON RHUDY

Like many businesses, when farmers need money for operating expenses, to purchase equipment, or lease or purchase land, they often look toward friends and family, banks, the Farm Service Agency (FSA), and sometimes even credit cards. But for many, particularly those from historically disadvantaged communities, raising money from friends and family is simply not an option, and conventional financing strategies rely on institutions that have a history of exploitive, extractive, and discriminatory practices, including predatory lending and redlining. Getting a loan from a bank or the FSA can be difficult because eligibility for these conventional financing options are primarily based on financial factors, such as cash flow, projected revenue, Schedule F income history, and credit scores. These eligibility standards help lenders avoid risk but don't account for the unique and non-financial circumstances of each farmer or enterprise.

So what if we took the power away from the large financial institutions and the government and gave it to our communities? What if

we harnessed the collective wealth of the community through strategies such as investment crowdfunding, and let the community decide who should get capital?

Donation crowdfunding, whereby farmers can raise money in the form of donations or in exchange for nominal rewards, has become increasingly popular, but it can be difficult to raise a significant amount of money solely through donations. For example, the average raise amount for projects on Barnraiser, a crowdfunding platform for food and farm-related projects, is about twelve thousand dollars. Investment crowdfunding takes the concept of raising small amounts of money from a large number of people a step further, but instead raises money from "investors" who either lend the enterprise money or buy an equity stake (i.e., become a shareholder) in the operation.

Raising capital in this way minimizes the risk for any particular investor and also allows a farmer to control the terms of the loans or structure the equity investments in ways that are more aligned with the practicalities of farming.

This could mean longer grace periods before repayment begins, more reasonable interest rates, or shares that have limited or no management rights, so farmers still maintain control over the business.

If this sounds familiar, it's because community supported agriculture (CSA) programs are, in some ways, a form of community capital raising. CSAs allow farmers to raise money directly from their customer community through pre-sales of produce or other products. But there is a limit to what individuals, restaurants, and other CSA customers can consume in any given period of time. One family can only eat so many carrots. Investment crowdfunding breaks through the carrot-consumption barrier, by creating another opportunity for customers, and other community members, to financially support farms and farmers in their region.

Another compelling reason for considering raising capital from the community is its potential as a locally-controlled, wealth-building tool. Most of the financial institutions that lend to farmers are not locally-owned, which means that the income they generate from lending to farmers pull valuable resources and money out of the community, compromising a local economy that may already be struggling. With investment crowdfunding, however, financial returns are largely retained locally, where they can recycle within, and sustain, the local economy.

Like with any investment—or farming for that matter—there's risk involved. And raising capital from the community may not be the best fit for every farm enterprise. It can require increased legal compliance, for example.[1] There is also a relational risk that comes with community investment. Loans can be defaulted on, businesses can fail. For farmers, the success or failure of the business is much more visible when the farmer knows their investors and their families. Of course, this could also have the positive impact of a higher sense of care and support from community investors who are

invested in not just the profit-making business, but in the people.

Community investors also need to make an assessment of the farm enterprise they are considering investing in to ultimately determine whether it's the right fit. Yet, investment crowdfunding allows people to take a more holistic approach to their assessment of risk and reward. In addition to factoring in financial considerations, community investors can also base their decisions on who grows the best organic tomatoes or grains, who raises pastured chickens, or who implements climate beneficial farming practices. Beyond the farm itself, local investment creates space for financial transactions to prioritize community benefits like increased access to healthy food and stability of farmers as dedicated stewards of land.

Despite the risks, strategies such as investment crowdfunding are worth considering along with conventional financing options. More and more people are asking and learning about where their food comes from, and we are discovering just how powerful a community can be when the decision-making power is taken out of the hands of large financial institutions and corporations. A food system full of farms that are both sustainable and locally-funded may be closer than we think.

NOTES

1. The amount of legal assistance required will depend on many factors, including whether the offering will be conducted using an online platform that facilitates investment crowdfunding (e.g., Wefunder, although there are many others). These platforms do charge a fee, but often provide templates for the documents that will be needed, which can reduce legal fees. On the other hand, online platforms have limitations and are not always the best fit, in which case more legal assistance may be necessary. Farmers can reduce their reliance on lawyers by learning some of the legal basics of raising community capital. Although the discussions on state law are specific to California, the Sustainable Economies Law Center's *Grassroots Financing Guide for California Farmers* (theselc.org/grassroots_finance_for_farmers) is a great place to start.

GARLIC PLANTING

M.T. SAMUEL

New moon to full
has found me
speaking
in a tongue that twirls with the facets of frost
left in the shadows
when the sun comes around,
late morning.

Words, newly acquainted,
have established themselves
residents
in my repertoire,
rooting me down
into rows
studded with granite stones
and garlic cloves.

Twilight magic silences the planting
and the sun settles peachy
at the edge
of the potato field.

Two sleeps on
and I'm pulled from the land
and put on the rails
with these words
small scuffs on my boots
and a crown of dried flowers.

Originally published in Abandon.

Frank Utpatel | At Day's End
Used with the permission of Stanton & Lee Publishers, Inc.

H-2A, THE US "GUEST WORKER" PROGRAM

The Farm Labor Dilemma

ELIZABETH HENDERSON

Petition for a Nonimmigrant Worker

Department of Homeland Security
U.S. Citizenship and Immigration Services

USCIS
Form I-129
OMB No. 1615-0009
Expires 12/31/2018

Farmers in need of workers often face a painful dilemma when hiring: revenues from farm products are too low to pay attractive wages, most local residents prefer jobs that are cleaner and less physically demanding, and many of the available skilled farm workers are undocumented immigrants. Knowingly hiring an undocumented worker is a felony. As the Trump administration's anti-immigrant policies continue to unfold, farmers and farm organizations have pressured to expand the H-2A "guest worker" program.

The program, which provides farmers a legal avenue to hire seasonal workers who are not US citizens, has its share of detractors.

Farmworker advocates question the program's fairness both to participating workers and to the undocumented workers, who make up fifty-five to seventy percent of current farm labor, and who may lose their jobs to workers who come to the United States through the program. In December 2016, United Nations Special Rapporteur Maria Grazia Giammarinaro came to the U.S. to observe the H-2A and B programs and reported that "the temporary visa for migrant workers, in the agriculture or in other sectors, which ties every worker to a particular employer, exposes them to the risk of exploitation and trafficking, as they are prevented from denouncing exploitation for fear of losing their job or their residence status." Farmers complain about the H-2A program's paperwork and bureaucratic snarls.

Despite the shortcomings of H-2A, some farmers have been using the program for many years and rely on it. Chris Cashen, of the Farm at Miller's Crossing in Hudson, New York, says, "We have been bringing back the same men for fourteen years. Hiring H-2A workers has enabled us to build our business on reliable labor." This is why he has gone to the trouble of using the H-2A program to hire workers from Guatemala for seven months each year.

The US Departments of Labor and Justice jointly instituted H-2A in 1952 to meet the need for seasonal and temporary labor on American farms, granting foreign nationals legal permission to work in the U.S. at specified jobs for up to ten months before returning home. Year-round workers are not covered, but in 2017, Virginia Congressman Bob Goodlatte, Chair of the House Judiciary Committee, proposed new legislation to replace the H-2A program with an H-2C program, which would allow three-year permits and weaken the few worker protections in the H-2A program. Goodlatte's bill has not passed.

Today, three federal agencies within the US Department of Homeland Security manage the H-2A program. The US Department of Labor (DoL) issues the H-2A labor certifications and oversees compliance with labor laws. The US Citizenship and Immigration Services (USCIS) processes the H-2A petitions from employers, and the US Department of State issues visas to workers at consulates in their countries. State departments of labor make on-farm inspections and state departments of health inspect the housing, which farmers must provide free of charge. Not surprisingly with so many agencies involved, miscommunications and delays occur. Until this year, applications had to be sent by snail mail. But there is no other legal program that allows farmers to bring agricultural workers from other countries to their farms.

As the number of H-2A workers has risen over the past decade—from fewer than 80,000 in 2006 to more than 200,000 by 2017—the complexities of the program have resulted in increasing delays in the processing of labor certifications, visa petitions and interviews for the final border crossing, and thus delays in worker arrivals at farms. Some farmers complain about workers not arriving in time for harvests and about the confusing amount of paperwork. But for farms like the Martins in Brockport, New York; Indian Ladders near Albany, New York; Zaid Kurdieh at Norwich Meadows Farm in Norwich, New York; and Chris Cashen's Farm at Miller's Crossing, the benefits outweigh the shortcomings.

The first year, Cashen says, he hired an agency to help with the paperwork, but found they were expensive and slow. So he and his wife, Katie, decided to do it themselves. The process begins in November when they post their job opening with the New York Department of Labor. The job is then available to domestic workers who have the experience required as defined in the work order. Then they post the job statewide, for the upcoming seasonal position. The next step requires authorization from the DoL, which can only be accomplished after moving through the New York state posting process. Once approved by the DoL, the next agency stop is Homeland Security, for identification of the business applying, and then onto USCIS. All of this takes ninety days.

Every year, the Cashens must prove that "there are not sufficient workers who are able, willing, qualified, and available." To do this, they must advertise the jobs in local papers and in two other states and keep a recruitment log. (Cashen remarked ruefully that he has only managed to recruit one worker with these ads, and that worker only stayed two years.) Once all necessary approvals are in hand, the Cashens can hire internationally, and choose which country the farm would like to hire workers from. In the Cashens' case, after USCIS gives the green light, their eight workers go for interviews at the US consulate in Guatemala. They must demonstrate that they have had no arrests and that they have solid attachments to their home country. It takes three to five days for them to get visas. Meanwhile, Cashen buys them round trip plane tickets.

Once the H-2A workers arrive at the Farm at Miller's Crossing, they work ten to twelve hour days. The farm has to comply with all federal and state labor regulations, and workers are covered by most US labor laws, including the Wage and Hour Act, Worker's Compensation, and the Affordable Care Act. The Cashens also have to purchase unemployment insurance for these workers—a mandate only required in the state of New York—even though non-citizens are not eligible for unemployment insurance payments. They must pay H-2A workers the higher of federal or state minimum wage, the applicable prevailing hourly wage rate, or the applicable adverse effect wage rate, which, in 2016, averaged ten to fourteen dollars an hour in New York. In addition, the Cashens must pay at least that rate to any local hires doing the same job. They also hire two to three locals who work year-round.

Before switching to H-2A, Cashen's farm hired interns. After a few experiences with interns who left after deciding that farming wasn't for them, Cashen gave up trying to both run a farm and be a teacher-mentor. He says he is comfortable with his current labor situation. The H-2A men live on the farm and, as Cashen puts it, "they are there for the farm," which has seventy acres of vegetables and a CSA with one thousand members, as well as wholesale markets. The men will work alongside him from dawn till dusk, and can each earn up to twenty-five thousand dollars in the seven- to eight-month season. Cashen says they are well worth it, as he has found them to be "professional, skilled, intelligent men, who do not drink, are religious and good citizens." In fourteen years, he has had no serious conflicts, and there have been few personality clashes among the men.

At Indian Ladders Farm, an apple farm that celebrated its centennial anniversary last year, Peter Ten Eyck says he is a "big fan" of H-2A, and has used the program for over thirty years. Ten workers come to his farm from Jamaica to harvest apples for six to seven weeks, most of them the same who have come year after year. A co-operative of Hudson Valley farms that Ten Eyck belongs to helps with some of the paperwork. Transportation and all the fees cost $1,090 per worker and their pay rate in 2017 was $11.76 per hour. The many inspectors to his farm wear deep ruts into the road—there is the Jamaican government, the county health department, and the Occupational Safety and Health Administration. H-2A housing must be up to code and, as Ten Eyck reports, there are plenty of inspections to ensure this.

Cathy Martin of Martin's Farm, a large vegetable farm in western New York, hires over ninety H-2A workers a year. She explained that it has been almost impossible to get enough

Clockwise: Latino migrant farm workers in California: 1936, 1972, and 2013

increasing focus on border security, employer compliance with immigration laws and curbing illegal immigration, agricultural employers have fewer and fewer options to find workers. The H-2A program, with its significant regulatory burdens and increased costs, is already a cumbersome program… It is imperative that the H-2A program is reformed or replaced..."

The National Sustainable Agriculture Coalition (NSAC) urges an approach that gives the workers a portable work visa and allows the labor market to function. The NSAC proposal envisages creating North American Agricultural Work Visas, "dual intent" visas that would allow guestworkers to change employers and to come and go across the border.[1]

To create a truly sustainable solution, the "dual intent" visas must be part of full immigration reform based on human rights. The same freedom to cross the border legally should be instituted for the eleven million undocumented people in the U.S., including the million or more undocumented farm workers. Improving the guestworker program does not go to the source of the problem: the U.S.'s cheap food system functions as long as there are sources of cheap labor. In a farming system worth sustaining, work as a professional farm worker will be a respected vocation that provides living wages with decent benefits. But while we are transforming the cheap food system, we can at least reduce the injustices farmers and workers must navigate in order to survive.

Elizabeth Henderson is lead author of Sharing the Harvest: A Citizen's Guide to Community Supported Agriculture *(Chelsea Green, 2007), and a member of the boards of the Northeast Organic Farming Association of NewYork and the Agricultural Justice Project.*

NOTES

1. See sustainableagriculture.net/blog/immigration-reform-principles/ for more detail and a link to a description of North American Agricultural Work Visas.

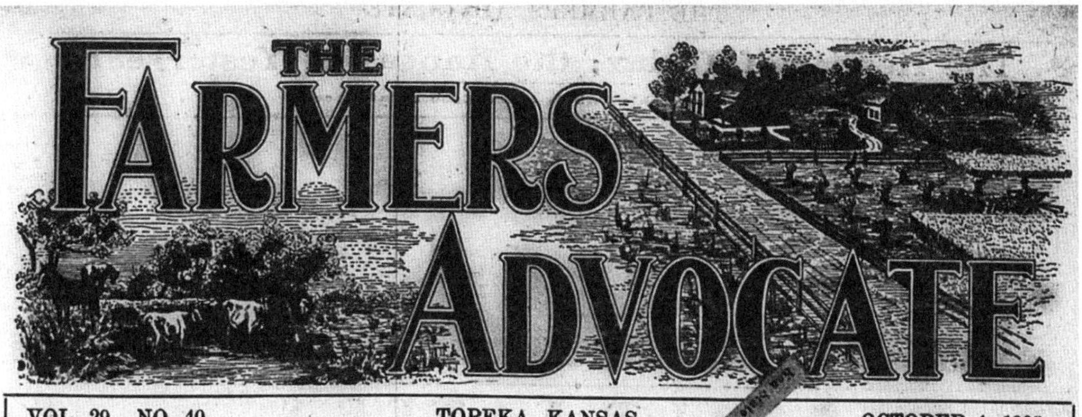

THE FARMERS ADVOCATE

VOL. 29. NO. 40. TOPEKA, KANSAS. OCTOBER 4. 1906.

MILLIONAIRES

Greatest Failures in Our Modern Life, Says William Allen White.

MILLIONAIRES are the greatest failures of modern life, but in spite of them and their influence the human race is nearer the "eternal justice" than ever before, according to William Allen White in the American Magazine for October.

In an article entitled "the Partnership of Society," Mr. White presents some relations between various classes of society calculated to show that "the battle line of the race has moved far forward."

"Today," he says, "we are fighting the world-old fight for the establishment of ideal justice between men. We know there is such justice; we know that it can come, and it must come before the contest ends. And we know that each century has brought ideal justice a little nearer reality. We know that something in our souls is seeking for the great harmony of life."

Mr. White then proceeds to deliver some blows against the selfish interests that have stood in the way of the steady progress he claims for society.

"As men in ancient days sought advantage over their fellows," he says, "so men today seek their special privileges in the partnership by money. * * Many a man bends his energies to make money that he may do good with it, not realizing that the same energy devoted to helping his fellows would accomplish much more good and leave him a better man.

"If, for instance, Jacob Riis had put off helping the poor of New York until he had made enough money to hire some one else to do it, he would have accomplished little.

Then Mr. White pays his respects to millionaires in this way:

"The greatest failures in our modern life are our millionaires. As a rule, they have accumulated money without giving society a just and equitable return for that money, they have acquired what seems to them a vast amount of power, without intelligence to use it, and they are going through life looking for joy and happiness, but finding only pleasure that burns out their souls and does not satisfy their hearts."

Of New York millionaires in particular, Mr. White says:

"We know America fairly well; it is probably as highly civilized as any other part of the globe. In New York city there are said to be 5,000 millionaires. Probably there are 10,000 or even let us say 20,000 men who are nearly millionaires, and 50,000 more who are living in the blessed hope of becoming millionaires reasonably soon.

"Let us say that there are 100,000 people who certainly are inspired by the love of money. These 100,000 people have killed the social instinct in their own hearts. They serve their fellows only for the money there is in it. They live parasitic existence. But what of the 3,000,000 other men and women in New York? Is the civilization of New York dependent upon the 100,000 parasites, or is it dependent upon the 3,000,000 people?

"Three million people are working day by day for money with which to buy the necessaries and comforts and luxuries of life. The 3,000,000 people devote eight hours every day to money-getting; but what of the other sixteen hours during the day? In the eight waking hours that are left what a vast amount of work is done for the love of it; and as we descend to those levels which are falsely called the lower levels of society—to the poor—what a vast amount of social work is done without the thought of pay."

Mr. White continues by declaring that it is love of money, legalized, that results in the injustices and immoral inequalities, "that deaden the consciences of preachers and allow them to take thousands from big thieves who are impenitent; that makes men take railroad rebates in secret which they confess in low-voiced shame in public on the stand; that makes our college presidents plead for a business system founded on deceit, chicanery and competitive malice.

"The whole of high finance has been shaken in the five years last past by exposures which showed how men in frock coats would steal and cheat and swear falsely—not for bread and butter, nor for the comforts of life, not even for the luxuries that money will bring—they might have had these things honestly—but for the lust of money!

"It is the problem of this generation and the next and the next to civilize the conscienless devil of malicious greed out of the hearts of men."

The Price of Being a Radical

By L. T. Breeling
Sanish, N. Dak.

Often we hear expressions of fear concerning the work of some one whom the world deems a radical.

Very little use to worry. The greater per cent of the human race do not want to think. The only thinking they do is when compelled to by the turbulance of some radical. The ripple soon passes from the surface and the majority settle back. We are especially fortunate in that we have a few who, when some one with independence enough to think for themselves steps out of the beaten track, will follow.

Very little progress would we make if we were not fortunate enough to have these radicals to take conventionalities in hand and shake the dust from them and expose their rotten vanities.

Whenever a person comes forth with stability, courage, and character enough to voice a new principle, the world looks aghast. A hue and cry is put forth that we are degenerating. We are becoming impious and sacrilegious. The change may appear at first to be retrogression, but it is only the step backward for the long running-jump ahead.

The triumph of the new truth which declared that the sun is our central orb, that the earth and the planets are a family of worlds, must be attributed to Galileo. He was a radical. What price did he pay? He was assailed from all sides with malice, ignorance, and ridicule.

He was imprisoned in a cell in the Inquisition. He was compelled to go down on his knees and renounce his teachings. Upon rising, he is said to have uttered in an undertone, "The sun moves, for all that."

It is easy enough to trail along with the noted. It takes energy, and nobility of purpose to oppose. The battle of the radical, shaking up conventionalities has lasted through centuries in the past, so it will last through other centuries in the future. These radicals, breakers of idols, have done more for the world's progress than all the hereditary or self appointed rulers that have ever lived on the earth.

We, in our own time have never given full credit to our greatest benefactors.

Thomas Paine was belittled, even called "a filthy little atheist". He was neither atheist, agnos, nor infidel. He attacked theology, but was a firm believer in God. He said, "The world is my country; to do good is my religion."

In January, 1776 when Paine published his first political argument, "Common Sense", it fell on the divided, undecided colonists like a bomb. But it, and it alone was given full credit by Washington, Jefferson, and Adams for having decided the issue. If any one person is entitled to credit for the Declaration of Independence, it is Thomas Paine. He it was who had courage and nerve to mold public opinion to that end.

When Washington's army had dwindled from eighteen thousand to a scant one thousand men, it was Thomas Paine who seized a drum and using the head for a desk, with his pen dripping fire, wrote "The Crisis". All that winter Washington ordered Paine's Crisis and other papers to be read at the head of each regiment.

As a nation we owe more to Thomas Paine than to any other human being. Every man who reads history knows we could not have won without Washington, and Washington could not have won without an organized public opinion back of him, and Paine is the one who organized it.

He, like other rebels and "breakers of idols," have their place— they have made men think and have spurred them to action. He too, paid the price.

People who are fearful of new ideas are always advocating suppression of the radical. Whenever we suppress, it is simply because the views expressed are not in accord with the views of the majority.

Look back to war times. The majority, we say, were in favor of war. Were they? (We had just elected Woodrow Wilson President because, "he kept us out of war".) Well, the dictating power was in favor of war. Numerous men and women were tried, convicted and imprisoned because they dared speak or write against war. Did they speak or write anything more bitter, more opposing, more radical than the pacifist writes today? No—but then they were opposing—were Radicals. Now, the pacifist writes the same propaganda. Now, it is safe and sane, because the majority are talking peace and the Peace Pacts must be advocated.

In the presidential campaign of 1896 W. J. Bryan was one of those ranting radicals. Where will we find a more strenuous speech than his "Cross of Gold"? The conservatives said that if Bryan's principles were put in force in the United States, the nation was doomed. But lo, before a decade had passed over our heads, we had accepted all of those direful principles except the ratio of sixteen to one.

In Los Angeles Max Rosenstein has been denied his diploma from high school because he is a communist. Think of it! In the twentieth century, asking a student concerning his politics or religion, when attending a public school, supported by public tax. As well ask whether his grandparents ate light or dark bread, or whether his mother does the family wash or sends it to the laundry.

If communist ideas are wrong, if they are base and vile, worry not. Give all the communist students diplomas, for by so doing you have given them an opportunity to learn other theories by attending high school. If wrong, their theories will die. Trying to oppress will only antagonize and make the communist spirit rise higher and spread farther. If their theories and dogmas are bad, they need not be feared. They will burn by their own fuel, die their own death, and bury themselves in an ignominious grave.

For the time being it is a sin and a crime to have certain ideas. Tomorrow they may be in popular favor. The crime of our age has more than once become the glory of the ages which followed.

Thirty-five years ago what is organized labor today were the radicals. Now they are organized. Now farmers are radicals because they are organizing.

Coming home from the Convention at Bismarck, I heard an elevator man tell a farmer that wheat might just as well keep on going down in price because the farmers are never satisfied anyway. Well, being a mere woman, I didn't offer any information, but I think he may soon learn, when we are going to be satisfied.

As a class we are showing our radical views, and we are blessed with enough so called radical leaders to put the program across. We are in disfavor today, asking too much, but when the equilibrium has been reached, and we are given a market that isn't less than cost, our preposterous ideas accepted, some one else will have to be the goat. We will cease to be radical.

Often times the theories which may benefit the future, unsettle the present. There must be an upheaval to liberate and free the minds; get them out of their groove so they may expand.

Look not for praise and glory for any ideas and principles put forth unless you are walking closely, very closely with the crowd. Rather listen for a wail of horror and prepare to be cast out with the cry of Radical hurled at you.

Where in the history of the world do you find the gratitude of a people given to their great reformers while those radicals are living? Men and women who have taught the world to think have gone down to defeat, as far as they knew of the world accepting their views. The monuments and statues we raise after they are dead.

BURY ME IN A FREE LAND

FRANCES ELLEN WATKINS HARPER

Make me a grave where'er you will,
In a lowly plain, or a lofty hill;
Make it among earth's humblest graves,
But not in a land where men are slaves.

I could not rest if around my grave
I heard the steps of a trembling slave;
His shadow above my silent tomb
Would make it a place of fearful gloom.

I could not rest if I heard the tread
Of a coffle gang to the shambles led,
And the mother's shriek of wild despair
Rise like a curse on the trembling air.

I could not sleep if I saw the lash
Drinking her blood at each fearful gash,
And I saw her babes torn from her breast,
Like trembling doves from their parent nest.

I'd shudder and start if I heard the bay
Of bloodhounds seizing their human prey,
And I heard the captive plead in vain
As they bound afresh his galling chain.

If I saw young girls from their mother's arms
Bartered and sold for their youthful charms,
My eye would flash with a mournful flame,
My death-paled cheek grow red with shame.

I would sleep, dear friends, where bloated might
Can rob no man of his dearest right;
My rest shall be calm in any grave
Where none can call his brother a slave.

I ask no monument, proud and high,
To arrest the gaze of the passers-by;
All that my yearning spirit craves,
Is bury me not in a land of slaves.

Lägerplats i Patagonien (Bagualesområdet).

SOME PASTURES

A Cowboy's Almanac

SAM RYERSON

July, Shipes Pasture—Monida, Montana

We moved the cows and calves in my group up to the Shipes Pasture in July, after Joe helped me sort the bulls out. It was the highest pasture on my side of the ranch, right along the Continental Divide and the Idaho state line on the south. I had gone around the fences in June. Up there on the Divide, a lot of it was let-down fence, four strands of barbed wire that you let down on the ground in the winter so the snow and wind and elk didn't destroy it. In June I'd picked it back up, out of the new grass and the remaining snowdrifts, and hooked it into the staples or wires that held the fence on the posts.

They were big steep open ridges up there. Not many cows ranged right up to the top, but Wally's Scottish Highlander cows did. They were shaggy old cows with more hair and horns than beef, but they used the country like goats while the big pretty Black Angus cows lay along the creeks in the shade. All the calves the wolves killed belonged to those black cows. They killed a few and a few more got sick and died, of pneumonia or high-altitude disease. I was working for a grazing association of about twenty member ranches that all ran cattle together in the summer, and owned shares in the association cooperatively and paid the summer crew. We had about four thousand pairs of cows and calves that summer. There were about nine hundred in my group. We sorted the cattle by their brands in the fall and shipped the calves to sell and sent the cows home to eat hay all winter.

Up there in the Shipes in the middle of summer was the perfect riding job. I rode every day from Middle Camp, about five miles up Beaver Creek, to the gate. I had some colts to ride and two good broke geldings. One big blue roan colt I was riding for a woman from town was so tall and narrow, it felt like he might tip over in the wind along the top of the Divide, and I got vertigo when I rode out on him to check that fence. I watched the calves for health and doctored a few and pushed the black cows up out of the creek bottoms to graze higher on the ridges. Working alone with my dogs all week was good, but I was glad when I ran into Scott on the road to his Forest Camp, when Wes came up from Headquarters to check on me, and on weekends when Laura Jean came up to visit in the old orange car.

ABSENCE MAKES THE HEART GROW FONDER

Mirroring Crop Diversity on a Restaurant Menu

LUCIA JAZAYERI, WITH PHOTOS BY JENNY ZHAO

They arrived from North Carolina in May. Twenty-five thousand sweet potato "slips." A slip is the green shoot grown from a mature sweet potato that's used for planting. Farmer Ray Young and the crew at Next Barn Over Farm planted the slips in two acres of earth in the small farming community of Hadley, Massachusetts.

"They look horrible at that point," Ray says. "We have a joke about ... don't even go out and look at them for the first week after planting because you'll just feel demoralized." I imagine little gray demons sprouting from the stems, but I think what Ray means is they look like they're not going to survive. Yet with the right growing conditions, each slip will grow roots and sprawling greens and pounds of sweet potatoes below the soil. If the season is especially good, each slip can yield two to three pounds of sweet potatoes.

Four months later, with the first leaves on trees in the distance turning yellow, my coworkers and I watch a crew of eight farmers pulling thousands of sweet potatoes out of the ground. It is September. The sweet potato season was a good one. The sweet potatoes came early and bountiful, about 125 percent more than expected. When the farmers pull them up, they make a satisfying sound, roots ripping. They are bigger than I could have thought possible. Some are the size of a large trout. One farmer shows me a picture of a recent sweet potato catch with a human forearm in the picture for scale.

There's something about sweet potatoes and the soil in Hadley. Located on a floodplain of the Connecticut River, the soil is sandy, which sweet potatoes love. It's also full of organic matter, the result of seasons of careful crop rotation. The farmers' hands are covered in light gray; the farmers call their land "sugar sand" because it grows such sweet vegetables.

"Do you mind if they take some photos?" Ray says to the crew, gesturing at us and laughing as he adds, "These guys own most of the sweet potatoes in this field." We work for Clover Food Lab, a mini empire of locavore restaurants in Boston conceived in 2008 as an experiment in serving vegetables to carnivores. When we put a sweet potato sandwich on our menu, we're not just putting sweet potatoes on the menu because it seems like a good idea. We're strictly following the ups and downs of a season on a farm.

We are not farmers, but we love farmers and we believe that the relationship between farmers and restaurants should be closer. "Farm fresh" items should not be a garnish. We change our menu to accommodate the produce coming out of this field and other fields in Massachusetts. And somehow we've convinced scores of customers to follow us on this journey. Ten years and forty seasons later, Bostonians are eating parsnips and sweet potatoes and pawpaws and brussels sprouts raised by the network of farmers we've built, mostly through word of mouth.

With fourteen years of experience in organic farming, Ray grows forty crops on fifty acres, and pays special attention to crop diversity. The two acres where he and his team raised sweet potatoes this year grew salad greens last year and were kept fallow the year before that. Unlike conventional farms, which often focus on one crop, Ray sows many crops and rotates them year to year, even planting crops like oats, peas, and rye that fetch little to no profit and are just turned under the soil to invigorate it with added organic matter. The following year's crop is more productive, less susceptible to pests and disease, and sweeter.

None of that can protect the farm from the forces of a rainy New England summer. Although the farm is buzzing with activity today, one month before, things were looking dismal. This summer was the worst in ten years for small farming in western Massachusetts. Heavy rains started one day and didn't stop, saturating the fields and making ponds. Tomatoes, zucchini, and peppers rotted on the vine. Vegetables grew to the right size but not the right color. If you picked up a squash it would disintegrate, filling your hand with mush.

Where Ray practices diversity in the field, we practice diversity on the menu. It's a symbiotic relationship, one that is largely unheard of in restaurants our size (we have twelve restaurants and feed thousands of Bostonians breakfast, lunch, and dinner each day).

When a crop is good, you'll see it all over the menu at each restaurant. When a crop is not good, we don't serve it at all. The summer rains this year disappointed lovers of the Heirloom Tomato Sandwich (an extremely simple but popular sandwich made with a thick steak of heirloom tomato, mayo, cheddar, and lettuce). Instead of six weeks, we ran it for two weeks. Some years we have strawberries, some years we do not.

Necessity is the mother of invention, and for us, narrow constraints help inspire us to create better food. Our founder, environmentalist and MIT grad Ayr Muir, grew up in western Massachusetts, and he's led us in creating a database of hundreds of recipes based on New England's fickle but beautiful growing seasons. When we were unable to get good lettuce in the winter for a BLT, we turned to seaweed harvested off the coast of Maine, fried until it tastes even better than the bacon or lettuce in the classic version of the sandwich. In 2014, when Ray had as good a season for sweet potatoes as this year, he called us up and asked if we wanted to buy the extra five thousand pounds. We said yes, and my friend Enzo, a chef at Clover, started working on a sandwich to use them.

The Japanese Sweet Potato Sandwich, or JSP, as we call it, was born. To make the sandwich, we roast sweet potatoes whole with the skin on (you can do that when you know your sweet potatoes were never sprayed with pesticides). We slice them and fry them so the outside is crisp. Then we add a salad made with raw scallions, raw cabbage, raw shiso, and a

Sweet potato harvest at Next Barn Over Farm

honey miso dressing. The final flourish: handfuls of black sesame seeds, poured into a liquidy tempura batter. That whole mess is dropped into a deep fryer and fried until it resembles a many-eyed sea monster. We break the fried seeds apart and sprinkle them liberally on top.

The JSP is so popular that this year it has its own fan club. The inaugural meeting was October 3, 2018, and forty people crowded around a table to eat the sandwich. That initial bumper crop has turned into a standing order, and we are now buying thirty-three thousand pounds of sweet potatoes from Ray every fall. We've gone sweet potato–crazy every year: sweet potato soda, sweet potato lemonade, sweet potato salad.

Every year around this time, a woman arrives at our Harvard Square restaurant asking if the JSP is back. When it arrives, she orders five each day. We suspect she is freezing the spares. We get hundreds of comments every year asking why this sandwich can't be on the menu year-round. We tell them that they wouldn't like it as much. Ray's experience, the magic of sugar sand, the two weeks of careful curing in a barn—all of this makes a difference. We imagine a world where not everything is available year-round, and everything tastes better because of that. Absence makes the heart grow fonder, but only when the love is sweet to begin with.

2017 COMMENCEMENT ADDRESS TO CENTER FOR LAND-BASED LEARNING'S CALIFORNIA FARM ACADEMY

TOM WILLEY

I landed my first career-track agriculture job in 1975, working under a man named Lester Travis, who superintended Newhall Land and Farming Company's ten-thousand-acre El Nido row crop farm, south of Merced on the San Joaquin River. Les's Ichabod Crane–like spindle neck earned him the moniker El Pescuezon amongst field hands on the ranch. If you will work in agriculture, be forewarned: political correctness has made little impact on the farmworker community to present day. When body types suggest it, people are nicknamed fatso, skinny, whitey, and darkie. Don't take it to heart.

As a World War II veteran, Les sought solace for a wounded soul in the soil. In sympathy with my fish-out-of-water bewilderment on that vast Newhall acreage, Les shared his post-war urban-rural transition experience, telling how he'd once acquired a piece of land with no knowledge of how to work it.

"I drove all the surrounding neighborhood each morning, spying on farmers who appeared to know what they were doing," recounted Les. "Then, returning to my place, I performed those same tasks." Some years later, I adopted Les's neighbor mimicry with some success, and I recommend it to you.

Under superintendent Lester Travis served three foremen: George Uyeda, Lloyd Smith, and Roy Spears, who could all be described in one important sense as failed farmers. That's not to imply these gentlemen could not grow fine, high-yielding crops—had that been the case, Newhall Land and Farming would never have entrusted them with ten thousand acres of canning tomatoes, sugar beets, and alfalfa. What that trio failed to do on each of their family's generational farms was to earn profits enough to remain in business. They shared this fate with countless farmers, whose numbers have declined, over the last century, from three of every ten Americans to fewer than one in a hundred. Some of you graduates are of a mind to tilt that windmill and beat those odds, by establishing your own successful farm businesses. I'm here to witness that it can be done, but also to warn, it is very difficult.

George Uyeda, who managed Newhall's tomato canning production, enjoyed a somewhat deserved reputation among El Nido farmhands for being "so smart that he was stupid." George's decision to hire me for an assistant was judged by fellow foremen and many farm employees to be one of his stupidest moves. Unlike his colleagues, George valued book learning. Several nights a week, after a long day's work, he would drive fifty miles each way to attend Fresno State agriculture classes, ultimately earning the advanced degree he sought. At the time he hired me, all I had to recommend me was one year's book learning in botany, chemistry, irrigation, soils, entomology, plant disease, and tending student grape and cotton plots on Fresno State's campus farm. George took that chance.

Hardly able to identify a tractor's front end from its rear, I spent the first year at Newhall in mortal terror of killing myself or someone else. By year four, mentor George had moved on, and I was managing El Nido's entire seven-hundred-acre tomato crop.

So, hang in there.

George's dubious reputation owed in part to a fertile mind's fondness for experimentation, resulting in his rarely applying the same procedures to the accomplishment of any task more than once. I learned from George's example, and default, to repeat methods that demonstrate success, and to engage in experimentation only when they don't. Other precepts gleaned from my association with George were imprinted on the day I approached him with every intention of resigning. Sensing the clouds of doom that enveloped me, the shrewd mentor launched into a fatherly lecture before I could mouth one word.

"If working the land is not a deep commitment and love within you," George said, "it is the worst occupation in the world, a torture even." And a bit further into the lesson, he shared something he'd learned from bitter experience: "If you don't know where and how you're going to sell a crop for more than what it costs to grow it, all the resources, time and effort that go into its production are a complete waste."

I stayed on the job.

I never learned just how Lester Travis chose the land he took to cultivating after the war. When I shifted careers from criminal reform to farming, the San Joaquin Valley's reputation as the planet's most productive agricultural real estate lured me down there. At twenty-five years of age, with a sociology degree already under my belt, I had little patience for eighteen-year-old clod-kicking beer drinkers, sleeping off hangovers in class. I figured a year's worth of cramming on the most relevant agricultural curriculum I could cobble together just might land me a job on a real production farm, where I could truly begin the learning process.

Shaking my severe allergy to anything scientific proved a challenge in old Doc Arce's botany class, but what I learned from that five-foot-tall Cuban firebrand became foundational to my success as an agriculturist. We farmers are simply managers of photosynthesis: that singular, miraculous photochemical reaction that energizes all earthly life. Acquiring a

The lion gives advice

deeper understanding of how that process self-manages in highly productive natural systems would ideally be every farmer's intellectual quest. Incorporating mimicry of those elegant natural systems into the management of a farm's agroecology brings one nearer to that elusive aspiration for sustainable agriculture.

After climate, the most fateful choice anyone makes in locating a farm is soil, and another professor, Jim Brownell, taught a soils course at Fresno State that proved invaluable in establishing four iterations of T&D Willey Farms over thirty-five years, even when only the last represented anything near a prime quality soil.

The ingenious Storie Index, dating from 1933, is a method for rating the agricultural value of soils that is particularly well adapted to California's irrigated agriculture. University of California (UC) Berkeley professor and pioneering soil surveyor Earl Storie continued to refine the index up until his death in 1981. Storie Index ratings consider soil depth and texture (sand, silt, clay), water permeability and holding capacity, fertility related chemical characteristics, internal drainage, slope, and surface runoff, scoring a soil from zero to one hundred for suitability to produce crops common to its climatic region.

Detailed county-by-county soil surveys, conducted cooperatively between the USDA's Soil Conservation Service (NRCS) and UC's Agricultural Experiment Station, commenced in 1900, focusing on our state's important production regions as each developed historically. Most of the now classic surveys were published by the 1970s, while less prominent regions' surveys were not completed until the century's end. This priceless legacy, which represents countless hours spent characterizing gridded soil cores from every nook and cranny of the Golden State, belongs to all of us.

I've seen too many novice farmers choose a farm's location for lifestyle reasons over its soil quality. A soil's inherent productivity, its potential to sponsor photosynthesis, is what will pay a farm's mortgage, not the viewshed. Locating a farm before one has acquired adequate soil judging skills has proven to be treacherous and costly in many cases. Such mistakes can be avoided, and were in my case, by engaging Earl Storie as a partner. His Index was revised in 2005 to generate ratings digitally from NRCS' National Soil Information System, where all US soil survey data has now been compiled. In this techie upgrade, original hand-evaluated ratings were considered subjective because no single person generated them for the whole state. Well and good, but you also want to get your hands on those original print-published versions, if possible.

Unless you fancy dryland farming, California's most fertile soils will sponsor little photosynthesis over our seven-month dry season without adequate access to irrigation. Another error common to neophyte cultivators is establishing farms where too little water is available to maximize the productive potential of their land.

California's surface and groundwater hydrology is a vast mosaic, the complexity of which rivals the diversity of her soils. Access to water, legally and physically, can vary enormously from neighborhood to neighborhood. During the current—or is it past?—drought in our San Joaquin Valley region, disparity could be so stark as to find one farmer enjoying an inexpensive water source so generous that he can grow rice, while his neighbor just across the road helplessly watches an almond orchard wither for lack of irrigation. California recently passed the Sustainable Groundwater Management Act, which will soon impact every farmer's ability to extract groundwater at will, as has been the privilege of agriculturists since the Gold Rush era. Per acre sales prices of California agricultural land will always reflect the value of the land's water supply, quantitatively and qualitatively. Rents will do the same. Time and effort spent studying your region's hydrology, as well as its soils, will pay generously.

That returns us to George Uyeda's second admonition: to have a solid plan for earning more from a crop than you spend growing it, which his own failure to heed had transformed him from a freeholder into a wage laborer on Newhall's corporate spread.

You're likely familiar with the quip, "Farmers are price takers, not price makers."

At the dawn of the Holocene, as humans began growing food instead of gathering it from the wild (a practice of questionable wisdom), clever schemers among them figured out how to relieve cultivators of their surplus production for less than its intrinsic value. Agriculture—the origin of human social and economic inequality—has also extracted a heavy toll on earth's natural resource base, particularly now that forty percent of the planet's entire dryland surface is under human management for food production.

Some forty years ago, a loose aggregation of back-to-the-landers conceptualized a natural systems-based agriculture that might go easier on soil resources while providing higher quality, less toxic food for its cultivators and community members alike. The organic movement has made modest but significant advances toward those goals over the ensuing four decades, thus providing some economic refuge for family-scale farmers, a demographic otherwise headed for extinction. Certified organic fruits and vegetables will likely represent fifteen percent of the nation's entire produce sales this year. Most of this gain, particularly early on, was achieved through face to face relationships on farms or in farmers markets, and through dedicated mom-and-pop retailers who convinced customers, one by one, to pay higher prices for food that was better for them, the people who cultivated it, and the environment in which it was grown.

Organic market growth has now reached a tipping point, at which its innovators are pushed aside by large concerns that scale up the new wrinkle. This trajectory, well underway as Walmart and Costco vied to become the nation's largest retailer of organics, was recently punctuated by Amazon's takeover of Whole Foods. Where does that leave family-scale farmers like me, or people aspiring to become one, like many of you? Advanced age, retail consolidation, food safety issues, water shortages, labor's scarcity and rising cost, all afforded me excuses to retire from the madness. I can hardly suggest that to you. Because we no longer grow our own food, Denesse and I have been eating a lot of industrial organic produce, and it's a big disappointment. Input substitution and monoculture, legally Certified Organic or not, does not satisfy discriminating customers who seek flavorful and nutrient-dense food.

There lies your opportunity, and your potential for success, which, unlike McDonald's and Walmart's, does not depend on "Billions and Billions Served." You need, perhaps, a few hundred families for whom you can become a personal farmer. Perform that responsibility well, educate loyal customers to value your food and service as they should, and, while it may not bring you great wealth and fame, you may enjoy a life of integrity and modest economic success. Favorite business guru and *In Search of Excellence* author Tom Peters advised, "No matter how fierce the competition, there's always room at the top." Go there!

Imagine a farm that covers 3/4 of the State of California, and uses as much water as California, Ohio, and Texas combined. When you harvest that farm, it is enough food to fill a tractor trailer every 20 seconds, and then it drives all over the country, except instead of going to people to eat it, it goes straight to the landfill. That is essentially what we are doing today. In fact, food is the number one product entering our landfills today.

--DANA GUNDERS, SENIOR SCIENTIST, FOOD AND AGRICULTURE PROGRAM, NATURAL RESOURCES DEFENSE COUNCIL

From Committee on Agriculture House of Representatives, 114th Congress, 2nd Session. "Food Waste from Field to Table," May 25, 2016. Serial No. 114-52, US Government Published Office, Washington: 2016.

CAUCASIANS.

AUGUST

Reparations

BEHOLD THE LAND

W. E. B. DU BOIS

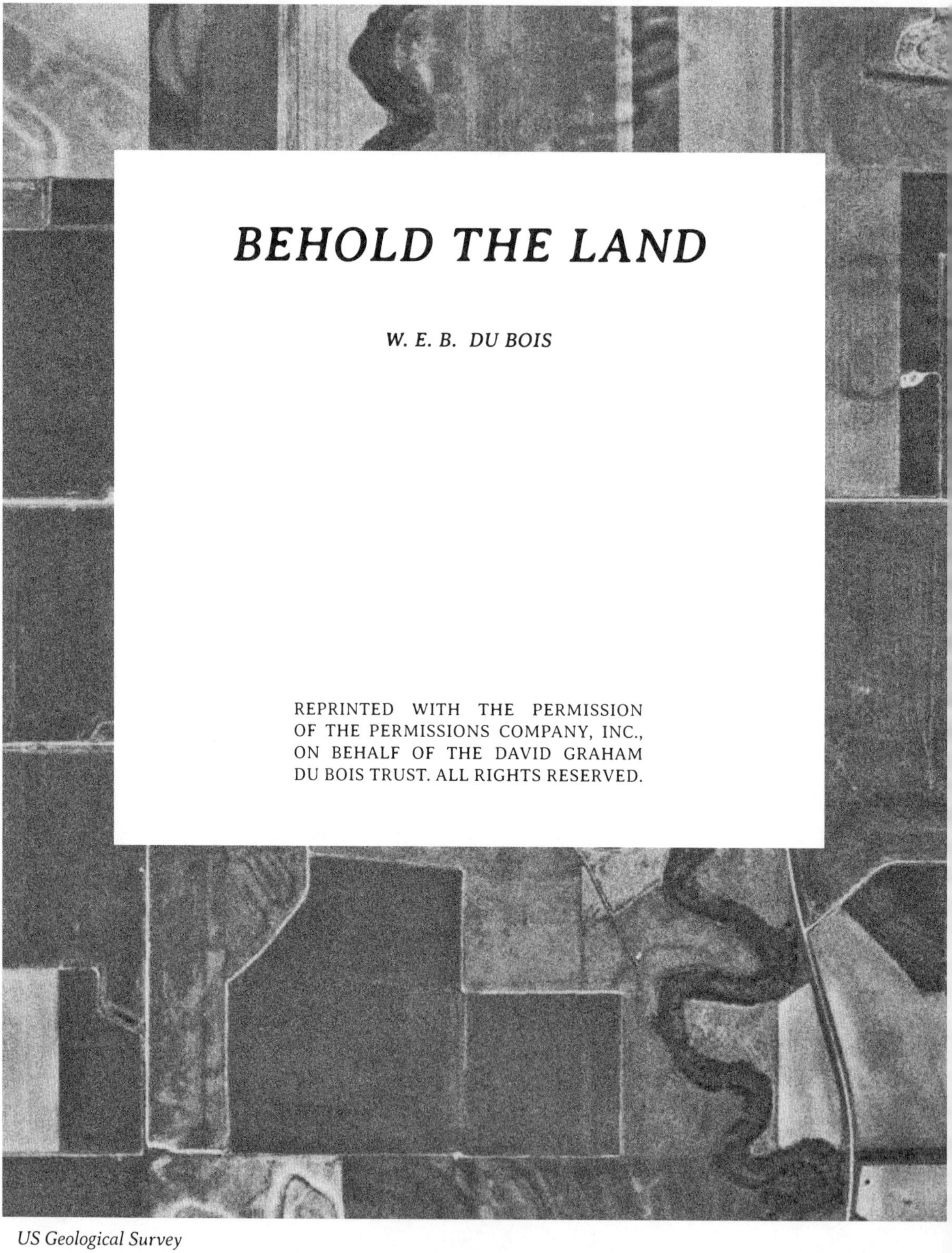

US Geological Survey

To the Southern Negro Youth Congress, October 20, 1946

The future of American Negroes is in the South. Here three hundred and twenty-seven years ago, they began to enter what is now the United States of America; here they have made their greatest contribution to American culture; and here they have suffered the damnation of slavery, the frustration of reconstruction and the lynching of emancipation. I trust then that an organization like yours is going to regard the South as the battle-ground of a great crusade. Here is the magnificent climate; here is the fruitful earth under the beauty of the Southern sun; and here if anywhere on earth, is the need of the thinker, the worker and the dreamer This is the firing line not simply for the emancipation of the American Negro but for the emancipation of the African Negro and the Negroes of the West Indies; for the emancipation of the colored races; and for the emancipation of the white slaves of modern capitalistic monopoly.

36-7-19

Remember here, too, that you do not stand alone. It may seem like a failing fight when the newspapers ignore you; when every effort is made by white people in the South to count you out of citizenship and to act as though you did not exist as human beings while all the time they are profiting by your labor; gleaning wealth from your sacrifices and trying to build a nation and a civilization upon your degradation. You must remember that despite all this, you have allies and allies even in the white South. First and greatest of these possible allies are the white working classes about you. The poort whites whom you have been taught to despise and who in turn have learned to fear and hate you. This must not deter you from efforts to make them understand, because in the past in their ignoranae and suffering they have been led foolishly to look upon you as the cause of most of their distress. You must remember that this attitude is hereditary from slavery and that it has been deliberately cultivated ever since emancipation.

Slowly but surely the working people of the South, white and

2

black, must come to remember that their emancipation depends upon their mutual cooperation; upon their acquaintanceship with each other; upon their friendship; upon their social intermingling. Unless this happens each is going to be made the football to break the heads and hearts of the other.

White youth in the South is peculiarly frustrated. There is not a single great ideal which they can express or aspire to, that does not bring them into flat contradiction with the Negro problem. The more they try to escape it, the more they land into hypocrisy, lying and double-dealing; the more they become, what they least wish to become, the oppressors and despisers of human beings. Some of them, in larger and larger numbers, are bound to turn toward the truth and to recognize you as brothers and sisters, as fellow travellers toward the dawn.

36-7-20

There has always been in the South that intellectual elite who saw the Negro problem clearly. They have always lacked and some still lack the courage to stand up for what they know is right. Nevertheless they can be depended on in the long run to follow their own clear thinking and their own decent choice. Finally even the politicans must eventually recongize the trend in the world, in this country, and in the South. James Byrnes, that favorite son of this commonwealth, and Secretary of State of the United States, is today occupying an indefensible and impossible position; and if he survives in the memory of men,he must begin to help establish in his own South Carolina something of that democracy which he has been recently so loudly preaching to Russia. He is the end of a long series of men whose eternal damnation is the fact that they looked TRUTH in the face and did not see it: John C. Calhoun, Wade Hampton, Ben Tillman are men whose names must ever be besmirched by the fact that they fought against freedom and democracy in a land which was founded upon Democracy and Freedom.

3

Eventually this class of men must yield to the writing in the stars. That great hypocrite, Jan Smuts, who today is talking of humanity and standing beside Byrnes for a United Nations, is at the same time, oppressing the black people of Africa to an extent which makes their two countries, South Africa and the Southern South, the most reactionary peoples on earth. Peoples whose exploitation of the poor and helpless, reahes the last degree of shame. They must in the long run yield to the forward march of civilization or die.

30-7-21

If now you young people instead of running away from the battle here in Carolina, Georgia, Alabama, Louisiana and Mississippi, instead of seeking freedom and opportunity in Chicago and New York which do spell opportunity - nevertheless grit your teeth and make up your minds to fight it out right here if it takes every day of your lives and the lives of your childrens' children. If you do this, you must in meetings like this ask yourselves what does the fight mean? How can it be carried on? What are the best tools, arms, and methods? And where does it lead?

I should be the last to insist that the uplift of mankind never calls for force and death. There are times, as both you and I know, when

"Tho' love repine and reason chafe,
There came a voice without reply,
'Tis man's perdition to be safe
When for the truth he ought to die.'

At the same time and even more clearly in a day like this, after the millions of mass murders that have been done in the world since 1914, we ought to be the last to believe that force is ever the final word. We cannot escape the clear fact that what is going to win in this world is reason if this ever becomes a reasonable world. The careful reasoning of the human mind backed by the facts of science is the one salvation of man. The world, if it resumes its march toward civilization, cannot ignore reason. This has been the tragedy of the South in the past; it is still its awful and unforgivable sin that it has set its face against reason and against the fact. It tried to build slavery upon freedom; it tried to build tyranny upon democracy; it tried to build mob violence on law and law on lynching and in all that despicable endeavor, the state of South Carolina has led the South for a century. It began not the Civil War - not the War between the States but the War to Preserve Slavery; it began mob violence and lynching and today it stands in the front rank of those defying the Supreme Court on disfranchisement.

36-7-22

Nevertheless reason can and will prevail; but of course it can only prevail with publicity - pitiless, blatant publicity. You have got to make the people of the United States and of the world know what is going on in the South. You have got to use every field of publicity to force the truth into their ears, and before their eyes. You have got to make it impossible for any human being to live in the South and not realize the barbarities that prevail here. You may be condemned for flamboyant methods; for calling a congress like this; for waving your grievances under the noses and in the faces of men. That makes no difference it is your duty to do it. It is your duty to do more of this sort of thing than you have done in the past. As a result of this you are going to be called upon for sacrifice. It is no easy thing for a young black man or a young black woman to live in the South today and to plan to continue to live here; to marry and raise children; to establish a home. They are in the midst of legal caste and customary insults; they are in continuous danger of mob violence; they are mistreated by the officers of the law and they have no hearing before the courts and the churches and public opinion commensurate with the attention which they ought to receive. But that sacrifice is only the Beginning of Battle, you must re-build this South.

There are enormous opportunities here for a new nation, a new Economy, a new culture in a South really new and not a mere renewal of

5

an old South of slavery, monopoly and race hate. There is a chance for a new cooperative agriculture on renewed land owned by the State with capital furnished by the State, mechanized zand coordinated with city life. There is chance for strong, virile Trade Unions without race discrimination, with high wage, closed shop and decent conditions of work to beat back and hold in check the swarm of landlords, monopolists and profiteers who are today sucking the blood out of this land. There is chance for cooperative industry, built on the cheap power of T.V.A. and its future extensions. There is opportunity to organize and mechanize domestic service with decent hours, and high wage and dignified training. There is a vast field for consumers cooperation, building business on public service and not on private profit as the main-spring of industry. There is chance for a broad, sunny, healthy home life, shorn of the fear of mobs and liquor, and rescued from lying, stealing polititions, who build their deviltry on race prejudice. Here in this South is the gateway to the colored millions of the West Indies, Central and South America. Here is the straight path to Africa, the Indies, China and the South Seas. Here is the Path to the Greater, Freer truer World. It would be shame and cowardice to surrender this glorious land and its opportunities for civilization and humanity to the thugs and lynchers, the mobs and profiteers, the monopolists and gamblers who today choke its soul and steal its resources. The oil and sulpher: the coal and iron; the cotton and corn; the lumber and cattle belong to you the workers, black and white, and not to the thieves who hold them and use them to enslave you. They can be rescued and restored to the people if you have the guts to strive for the real right to vote, the right to real education, the right to happiness and health and the total abolition of the father of these scourges of mankind, POVERTY.

6

"Behold the beautiful land which the Lord thy God hath given thee." Behold the land, the rich and resourceful land, from which for a hundred years its best elements have been running away, its youth and hope black and white, scurrying North because they are afraid of each other, and dare not face a future of equal, independent, upstanding human beings, in a real and not a sham democracy.

36-7-74

To rescue this land, in this way, calls for the Great Sacrifice; This is the thing that you are called upon to do because it is the right thing to do. Because you are embarked upon a great and holy crusade, the emancipation of mankind black and white; the upbuilding of democracy; the breaking down, particularly here in the South, of forces of evil represented by race prejudice in South Carolina; by Lynching in Georgia; by disfranchisement in Mississippi; by ignorance in Louisiana and by all these and monopoly of wealth in the whole South.

There could be no more splendid vocation beckoning to the youth of the twentieth century, after the flat failures of white civilization, after the flamboyant establishment of an industrial system which creates poverty and the children of poverty which are ignorance and disease and c crime; after the crazy boasting of a white culture that finally ended in wars which ruined civilization in the whole world; in the midst of allied peoples who have yelled about democracy and never practised it either in the British Empire or in the American Commonwealth or in South Carolina.

Here is the chance for young women and young men of devotion to lift again the banner of humanity and to walk toward a civilization which will be free and intelligent; which will be healthy and unafraid; and build in the world a culture led by black folk and joined by peoples of all colors and all races - without poverty, ignorance and disease!

7

Once a great German poet cried: "Selig der Er in Sieges Glänze findet"

"Happy man whom Death shall find in Victory's splendor"

But I know a happier one: he who fights in despair and in defeat still fights. Singing with Arna Bontemps the quiet, determined philosophy of undefeatable men:

36-7-25

"I thought I saw an angel flying low,
I thought I saw the flicker of a wing
Above the mulberry trees; but not again,
Bethesda sleeps. This ancient pool that healed
A Host of bearded Jews does not awake.
This pool that once the angels troubled does not move.
No angel stirs it now, no Saviour comes
With healing in His hands to raise the sick
and bid the lame man leap upon the ground.

The golden days are gone. Why do we wait
So long upon the marble steps, blood
Falling from our open wounds? and why
Do our black faces search the empty sky?
Is there something we have forgotten? Some precious thing
We have lost, wandering in strange lands?

There was a day, I remember now,
I beat my breast and cried, "Wash me God,"
Wash me with a wave of wind upon
The barley; O quiet one, draw near, draw near!
Walk upon the hills with lovely feet
And in the waterfall stand and speak!

W. E. B. Du Bois

WEALTH

FARMING FOR SOCIAL CHANGE

ANTHONY REYES

Sustainable agriculture is loosely defined as an approach to agriculture that strives to create environmentally sound, economically viable, and socially just food and agriculture systems. There are numerous resources that focus on how to establish economically viable and environmentally sound practices, but few that address how to establish farming practices that are socially just. How do we use the food and fiber that we grow as the vehicle that drives social change?

I would like to invite you to take a minute and ask yourself: What does social justice look like? And how do we know when we are pursuing social justice?

I've often pondered these questions while working in the field, coworkers and crew members side by side. Sometimes that's social justice. Sometimes it's awareness around seed choices and preserving culturally significant cultivars, or providing a fair wage for labor and incorporating a more equitable hiring process. What we need to acknowledge and interrogate at the core of all of our responses are the historical and institutionalized systems of oppression. Everything we know about colonial western agricultural practices is rooted in a deep history of racism, sexism, and classism, among many other prejudices. Our nation, its economic policies, and its agricultural systems were founded on the brutal exploitation of people and labor. We must learn to redefine our agricultural practices and food systems to acknowledge and address these power dynamics, and challenge and dismantle them at their roots, utilizing our strengths as farmers and those working in the food system to create a larger systemic change and, ultimately, a paradigm shift.

When thinking about the systems of oppression in the food system and beyond, we have to acknowledge historical contexts and the difference among approaches. Equality, equity, liberation, and radical inclusion are among these. We hear so much about the need to create an egalitarian or an equitable society—and while I agree that these are steps in the right direction, I believe we can go further. Equality and equity still work within the confines and constructs of an oppressive state. Equality is the idea and practice of giving everyone the same resources regardless of their privilege, and equity is the idea and practice of providing resources and opportunities based on privileges to try and create a "more just society." The two responses may acknowledge the oppressive framework in which they are set, but not that they work to create opportunities and resources accordingly in attempt to fix an unjust system. In that, they are very much so reactionary to these institutionalized systems.

Liberation and radical inclusion move beyond equality and equity to analyze and combat the injustices, taking more proactive steps towards dismantling systems of oppression. Liberation works to dismantle the existing discriminatory power dynamics; radical inclusion involves having everyone's voice present, equally represented in making decisions and setting the course of the society, space, and culture. In order to host and facilitate an inclusive environment, we need a liberated space. Otherwise, institutionalized power dynamics will continue to play out, creating an illusion of inclusivity and putting those in food and agriculture at risk of creating a space where marginalized voices are still tokenized.

How, then, do we create and envision a liberated space within a society and system that has so much historical and cultural oppression? Firstly, we need to practice prefigurative politics, a mode of organizing that strives to reflect the future society being sought by the group. So, in organizing and holding space that is bringing everyone to the table, we must model the social and cultural relationships that we strive to create. We also need to find ways to facilitate spaces and dialogues collaboratively and work to center the voices of those most impacted by systemic oppressive power structures.

Then, we can move further to reclaim our food system to reflect a liberated peoples and society, and reclaim the power and the culture of agriculture. We need to be more knowledgeable of our own varied places and narratives, our individual positions and privileges within these existing systems, and bring forward the voices of our indigenous brothers, sisters, and siblings, many of whom were forcibly removed from and/or robbed of the land that now produces the nation's agricultural bounty. Further, we need to interrogate how race, class, and privilege permeate most every social interaction and space. When met with prejudice, we must embrace curiosity and create dialogue.

Working at the intersection of agriculture and social justice means engaging with systems of oppression and finding means of breaking them down. It means recognizing historical contexts and the complexities of our social relationships and everyday institutionalized social dynamics, and how power is represented and held. Collectively, we can challenge and dismantle the systems of oppression and make proactive change in the food and agriculture system and beyond, but we need to be present and show up for one another. While we didn't create these systems, we can change them.

This writing is partially adapted from an interactive workshop I gave at the 2017 Tilth Producers Conference in collaboration with Nat Mengist. My inspiration is drawn from my work with Seattle Youth Garden Works, Tilth Alliance, Homeless Garden Project, and all the collaboration and friendships therein.

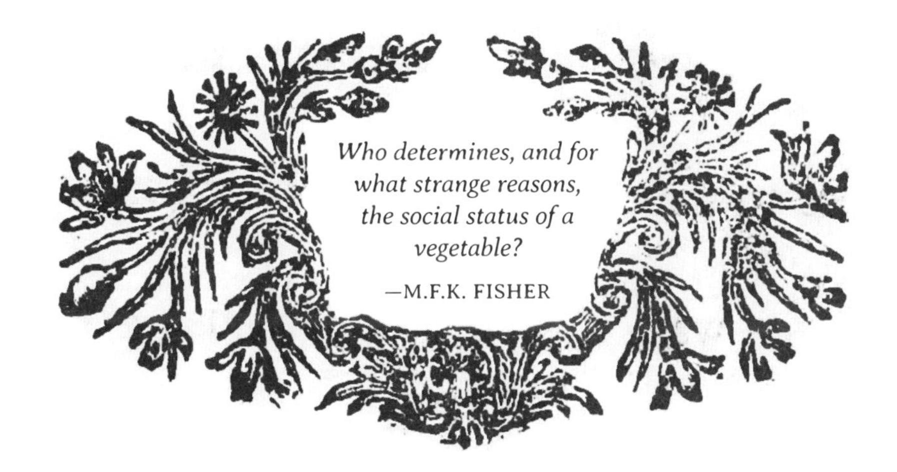
Who determines, and for what strange reasons, the social status of a vegetable?
—M.F.K. FISHER

HERE YOU THERE

AZURÉ KEAHI

There I was —
tip-toeing on a canopy of monstera,
tangled in the hau (Hibiscus tiliaceus)
with rhythmic squeals of to-be-slaughtered swine
ricocheting off stones in the nearby creek.
If a pig dies in the jungle, and I am the only one who hears it,
does it make a sound?

It was there —
I caught tadpoles,
watched them dance in my palm,
let infant frogs stand on my fingertips.
Where were the mothers?

There it grew —
the ovular mountain apple,
leaving tongue-tingles
through scarlet skin;
the rosy-pink flesh of guava,
taunting with seeds
and sunny rind.
Did we forage or simply feed?

I saw it there —
they stuck straws in coconuts
so the others could drink from them.
No need to yield a machete;
leave it to the fruit-stand-man.
He even keeps it ice cold.
Is it the same without the struggle?

We did that there —
we'd catch chicks and raise them to be hens
or cuddle cocks before the fight.
Just look at the plumes,
where do they lead you?

They took it all —
the goats grazed on the ginger
and the black dog growled
until they scurried,
then life was barren.
Where did they go?

I left them there —
I came to the city.
I looked for the forest snacks,
the passion fruit vines,
the loose, roaming roosters,
but no.
Are they still?

I found them there —
the toxic lots,
the needles, the feces, the Carhartt overalls,
the boots —
the boot we got.
We built paradise
on the colonizer's forgotten ground.
Now, will you remember?

RACIAL JUSTICE IN THE FOOD AND FARM SYSTEM REQUIRES LAND JUSTICE

NEIL THAPAR

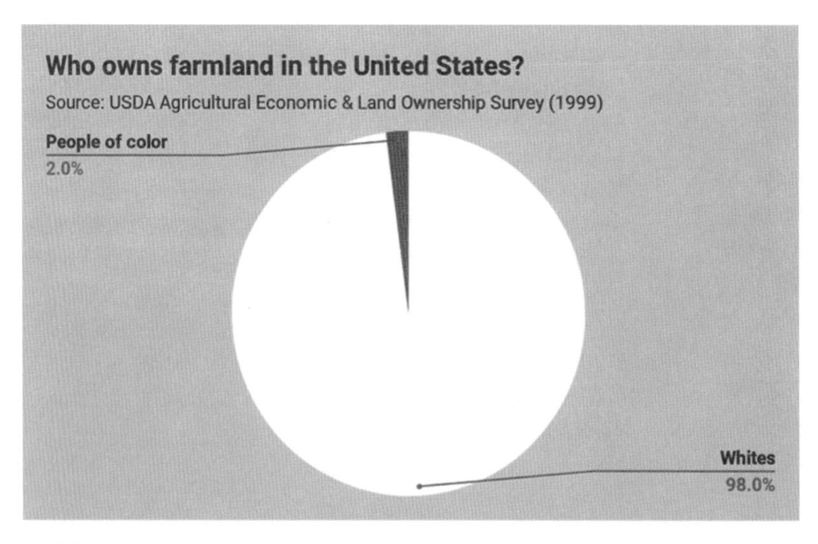

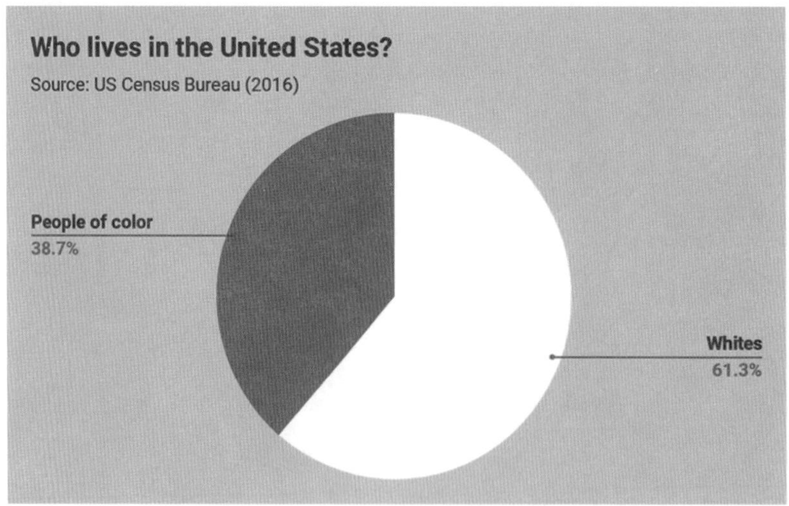

Charts by Neil Thapar

The last year that the United States Department of Agriculture (USDA) conducted a national survey of farmland ownership in the U.S. was 1999. At the time, the survey results, published in 2002, showed that ninety-six percent of farmland owners were white and ninety-seven percent of farmland value was owned by white people, as were ninety-eight percent of farm acres. In raw acreage, that year's data reported that white people owned 1.25 billion acres of farmland compared to the 25 million acres owned by all people of color included in the survey—Black people, American Indians, Hispanic people, Asians, and those grouped together under the government label of "Other."

That was roughly twenty years ago. A few farm bills, twelve free trade agreements, and one Great Recession later, I would guess the disparity is approaching its absolute form. When I think about how the current state of ownership relates to social justice, I can't help but compare the state of farmland ownership now to the last time nearly all the land was owned by white people. Back then, from the colonial period through the civil war, people with darker skin were enslaved and forcibly removed from their land. For them, there was no democracy, no social justice, no human rights.

The crises of the food and farm system as they exist in the U.S. today—overproduction and undernourishment, bigger farms and fewer farmers, higher tech and lower wages—are results of a powerful and deep erosion of democracy. Democracy, in this context, is the right to meaningfully participate in the circumstances of one's life.

In action, democracy is essentially the ability to make choices. On the farm, these choices include whether to save seeds, spray chemicals, pay workers fair wages, or rest the land after an especially hard season. These are the choices that advocates constantly lift up as necessary to create a more just, sustainable, and fair food and farm systems. Without land ownership, there is no choice, no possibility of democracy.

From the decades-old USDA survey results, it's clear to see that for farmers of color, whether they operate the business or work in the fields for an employer, there has never been such a choice. Farmers of color have been locked out of land ownership by way of theft, discrimination, and lack of opportunity, all reinforced by public policy and law. Yet, in states across the country, people of color are the fastest growing demographic of farm operators and already constitute large percentages of the labor force in agriculture. Racial justice in the food and farm system is not only about raising wages, improving access to healthy food, and comprehensive immigration reform. At its core, racial justice is about

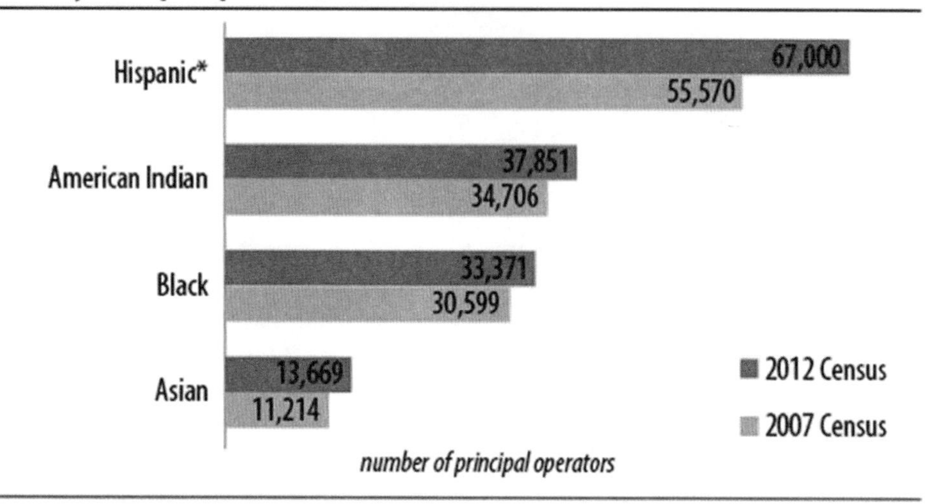

Source: USDA NASS, 2012 Census of Agriculture.

democratizing the ownership of land for people of color.

Being clear about the goal creates opportunities to be strategic in the approach, ensuring that each step is forward movement. This is important, because the challenge ahead is immense, requiring nothing less than a transformation of the normative notion of land as property, as an asset, a financial tool. For anyone—farmer, foodie, or advocate—working toward a more just, sustainable, and fair food and farm system, I encourage you to start asking difficult questions: What is the relationship between colonization and the invention of private property rights to the racial disparity in farmland ownership? What would it mean to reframe land ownership as a responsibility and privilege, instead of a right? What responsibility do landowners have to repatriate land to communities of color that has been acquired by theft or discrimination, perhaps generations before the current landowner's tenure?

Restructuring land ownership in this way will also create the conditions for achieving many of the downstream goals of food and farm advocates, including stronger rural economies, dignified and sustainable livelihoods for farmers, deeper environmental stewardship, and increased health equity. Here's why: when people of color own land, they have the ability to truly make their own choices, self-determine the kinds of farms they want to run, and create the futures they aspire to. The people most impacted by the social, economic, and environmental injustices in the food and farm system will finally be able to fully participate in creating the solutions to these injustices. In the end, a more racially just food and farm economy means a more democratic democracy. And true to the definition of the word itself, we all have a role to play.

TO AMERICA

JAMES WELDON JOHNSON

How would you have us, as we are?
Or sinking 'neath the load we bear?
Our eyes fixed forward on a star?
Or gazing empty at despair?

Rising or falling? Men or things?
With dragging pace or footsteps fleet?
Strong, willing sinews in your wings?
Or tightening chains about your feet?

HOW MAKING REPARATIONS CAN REMAKE THIS LAND

JEAN WILLOUGHBY AND DOUGLASS DECANDIA

Black refugees evicted from sharecropping, waiting by the roadside. Parkin, Arkansas, 1936

The cause of reparations is having a moment of resurgence in the United States. Author and journalist Ta-Nehisi Coates reinvigorated the idea in a sweeping and influential essay, "The Case for Reparations," published in *The Atlantic* in 2014. In 2016, after more than a decade of rigorous investigation, the United Nations' Working Group of Experts on People of African Descent determined that "past injustices and crimes against African Americans need to be addressed with reparatory justice."[1] In the United States, the term *reparations* is most associated with the idea of compensation for slavery, but the idea has also been explored by Coates and others as an integral part of redressing racial discrimination and injustice in our own times.

When the Movement for Black Lives released its first policy platform in 2016, it reflected a demonstrated consensus built among more than fifty organizations across the country. The platform is notable for its demand for reparations "for past and continuing harms" wrought by "government, responsible corporations and other institutions that have profited off of the harm they have inflicted on Black people—from colonialism to slavery through food and housing redlining, mass incarceration, and surveillance."[2] In a related vein, House Resolution 40, the Commission to Study Reparation Proposals for African Americans Act, has lately gained increasing attention from social justice groups and the media. Specifically, HR 40 would "examine slavery and discrimination in the colonies and the United States from 1619 to the present and recommend appropriate remedies."[3] Introduced in each session of Congress since 1989, the bill has never garnered enough votes to make it to the floor for debate. Both despite and because of this inertia, Coates described the bill as the vehicle for a national reckoning that is long overdue.[4] In their platform, the Movement for Black Lives calls for its immediate passage.

In 2018, a group of farmers of color and organizers based at Soul Fire Farm in New York launched the online Reparations Map for Black-Indigenous Farmers. Calling for "reparations of land and resources so that we can grow nourishing food and distribute it in our communities," the organizers assert that our "food system was built on the stolen land and stolen labor of Black, Indigenous, Latinx, Asian and people of color."[5] Their map shows projects of farmers of color across the country and lists the resources these farmers need for development and growth. The map's organizers encourage "people-to-people solidarity" in meeting these needs and view the endeavor not only as a method for increasing the self-determination of communities of color but also of providing an anchored source of connection and healing in people's everyday lives.

These efforts are powerful, instructive examples of individual and collective action. As advocates working for a just, equitable, and life-affirming food system, the authors of this essay have spent much of the past year reflecting on the power of reparations to remake this land that we can't help but call home. We come to this work as two white people who have each spent the past decade involved in sustainable agriculture. One of us works as an agricultural educator and grower with a food bank. The other works as an advocate and strategist in farmland preservation, developing models of land stewardship that expand on the principles and practices of community supported agriculture. Both of us enjoy connecting with other people in relation to the land and the sources of our daily sustenance. Antiracism work and community organizing also play strong roles in our professional and personal lives. We're both drawn to farms and gardens, fascinated by agriculture as a nexus between nature and humanity, and here we consider the rationale for reparations and the role they could play in our food system.

A BRIEF LOOK AT RACISM IN THE FOOD SYSTEM

Many of the ways that racism shapes our food system can be summarized in the word access—lack of access to food, to farms and farmland, to agricultural and business lending, and to the fundamental needs for clean air, water, and soil. Statistics on race and the food system mirror what we observe about racism across all systems in our country: that is, white people typically experience better outcomes than any other racial group, while black people experience some of the worst outcomes. In states with larger populations of Native Americans or Latinos, those groups typically experience some of the worst outcomes, but the pervasive pattern remains. Regarding food access and health, studies have shown that, on average, black households are significantly less likely to live near a supermarket than are white households and that supermarket inaccessibility corresponds with higher rates of obesity. Fresh, unprocessed food is out of reach for many black families whose neighborhoods are often surrounded by fast food chains and corner stores that offer cheap, calorie-dense meals and snacks.

Cheap food exacts a high price. The negative effects of unhealthy diets can be felt in all communities, but they are particularly pronounced and alarmingly fatal among groups who have been dispossessed and displaced from their land or community. Numerous scientific studies have found that high rates of diet-related illnesses such as heart disease and type-2 diabetes disproportionately affect black Americans—typically at rates two to three times higher than white Americans. Epidemiological research has also shown that displaced indigenous people, not only in the United States but around the world, often experience skyrocketing rates of these same illnesses in the years and generations following their dispossession and displacement.[6]

The taking of land from communities of color in this country of course does not end with indigenous people. In his book *Dispossession: Discrimination against African American Farmers in the Age of Civil Rights*, historian Pete Daniel gives a damning account of the discrimination experienced by black farmers at the hands of the US Department of Agriculture. Between 1940 and 1974, the number of black farmers plummeted, from 681,790 to 45,594, a 93 percent drop within a single generation.[7] Daniel notes that in the early 1970s, legal observers predicted that a fierce and protracted battle of lawsuits, lobbying, and organizing would be necessary to break the stranglehold of discrimination within the USDA. They were right.

Daniel's book provides more than ample explanation for why black farmers would ultimately turn to the courts. In 1999, the US government entered into an agreement with a group of black farmers in the *Pigford v. Glickman* case in what was and may still yet be the largest civil rights class-action settlement in US history. In short, the farmers sued the USDA for racial discrimination and won. The majority of the funds consisted of $50,000 payments to 33,256 individual black farmers or their heirs. Unfortunately, the payments received were typically not enough to restore the land and farms lost to them years before.

LAND OWNERSHIP AND THE RACIAL WEALTH GAP

Scholars and journalists rightly focus on the pivotal relationship between the federal government and the housing market in creating and maintaining the modern racial wealth gap. But wealth obtained from ownership of America's farmland is also a significant contributor. Today, between 95 and 98 percent of all farmland in the United States is owned by white families. This is no accident but rather the outcome of hundreds of years of public policy advantaging white people over people of color in both the right and the opportunity to own land.[8] Accounting for 40 to 50 percent of all the land within our borders, the farmland owned by white families is estimated to be worth more than $1 trillion. And while more than 90 percent of farm owners and operators are white, farm labor is another story. Surveys over the past decade have consistently found that around 80 percent of farmworkers identify as Hispanic or Latino, with the majority born in Mexico.[9] We have now an essentially racialized food production system comprised of white owners and brown workers.

According to recent research, the racial wealth gap is widening. The US Federal Reserve found that white families have the highest median net worth: $171,000. Black families' median net worth is less than 15 percent of that: $17,600. Hispanic and Latino families' median net worth is $20,700. The gap also persists across the lowest income levels: white households living near the poverty line nevertheless typically hold around $18,000 in wealth (often based on assets such as cars or real estate), while black households earning similarly low incomes typically own little to nothing that an accountant would classify as a valuable asset.[10] Furthermore, median black and Hispanic-Latino wealth appear to be declining. The nonprofits Prosperity Now and the Institute for Policy Studies have projected that by 2053 the median wealth of black families could hit zero.[11] Hispanic-Latino families are expected to drop to the same level by twenty years later.

In light of these alarming trends, a collaboration of scholars at Duke University's Samuel DuBois Cook Center on Social Equity and the Insight Center for Community Economic Development made a bold and decisive statement, declaring that there are "no actions that black Americans can take unilaterally that will have much of an effect on reducing the racial wealth gap." In their comprehensive report, they concluded that "addressing racial wealth inequality will require a major redistributive effort or another major public policy intervention."[12] To accomplish this, they suggest the creation of a reparations program to compensate black Americans for the intergenerational impact of slavery and legal discrimination or "the equivalent of a substantial trust fund for every wealth-poor American," noting that the two approaches need not be mutually exclusive.

MAKING AMENDS

Reparations can be simply defined as making amends for past wrongs. It's generally understood as a practice of making some form of payment with the goal of ameliorating the living conditions of those harmed. Payment is where the discussion over reparations tends to get stuck. Figuring the math of reparations can easily consume any discussion of the subject. For all their serious commitment to seeing the numbers work, those who focus the conversation on debating methods of calculation and distribution tend to miss its point.

Reparations has alternatively been discussed as a kind of moral accounting. Coates relies on this metaphor in describing America's "compounding moral debt." Our nation's moral "credit-card bill" keeps growing, he writes, and though we've "pledged to charge no more ... the effects of that balance, interest accruing daily, are all around us."[13] Though the true cost of righting past wrongs will forever remain incalculable, scholars insist that it is very possible to estimate the financial impact of discrimination against people of color, particularly black people, locked out of housing markets, education, health care, and other resources more readily available to white people. But the work of reparations encompasses far more than the practicalities of the redistribution of resources, the repayment of benefits, or the repatriation of land.

Countless people have justifiably questioned whether a price could ever be put on the defining tragedies of our nation's history: genocide, slavery, dispossession, discrimination. The answer, of course, is no. Nevertheless, the act of making reparations would constitute a profound and unprecedented acknowledgment of the lasting effect of past wrongs and a visionary commitment to an equitable society, to liberty, and to justice. It would speak to an understanding, a reckoning that stretches well beyond our usual reach of moral accounting.

Calculating the dollar amount owed to the people who have been wronged by our country will not necessarily compel us to give it. Indebtedness alone will not move us toward reconciliation, nor will morality alone. Knowing how much we're all still losing because of racism and our legacy of racial inequity might. To have the will to make reparations our reality, we will have to understand what our inequity is costing us, individually, in our communities, and as a nation.

In financial terms, we now know that the cost to the country is tremendous: nearly $2 trillion. A 2013 study from the W. K. Kellogg Foundation and researchers from Brandeis, Harvard, and Johns Hopkins Universities found that the United States loses trillions in "lost earnings, avoidable public expenditures, and lost economic output" due to the economic impact of racial inequity. They found that if the average wages of people of color were made equal to those earned by whites, overall earnings in the United States would increase by 12 percent, for a total of nearly $1 trillion. By closing this gap in earnings, the authors projected that the total value of our country's economic output would increase by $1.9 trillion.[14]

BUILDING TRUST

Many people distrust the idea of reparations. For some, that may have more to do with lack of trust in our government than the idea itself. Undoubtedly, many people are genuinely uncertain about the full basis for its claim and its potential implications. This makes the passage of a "study bill" like HR 40 critical for increasing public knowledge and confidence. In the broader context, it's also important to take into account the impact of race and racism on public trust. "Race is the life experience that has the biggest impact on trust," according to political scientist Eric M. Uslaner.[15] His work points to significant racial "trust gaps" that recall gaps in areas like wealth and education. Even where well-founded, a lack of trust can inculcate cynicism, apathy, and hopelessness in our interactions with government, institutions, and other people. A lack of trust weakens social bonds and can cause us to retreat from public life and political engagement. And, unsurprisingly, racism damages public trust in a multitude of ways. For our discussion, it's important to consider that a lack of public trust

Woman sharecropper picking cotton, 1939

can undermine support for publicly funded initiatives and make it less likely that the public will engage in efforts to redistribute power and resources.

In the past, calls for reparations that have managed to attain a popular audience haven't translated into sustained momentum. But there is reason to be hopeful. Millennials appear to be significantly more open to considering reparations than past generations. A Marist poll suggested that more than half of millennials are willing to consider the idea of reparations for slavery: 40 percent of millennials came out in favor of reparations while 11 percent were unsure. The conversation about reparations will continue to develop as the most diverse demographic in US history takes a place at the table.

Despite criticisms that reparations dwell on the distant past, the work of reparations is made on behalf of the present and the future. Disparities of wealth, land and property ownership, and access to resources are ultimately harmful to all of us and, if not reversed, will have a destructive impact on the world the next generation grows up in too. Making reparations is part of our work toward collective liberation from the social ills caused by inequity. In this context, sharing and redistributing resources is an act of solidarity, not simply an attempt to repay an unpayable debt. The call for reparations seems best understood not just as a request or demand for a compensatory payout, but as part of an ongoing, collaborative effort that moves us further toward ending racial inequity and racism.

Many groups presently organizing in support of reparations have embraced the need to build trust and relationships between and among people of color and white people. This is no small endeavor. Organizing cross-racially and building our solidarity on shared values allows us to model the world we want to create, cultivating a sense of belonging and connection. By supporting the cause of reparations, those who have gained an advantage from racism can take part in dismantling this unjust system.

REPARATIONS ON THE LAND

Agriculture is a profession in which our values and decisions are written on the landscape. As agrarian organizers, we want to cultivate an agriculture that reflects our commitments to sustainability, biodiversity, cooperation, and fairness—and that embraces creative and effective strategies to get us there. In rural America, reparations might find expression on the land though the legal restoration of land, bequests, gifts of land, grants, community supported farming initiatives that respond to the needs of farmers of color, and other reparatory efforts undertaken by landowners.

We're curious about how we can step up to support these kinds of efforts in farming communities. At present, we have more questions than answers: How can communities organize and come to a consensus on taking action for reparations? How can institutional landowners be engaged and involved? What are some of the ways that entities like towns and cities, religious institutions, companies, nonprofits, land trusts, and universities can participate in dialogue with their constituents and communities around these questions? And how can we, as gatekeepers within our groups and institutions, achieve accountability and reparatory justice in the ways resources such as land and space are allocated and accessed?

As community organizers, we want to ensure that dialogue and inquiry lead to action and reconciliation. By collaborating with reparations efforts nationally, it's possible to share what's working and gain exposure to different approaches. As relevant initiatives continue to emerge, many seem designed with the next generation of reparationists in mind. Innovative tools like the online reparations map and the Native Land app are bringing heightened visibility and the power of the internet to centuries-old struggles. The many reparatory justice groups active on social media are connecting people who wish to practice reparations through posts of "requests" and "offerings" that can be shared or amplified by others. Social justice groups that organize among young people with wealth are fundraising for organizations led by people of color both as a form of reparations and as part of their long-term strategy to end racial inequity. With each advance, activists and organizers are developing the resources and capacity to write a new chapter in the long history of the movement for reparations. It may be that the liberation we seek is already being traced on the land.

NOTES

1. United Nations Office of the High Commissioner for Human Rights, "Statement to the Media by the United Nations' Working Group of Experts on People of African Descent, on the Conclusion of its Official Visit to USA," January 2016. ohchr.org/EN/NewsEvents/Pages/DisplayNews.aspx?NewsID=17000.
2. The Movement for Black Lives, "Reparations," policy.m4bl.org/reparations.
3. US Congress. House. *Commission to Study and Develop Reparation Proposals for African-Americans Act*. 115th Congress. Introduced January 3, 2017.
4. Ta-Nehisi Coates, "The Case for Reparations," *The Atlantic*, June 2014.
5. Soul Fire Farm, "Reparations: Reparations Map for Black-Indigenous Farmers," soulfirefarm.org/support/reparations.
6. California Newsreel, *Unnatural Causes: Is Inequality Making Us Sick?* San Francisco: California Newsreel, 2008. unnaturalcauses.org.
7. Pete Daniel, *Dispossession: Discrimination against African American Farmers in the Age of Civil Rights*. (Chapel Hill, NC: University of North Carolina Press, 2013).
8. Howard Zinn. *A People's History of the United States*. (New York: Harper & Row, 1990).
9. Trish Hernandez, Susan Gabbard, and Daniel Carroll. "Findings from the National Agricultural Workers Survey 2013-2014: A Demographic and Employment Profile of United States Farmworkers." U.S. Department of Labor Employment and Training Administration Office of Policy Development and Research, December 2016, doleta.gov/agworker/pdf/NAWS_Research_Report_12_Final_508_Compliant.pdf.
10. Lisa J. Dettling, Joanne W. Hsu, Lindsay Jacobs, Kevin B. Moore, and Jeffrey P. Thompson. "Recent Trends in Wealth-Holding by Race and Ethnicity: Evidence from the Survey of Consumer Finances," *FEDS Notes*. Washington: Board of Governors of the Federal Reserve System, September 27, 2017, doi.org/10.17016/2380-7172.2083.
11. Chuck Collins, Dedrick Asante-Muhammed, Emanuel Nieves, and Josh Hoxie. "The Road to Zero Wealth: How the Racial Wealth Divide Is Hollowing Out America's Middle Class." *Prosperity Now*, September 11, 2017.
12. William Darity Jr., Darrick Hamilton, Mark Paul, Alan Aja, Anne Price, Antonio Moore, and Caterina Chiopris, "What We Get Wrong About Closing the Racial Wealth Gap," Samuel DuBois Cook Center on Social Equity and Insight Center for Community Economic Development, April 2018, goo.gl/SJJyLo.
13. Coates, "The Case for Reparations."
14. Ani Turner, Dolores Acevedo-Garcia, Darrell Gaskin, Thomas LaVeist, David R. Williams, Laura Segal, and George Miller. "The Business Case for Racial Equity." W.K. Kellogg Foundation and Altarum Institute, October 2013, wkkf.org/resource-directory/resource/2013/10/the-business-case-for-racial-equity.
15. Eric M. Uslaner, *The Moral Foundations of Trust*. (Cambridge, UK: Cambridge University Press, 2002).

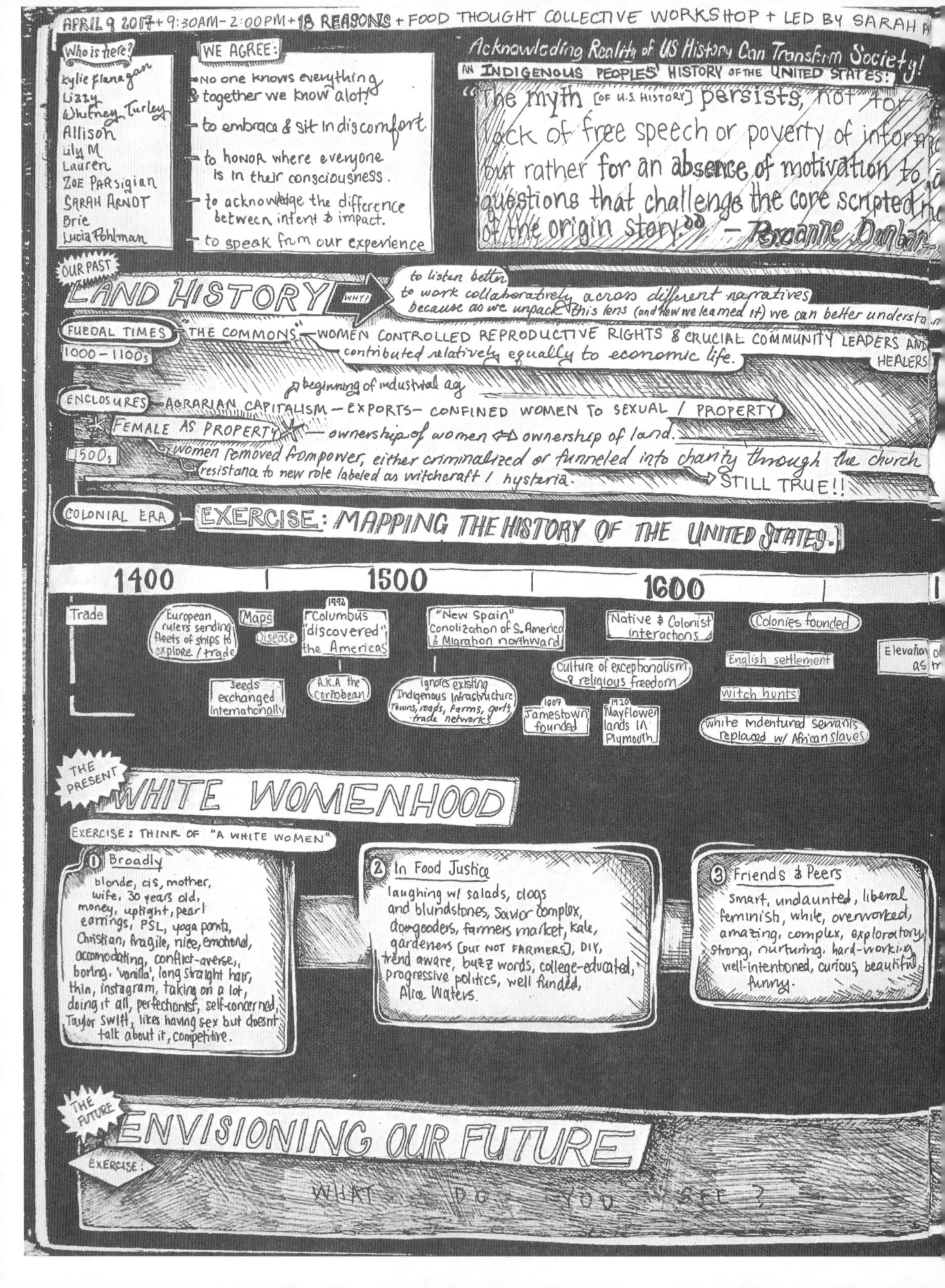
APRIL 9 2017 + 9:30AM-2:00PM + 18 REASONS + FOOD THOUGHT COLLECTIVE WORKSHOP + LED BY SARAH A
Who is there?
Kylie Flanagan
Lizzy
Whitney Turley
Allison
Lily M
Lauren
ZOE PARSIGIAN
SARAH ARNDT
Brie
Lucia Pohlman
WE AGREE:
No one knows everything & together we know alot!
to embrace & sit in discomfort
to honor where everyone is in their consciousness.
to acknowledge the difference between intent & impact.
to speak from our experience
Acknowleding Reality of US History Can Transform Society!
AN INDIGENOUS PEOPLES' HISTORY OF THE UNITED STATES:
"The myth [OF U.S. HISTORY] persists, not for lack of free speech or poverty of inform but rather for an absence of motivation to questions that challenge the core scripted of the origin story." — Roxanne Dunbar
OUR PAST
LAND HISTORY
WHY?
to listen better
to work collaboratively across different narratives
because as we unpack this lens (and how we learned it) we can better understa
FUEDAL TIMES
1000-1100s
"THE COMMONS" — WOMEN CONTROLLED REPRODUCTIVE RIGHTS & CRUCIAL COMMUNITY LEADERS AND HEALERS
contributed relatively equally to economic life.
beginning of industrial ag
ENCLOSURES — AGRARIAN CAPITALISM — EXPORTS — CONFINED WOMEN TO SEXUAL / PROPERTY
FEMALE AS PROPERTY — ownership of women ⇔ ownership of land.
1500s
women removed from power, either criminalized or funneled into charity through the church
resistance to new role labeled as witchcraft / hysteria.
STILL TRUE!!
COLONIAL ERA
EXERCISE: MAPPING THE HISTORY OF THE UNITED STATES.
1400
1500
1600
Trade
European rulers sending fleets of ships to explore / trade
Maps
Disease
Seeds exchanged internationally
1492
Columbus "discovered" the Americas
A.K.A the Caribbean
"New Spain" Conolization of S. America & Migration northward
ignores existing Indigenous infrastructure Towns, roads, farms, gov'ts trade networks
Native & Colonist interactions
Culture of exceptionalism & religious freedom
1607
Jamestown founded
1620
Mayflower lands in Plymouth
Colonies founded
English settlement
Witch hunts
white indentured servants replaced w/ African slaves
Elevation of
as
THE PRESENT
WHITE WOMENHOOD
EXERCISE: THINK OF "A WHITE WOMEN"
1 Broadly
blonde, cis, mother, wife, 30 years old, money, uptight, pearl earrings, PSL, yoga pants, Christian, fragile, nice, emotional, accomodating, conflict-averse, boring, 'vanilla', long straight hair, thin, instagram, taking on a lot, doing it all, perfectionist, self-concerned, Taylor Swift, likes having sex but doesn't talk about it, competitive.
2 In Food Justice
laughing w/ salads, clogs and blundstones, savior complex, do-gooders, farmers market, kale, gardeners [BUT NOT FARMERS], DIY, trend aware, buzz words, college-educated, progressive politics, well funded, Alice Waters.
3 Friends & Peers
smart, undaunted, liberal feminish, white, overworked, amazing, complex, exploratory, strong, nurturing, hard-working, well-intentioned, curious, beautiful, funny.
THE FUTURE
ENVISIONING OUR FUTURE
EXERCISE:
WHAT DO YOU SEE?

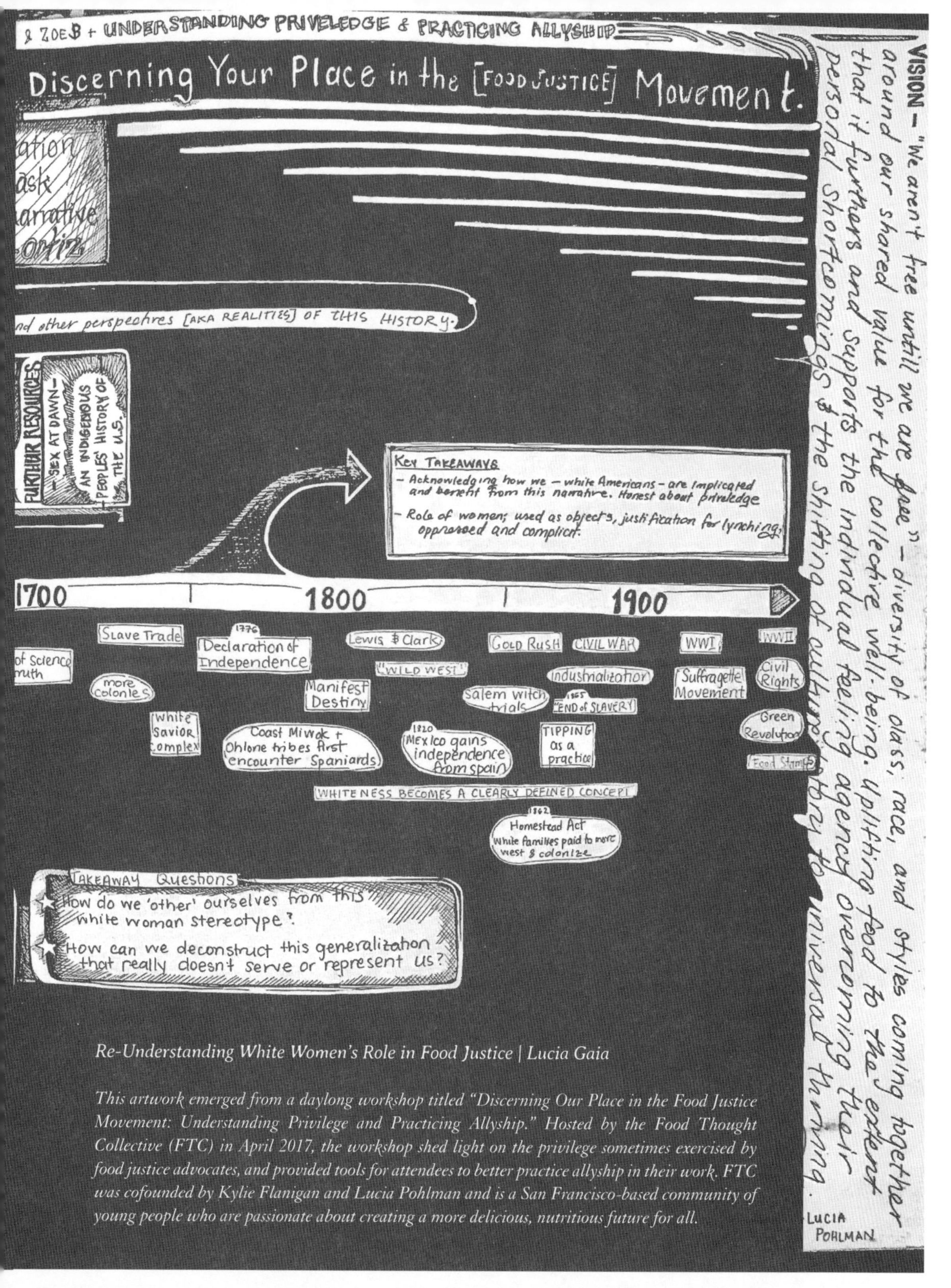

Re-Understanding White Women's Role in Food Justice | Lucia Gaia

This artwork emerged from a daylong workshop titled "Discerning Our Place in the Food Justice Movement: Understanding Privilege and Practicing Allyship." Hosted by the Food Thought Collective (FTC) in April 2017, the workshop shed light on the privilege sometimes exercised by food justice advocates, and provided tools for attendees to better practice allyship in their work. FTC was cofounded by Kylie Flanigan and Lucia Pohlman and is a San Francisco-based community of young people who are passionate about creating a more delicious, nutritious future for all.

Members of the Black-owned Bluffton Oyster Cooperative, crabbing in the off-season
South Carolina, 1937

BLACK COOPERATIVES IN THE UNITED STATES

An Abridged History

JESSICA GORDON-NEMBHARD

Black scholar W. E. B. Du Bois was one of several African Americans to view cooperative economics as a promising antidote to persistent racial economic inequality. In 1907 Du Bois wrote a monograph as part of his Atlanta University series on the Negro, *Economic Cooperation among Negro Americans*, a topic he'd first explored in his 1898 volume, *Some Efforts of American Negroes for their Own Social Betterment*. Blacks pooled money in order to help each other buy themselves out of enslavement. In addition, fugitives from enslavement formed their own communities, often isolated Maroons, where they eluded or fought off bounty hunters, and lived collective existence in relative isolation. Immediately after the Civil War, some Blacks organized themselves (or were organized) into intentional communities and communes, where they could live and develop under their own leadership, creating their own economy.[1]

Du Bois's 1907 monograph is a comprehensive study of cooperative activities among African Americans from the 1800s to 1907. It is less a theoretical study of cooperative economic development and more an analysis of a variety of ways African Americans were cooperating economically, and a listing of Black-owned cooperative businesses, organizations, and projects. Du Bois explains that Blacks pooled resources through churches, mutual aid societies, fraternal organizations, and jointly owned businesses. He documented hundreds of mutual aid societies and cooperative projects through religious and benevolence institutions, beneficial and insurance societies, secret societies, schools, and financial institutions. Mutual Aid Societies and Beneficial Societies provided joint purchasing and marketing; revolving loan funds; and sickness, widow and orphan, and death benefits. They often operated informally through Black religious organizations and Black independent schools. Many were founded and headed by Black women. These were the precursors to the African American–owned cooperatives of the 20th century.

Often white landlords, insurance agents, banks, and even the federal government created barriers to thwart the success of these businesses by raising the rent, refusing a line of credit, withdrawing an insurance policy, or even accusing the company of fraud. Cooperative activity among the Knights of Labor (KoL) was controversial; there was opposition within and outside the trade union movement. Many Black organizers had to operate from "underground" so not to be beaten or killed, and sometimes white vigilantes burned down the co-op stores and factories.[2]

1

1880S: COLORED FARMERS' ALLIANCE

The Colored Farmers' National Alliance and Cooperative Union (CFACU) grew out of efforts with organized labor and other populist organizations. The Knights of Labor organized integrated cooperatives and labor unions in the mid to late 1800s and early 1900s. Some Southern KoL chapters were all Black; by 1887, between sixty and ninety thousand African Americans were members. The CFACU formed to establish African American leadership and control over racial agrarian reform in the south in the 1880s, and operated under fear and harassment by the white plantation bloc, but managed to operate several cooperatives in the late 19th century before having to disband.[3] Members shared agricultural techniques and innovations, and coordinated cooperative efforts for planting and harvesting.[3] The CFACU promoted alliances between farmers and laborers and was active in local and regional politics, striving to maintain rights for African Americans after Reconstruction. Best estimates suggest the CFACU had over one million members and was the largest Black organization of its time.

2

1919: CITIZENS' CO-OPERATIVE STORES, MEMPHIS

Citizens' co-op stores were established in direct response to the Negro Cooperative Guild meeting in August 1918. A "Mr. Ruddy" returned home to Memphis and organized a study group. In February 1919, the Memphis group incorporated as the Citizens' Co-operative Stores to operate cooperative meat markets. According to an article in the *Crisis*, the cooperative sold double the amount of the original shares they offered. By August 1919, five stores were in operation in Memphis, serving about seventy-five thousand people. Members of the local guilds associated with each store met monthly to study cooperatives and discuss issues.

3

1927: COLORED MERCHANTS SOCIETY, MONTGOMERY, ALABAMA

The Colored Merchants Society was founded by the National Negro Business League in Montgomery, Alabama, in 1927. The society was an association of independent grocers organized into a buying and advertising cooperative, created to support independent Black grocery stores with mutual support and collective marketing in a harsh market dominated by chain stores. Chapters were organized in cities with ten or more stores. Dues were $5 per month per store.[4] By 1930, more than two hundred fifty stores were part of the society's network, including thirty-two stores in Tulsa, twenty-five in Dallas, twenty-five in New York City's Manhattan borough, and ten in Omaha.[5] The society was relatively successful for several years, but the major chain grocery stores gave fierce competition and often undercut the prices of the society's stores.

1930S AND 1940S: EASTERN CAROLINA COUNCIL FEDERATION OF NORTH CAROLINIAN COOPERATIVES

During the 1930s and 1940s, there was extensive cooperative activity through two Black independent schools: the Bricks Rural Life School and the Tyrrell County Training School. These schools promoted cooperative education and development among their students and with the students' families. For example, Tyrrell established study groups on cooperative economics. By 1939, twenty-five neighbors had established a credit union. In the first year, membership increased to nearly two hundred, and the credit union started a student savings account program. The credit union helped several families save their farms from foreclosure or to purchase a farm, and it financed group purchases of farm equipment. Buying clubs and machinery cooperatives (purchasing co-ops) were established through 1945.[6]

In 1939, Bricks and Tyrrell County school leaders helped to establish the Eastern Carolina Council, a federation of African American cooperatives in North Carolina.

In 1945, the Eastern Carolina Council helped to form the North Carolina Council for Credit Unions and Associates, shortened to the North Carolina Council. By 1948 the North Carolina Council had established ninety-eight Black-owned credit unions and forty-eight additional co-op enterprises: nine consumer stores, thirty-two machinery co-ops, four curb markets, two health associations, and one housing project.[7] Pitts's is the only study about this, so we do not know what happened to those cooperatives after the 1950s. It is also the only statewide study of Black cooperatives, so we do not have this kind of information about any other state.

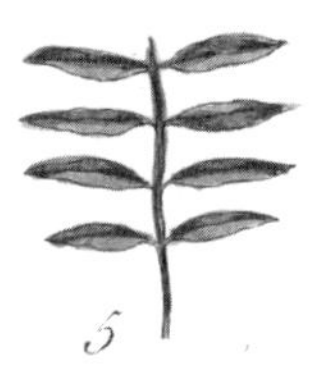

1966: FREEDOM QUILTING BEE, ALBERTA, ALABAMA

The Freedom Quilting Bee (FQB) is a handicraft cooperative and a member of the Federation of Southern Cooperatives/Land Assistance Fund (FSC/LAF). It was founded by women in sharecropping families looking to increase and stabilize their incomes. Since sharecropping is essentially debt peonage, their families were very poor, and the women began selling quilts to supplement their families' farm incomes. The seed money for the cooperative came from an initial sale of one hundred quilts sold for them in New York by an Episcopalian minister, Francis Walters, who wanted to support the effort.[8] Cofounder Estelle Witherspoon was FQB's first president, with her mother, Mama Willie Abrams, who was also a founding member of the

Federation of Southern Cooperatives.

In 1968 the cooperative bought twenty-three acres of land. This was an important acquisition because it allowed them to build the sewing factory and to increase Black land ownership. They sold eight lots to families who had been evicted from their homes for registering to vote or attending a speech by Martin Luther King Jr.[9] The quilters also began using other entrepreneurial strategies to increase the economic activity under their control—making pot holders and conference canvass bags, for example. They housed a day care and an after-school program in the sewing factory. Some of the quilters were highlighted in the Smithsonian *Gees Bend Quilters* exhibit. The cooperative continues to exist, but has suffered from the death or retirement of many of its original members and lack of new members and resources.

1967: FEDERATION OF SOUTHERN COOPERATIVES

The Federation of Southern Cooperatives is a nonprofit organization of state associations to support predominantly Black cooperatives in southern states. Founded in 1967, it later merged with the Land Emergency Fund to become the Federation of Southern Cooperatives/Land Assistance Fund. Member cooperatives engage in organic farming, marketing, agricultural processing, fishing, sewing, handicrafts, land buying, grocery cooperatives, and credit unions. The organization also engages in state and federal policy advocacy and provides technical assistance to protect Black-owned land and maintain Black land ownership. The Federation provides emergency services to its members during times of natural disaster. Since its founding, the organization has helped to create or support more than two hundred cooperatives and credit unions, mostly in the states where it operates: Alabama, Arkansas, Florida, Georgia, Kentucky, Louisiana, Missouri, Mississippi, South Carolina, Tennessee, Texas, and the Virgin Islands.[10]

The federation owns and runs a rural training and research center in Epes, Alabama, that showcases sustainable forestry, provides co-op education, and helps to develop Black youth-run co-ops, such as the Sankofa Youth Cooperative. Its headquarters is in East Point, Georgia. The FSC/ LAF also engages in cooperative development in Africa and the Caribbean. The organization has an important reach throughout the South, is connected to the larger US cooperative movement, and has successfully advocated for important measures in US farm bills to support Black farmers, Black land ownership, and Black co-op development.

1997: DAWSON WORKER-OWNED COOPERATIVE, DAWSON, GEORGIA

The Dawson Cooperative was created by workers from the abandoned Almark Mills fabric cutting and sewing plant, with help from a local business developer. The president and general manager of Almark Mills worked with the mayor of Dawson and former employees of the Mill to create a worker-owned sewing factory. In December 1997, seventy members who were former employees started work at the old plant, now as worker-owners in a new cooperative. The majority (seventy-six percent) of the mill's workforce was female, one-third of them single mothers, and most were Black.[11]

Almark Mills had been the largest employer of women in Terrell County, Georgia, and there were no other textile jobs within fifty miles. The worker-owners used their union fund from years of paying union dues, now available because the union had been dissolved with the closing of Almark Mills, as their equity investment in the new cooperative. Ownership shares were also paid in installments of weekly payroll deductions of $7.16 over four years. The co-op was fortunate to be able to access targeted government funds. The Clinton Administration's Community Adjustment and Investment Program authorized and funded the Department of Agriculture to make loans to businesses in up to fifty rural communities adversely impacted by NAFTA through the Business And Industry Loan Guarantee Program operated by the Rural Business Cooperative Service. Their county qualified, enabling the co-op to pay off the bank debts and to expand and hire one hundred more workers. Dawson Cooperative sales in 1998 were almost $5 million.

The mayor of Dawson, Robert Albritten, told Catherine Merlo that "Persons in this community doubted that women and minorities could make this work where it hadn't succeeded before, but we've made believers out of them."[12] Board chair Dianne Williams remarked that it changed the workers' lives, made them more optimistic and hopeful, and changed the way business was done in the factory. Jobs were now more secure and communication was now more open and transparent.

2009: MANDELA FOODS COOPERATIVE, WEST OAKLAND, CALIFORNIA

The Mandela Foods Cooperative is incorporated under California law as a for-profit cooperative. The co-op started in June 2009 with a group of local activists working on food security issues in Oakland. The store opened with eight worker-owners who operate in a nonhierarchical management structure by committee. The co-op partners with local farmers to increase the financial sustainability of family farms, and to increase access to fresh foods; it partners with its incubator, Mandela Marketplace, to support cooperative development in West Oakland. The co-op allows low-income residents in the neighborhood to provide services and to be involved in the cooperative as worker-owners.

An excerpt from "Black Cooperatives in the United States: An Excerpted History," prepared for the Babson-Equal Exchange Cooperative Curriculum. A version of this was also published on the Federation of Southern Cooperatives's website: federationsoutherncoop.com/coopinfo/Black%20 co-ops.pdf.

NOTES

1. William H. Pease and Jane H. Pease, *Black Utopia: Negro Communal Experiments in America* (Madison, WI: State Historical Society of Wisconsin, 1963); James DeFilippis, *Unmaking Goliath: Community Control in the Face of Global Capital* (New York: Routledge, 2004).
2. William F. Holmes, "The Laflore County Massacre and the Demise of the Colored Farmers' Alliance." *Phylon* 34:3 (1973), 267–47.
3. Bruce J. Reynolds, "Black Farmers in America, 1865–2000: The Pursuit of Independent Farming and the Role of Cooperatives. *Rural Business–Cooperative Service RBS Research Report* 194 (October 2002), US Department of Agriculture;
4. Omar H. Ali, "Black Populism in the New South, 1886–1898," PhD dissertation, Columbia University, 2003, p. 77; Clyde Woods, *Development Arrested: The Blues and Plantation Power in the Mississippi Delta* (London: Verso Press, 1998); Holmes, "Laflore County Massacre."
5. "Business: Negro Chain," *Time*, May 12, 1930.
6. Lisa Tolbert, "Colored Merchants Association: Challenging the Chain Stores," *Tar Heel Junior Historian* (Spring 2007).
7. Nathan Alvin Pitts, "The Cooperative Movement in Negro Communities of North Carolina: A Dissertation," *Studies in Sociology 33* (1950), Catholic University of America Press, 27–30.
8. Ibid.
9. Federation of Southern Cooperatives/Land Assistance Fund, "Freedom Quilting Bee of Alberta Alabama: A Southerns Legend," *25th Anniversary Annual Report, 1967–1992* (East Point, GA: Federation of Southern Cooperatives/Land Assistance Fund, 1992).
10. Freedom Quilting Bee, "History, Activities, Plans," Rural Development website, n.d., ruraldevelopment.org/FQBhistory.html.
11. "A Tribute to the Federation of Southern Cooperative/Land Assistance Fund," *Journal of Cooperative Development* 2:2 (Spring 2000). Examples of cooperatives in the Federation are: Freedom Quilting Bee (AL), North Bolivar County Farm Cooperative (MS), Panola Land Buyers Association Housing Development Corp. (AL), Southern Alternatives Cooperative (GA), Southwest Georgia Farmers Cooperatives (GA), Indian Springs Farmers Association Inc. (MS), Beat 4 Farms Cooperative (MS), South Plaquemine United Fisheries Cooperative (LA), South Carolina Sea Island Farmers Coop (SC), People's Cooperative (SC), Farmers Cooperative (Virgin Islands), Demopolis Citizens Federal Credit Union (AL), First Delta Federal Credit Union (MS), and Shreveport Federal Credit Union (MS).
12. Catherine Merlo, "The Triumph of Dawson's Textile Workers," *RBS Rural Cooperative Magazine*, March–April 1998.
13. Ibid.

RUMINATION ON THE FUTURE OF YOUNG BLACK FARMERS

SAVI HORNE

If we take a riff from W.E.B. Du Bois and posit that the question of the 21st century is the "land question," just as the question of the 20th century was the "color question," then we have witnessed the greatest decline in African-American land ownership since its height just before the Great Depression.

With our tenuous hold on less than seven million acres from upwards of twenty million, we are facing a future of land losses that confounds the collective memories of family farmland ownership. In an era where next-generation farmers are on the rise, this lack of access to farmland blunts the ambition of many future producers. According to Dr. Spencer Wood, "There is a concern that the robustness and resiliency of the rising movement of small producers that is ushering in a new era of food democracy is resting on uncertain footing, and that control of the land is not secure and efforts to shore it up are routinely met with obstruction by local governing bodies."

Across the southern United States, rural communities are looking to strengthen the local food economy as a means to improve health by expanding access to fresh and affordable produce. These communities are impeded by a lack of access to land that was previously available but has been lost through structural racism in the farm economy. Without agrarian reform in national farm policy and new land models that meet the needs of this emerging movement in the food rights space, the growth of African American farmers in the 21st century will be stymied and once again become a dream deferred, to the detriment of healthy Black bodies and healthy Black livelihoods.

SOME PASTURES

A Cowboy's Almanac

SAM RYERSON

August, Bench Pasture—Fishtail, Montana

Another summer, we helped Brandon, running goats for weed-control projects on ranches across Montana. On one of them, up on the big bench above the West Rosebud River outside Fishtail, all the old coulees were infested with leafy spurge. It was a noxious weed and made a sea of bright yellow. Some people said spurge was toxic to cattle. I never saw a cow sick from it, but I never saw one eat much of it. It could take over the most productive soils and choke out the native grasses. The ranch owners hired us to bring the goats and clean it up, and the goats loved it. We took them up from the headquarters and moved them through a series of temporary paddocks. They devoured the spurge. We'd left the guard dogs behind with Brandon for his other herd when we shipped out of Two Dot. We thought if we were always with the goats or had them inside the electric fence, we wouldn't have trouble with bears or coyotes.

One morning I went up from our camp down by the creek and found a dead goat inside the fence with her belly ripped open and her skin peeled back. I found bear sign around her and a six-inch gap in the fence between two sections, where one of us forgot to connect the clip that carried the charge. The bear had pulled the fence apart, crawled in, and got himself a goat. The next few nights I took my bedroll up and slept outside the goat pasture with my dogs, and a pistol. I have done the same where we had wolves near our cattle. If the weather was nice, it was fun camping out like that, but I doubt it made much difference for the cows or the wolves, except that I would know sooner when there was a problem. I have worked with some environmental groups on programs to try to prevent predator conflicts and served on the board of a coexistence council to help ranchers working in the presence of wolves. I've shot at some coyotes when they came right in by the house and were killing chickens. I never liked it. It always seemed like it was a better use of our time to manage our livestock than to hunt the predators. If a cow loses her calf to a coyote, I'll usually sell the cow before I hunt the coyote. And if I had kept the fence hot, I wouldn't have had trouble with that bear.

An American having struck a Bear but not killed him, escapes into a Tree.

issue 1 call

The JOURNAL of JUDGMENT, MEASUREMENT, 'N TECHNOLOGY

our hypothesis:
the way we measure our world is the way we talk about our world is the way we make decisions and act in our world. measurements are part of our shared grammar + consciousness.

are we placing too much emphasis on certain measures? is there a measure we are missing? how does our tech affect our measurements?

carbon footprint is a relatively new measure that has become prominent in the last 40 years. what else do we want to measure + converse about in the next 20-50 years?

01:30
00:40
Faster!

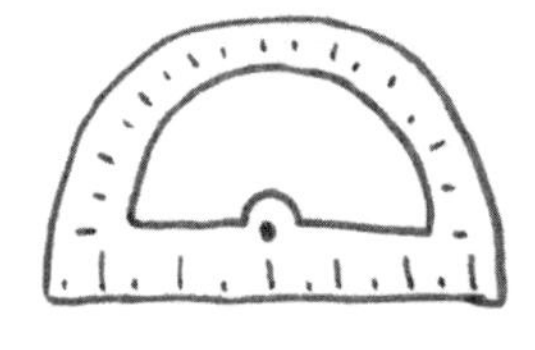

write to us:
JJMNT c/o danjo
12 Dunning St
Brunswick ME 04011

and miles to go before we sleep...

43°54'2"N 69°57'34"W, 29°F, -1°C, kWHrs, Calories, FGP, PPG, RBI, GDP,
GPA, MPG, CPI, DOW JONES, APR, PPM, PPB, lbs, in, kg, L, mph, Nm, Gauss,
unemployment rate, incarceration rate, LDL, HDL, Richter Scale, acres, hectares

Danjo Paluska

jjmnt

brief examples

AY - The acre year

The acre year is a unit of energy like a calorie or joule or KWHr. In the case of an acre year, the energy unit is in reference to biological growth on the land. Old timers in New England say you can harvest a cord of wood from an acre every year + retain a full forest. So one acre year is one cord of wood which is about 100-150 gallons of fuel oil in energy. How many acre years to heat your home each year?

CTI - The compost time index

Like the richter scale, CTI is a logarithmic measure. 0 means 1 year, 1 means 10 years, 2 means 100 ... 5 is 100,000, etc. CTI measures decomposition time in a compost pile. Food scraps are mostly CTI 0, while plastics are in the 3-5 range. Woods, metals, ceramics, etc are somewhere in between. Does CTI help me remember to bring my own bag?

TSI - Topsoil Spread Index

This measure is motivated by all the junk we see left behind on the land. An old barn collapsing is not so bad but what about lead paint or an old battery? or a computer? are there heavy metals somewhere? For any given product, how much area is it capable of polluting if it is dissolved into the topsoil? Could this number help us to be more conscious in our waste stream? Or add deposits to certain products?

scoville units, liters per flush, ROI, NPK, mg of sodium, 120/80, 98.6°F, happiness index, 0.8mm, 6×8", barometric pressure, psi, mm of mercury, drams, cups, tsp, TBSP, a pinch, 440 Hz, 20 Mbps DL

BOOK REVIEW

Land Justice: Re-imagining Land, Food and the Commons in the United States

ELIZABETH HENDERSON

If you are looking for a deeper understanding of the ills in our food system and a strategy to address them other than "vote with your fork," a book you should read as soon as possible is *Land Justice: Re-imagining Land, Food and the Commons in the United States*, which was edited by Justine M. Williams and Eric Holt-Gimenez and published by *Food First* in 2017.

Land Justice stands in contrast with so many food movement books that never question the basic premise that with a few adjustments, we can correct the excesses of the capitalist marketplace. Holt-Gimenez lays out the book's basic premise, writing, "Racial injustice and the stark inequities in property and wealth in the U.S. countryside aren't just a quirk of history, but a structural feature of capitalist agriculture. This means that in order to succeed in building an alternative agrarian future, today's social movements will have to dismantle those structures. It is the relationships in the food system, and how we govern them, that really matter."

A collection of essays, *Land Justice* brings together stories of old injustices and ongoing ones. We learn about the Gulluh Geechee farmers of the Atlantic coast, George Washington Carver and Booker T. Whatley, the Republic of New Africa, the Land Loss Prevention Fund, white and black women farmers, the Acequia Communities of New Mexico, the Mashpee Wampanoag Tribe of Massachusetts, Rosalinda Guillen and farm worker organizing in Washington state, the Black Community Food Security Network in Detroit, Michigan, and the People's Community Market in Oakland and students taking action in Occupy the Farm in Berkeley, California.

Three voices open *Land Justice*: Winona LaDuke, a woman and registered member of the Ojibwe Tribe; LaDonna Redmond, an African American woman; and George Naylor (who is white), a family-scale farmer. Often marginalized in our food system, these writers represent the voices that must be heard if we are to sort out the history of the "land problem" in this country and imagine a way forward.

LaDuke contrasts mainstream industrial, monocrop agriculture with the indigenous approach "based on biodiversity and the use of multiple locally adapted crops." Plants, LaDuke tells us, are magical and "provide complex nutrients, medicinal values, cultural and spiritual connections, and they feed the soil." She recounts the struggle of Native Americans for control of their land culminating in the successful class action suit, *Keepseagle vs. Vilsack* which, in 1999, won $680 million in reparations. While this award is far from adequate, it marks the resurgence and recovery of indigenous farming that is underway. LaDuke declares that it is time for decolonization.

Taking as her chant "This Land is Contested," Redmond laments the removal of Native American peoples that preceded the importation of slave labor. "The holocaust of the indigenous set the stage in the U.S. for the rise of capitalism," she writes, as the free labor of twelve million enslaved Africans on stolen land that followed "is what built the wealth of the so-called New World." Redmond urges solidarity between black people and Native Americans, "the water and land protectors," and calls for unity against "corporate oligarchy and federal imperialism."

Rejecting the "vote with your fork" analogy, Iowa farmer George Naylor declares that it's time to "recognize how market forces affect farmers, the land, and consumer behavior, and demand policy solutions to achieve a sustainable future." Naylor insists that we—citizens, consumers, and farmers, all—need to "de-commercialize food and land," proposing that we replace the cheap food policy that has enabled corporate dominance with a system based on parity like the New Deal farm programs involving "conservation-supply management to avoid wasteful, polluting overproduction; a price support that actually set a floor under the market prices rather than sending out government payments; grain reserves to avoid food shortages and food price spikes; and a quota system that was fair to all farmers and changed the incentives of production."

Despite the recitals of inhuman cruelty and brutal greed, *Land Justice* leaves the reader energized and inspired by the courage and determination of its contributing authors. Together, they show us how the future can be built: through alliances and collaboration with marginalized communities to "change the politics of property." This book is a major contribution to a radical and coherent program for transformative change. As Holt-Gimenez concludes his incisive introduction, the authors are "in a struggle to remake society." It is up to us to harken to their passionate words and to take land justice as "both a vision and a clarion call."

Originally published in The Natural Farmer *(Winter 2017). Reprinted with permission.*

Preparation of hemp beer

SEPTEMBER

Marijuana Culture

IN A COUNTRY FOUNDED ON SMALL FARMS

CASEY O'NEILL

In a country founded on small farms, it can be a struggle for us to survive in the modern industrial paradigm. We have to be excellent farmers, good at our craft, and willing to work long, hard hours. As farmers, we need the support of government systems that honor our contributions to the economy, to the social fabric of our communities, and to the environment we steward. Small farmers need every potential option available to us as producers of plants and participants in animal husbandry. The deck is often stacked against us by federal programs that provide support to large agribusiness operations, resulting in a skewed accounting of food costs and appropriate consumption patterns.

Diversified farms are the answer to the many externalized costs of industrial agriculture; it is my argument that cannabis and hemp can serve as important crops for many of these operations. Maintaining cash flow is one of the biggest issues for small operations. Creating a system of regulation that authorizes cannabis as a cash crop for small producers would go a long way toward leveling some of the disparities in agricultural production. It has become clear that Earl Butz's "Plow it fence row to fence row" has failed our nation. As farmers, we must always be assessing and reassessing our systems of agriculture.

Why do I advocate for cannabis to be held for small farms? Because the last hundred years are one long cascade of industry consolidation that has led us to today. We have a once-in-a-civilization opportunity to bring a major agricultural crop online under the auspices of supporting small farmers, while giving us the opportunity to shine as quality producers. By a cold economic reckoning, agriculture trends larger-scale to achieve "efficiencies of scale" that often rob the land and those who tend it. If true-cost accounting is used to measure the well-being of land, community, and economy, small farms are far and away the most efficient producers.

Engagement in the democratic process is a time-honored tradition for farmers. As the 21st century approaches the beginning of its third decade, cannabis is making a transition out of decades of Prohibition into the light of a new day. In many places, farmers are still denied the ability to cultivate both hemp and psychoactive varietals. There needs to be a robust dialogue about how cannabis should be reintroduced as a crop for American farmers.

Farming is the only industry in which producers buy everything at retail, sell everything at wholesale, and pay for transportation on both ends. Over the last hundred years, there is a disturbing parallel between the declining number of small farms and the declining diversity of crops, including the removal of hemp and cannabis from the small-farm portfolio. Farmers have been convinced to produce fewer types of crops in favor of larger volumes of more specialized production. This has been detrimental to the ability of small farms to survive.

There was a time, not too long ago, when hemp and medicinal cannabis varieties were commonplace in America. We were a nation of transplants, people who came from all over the world, bringing their favorite cultivars. America was known for the superior ropes we produced from hemp crops that arose out of the melting pot of plant breeding that occurred through the 19th and early 20th centuries.

When Henry Ford designed his first vehicles, he envisioned an engine capable of running on hemp-seed oil. He expected that farmers would purchase the Model T's because they would be able to produce fuel to power them. It is amazing to see hemp reemerging as a crop for farmers as society begins to shake off the blinders put on by decades of false propaganda.

It is empowering to read about the many modern uses of hemp, from BMW door panels to hempcrete to paper, clothing, fuel, and food. Hempcrete is less energy-intensive than concrete and is carbon-negative, sucking carbon out of the atmosphere for years after it is put in place. Hemp can be used to remediate soil and can sequester carbon from the atmosphere while producing large quantities of biomass that can be used for many applications.

When we turn to look toward the medicinal cultivars, it is instructive to look back to a time before modern advertising and propaganda skewed belief in herbal medicine. The medical pharmacopoeia from the 19th century is rife with the use of cannabis for all manner of human illnesses, and it has been used as traditional medicine and spiritual upliftment for eons.

It is worth considering what is known as "the stoned monkey hypothesis," which argues that humanity evolved through consumption of mind-altering substances that enabled expansion of consciousness. It is possible that humans learned how to be human with the use of plants to open windows of experience that alter perception. This may have created opportunity for cognitive leaps, generating new ideas and processes.

Cannabis has been used in ritual ceremonies by many religions for eons; there is a vast human history that includes use of this plant, and it behooves us to take a close look at it through the lens of modern technology and understanding. There is much for us to learn, but we must first engage in the process of renormalization, the need for which has been created by decades of prohibition.

The change is underway, but the process is already threatened by big business, which sees just another opportunity for plunder. We need to be thoughtful about this process, considering what is best for us as a country. In this time of great change and uncertainty, we could do no better than to strive to support small farmers. Cannabis can help.

 Previous and opposite photographs by Casey O'Neill

THE GREEN RUSH: ALL THAT GLITTERS IS NOT GOLD

Cannabis Farming in Southwest Oregon

DON TIPPING

Gold miners in California, circa 1850

Southern Oregon is the northern extension of the prime climate zone for growing outdoor cannabis. While not as well known as the famed Emerald Triangle of Humboldt, Mendocino, and Sonoma counties in California, southern Oregon has a similar climate, and as a result of more relaxed regulations in Jackson and Josephine counties, many growers from the Emerald Triangle have relocated to the area, which has led to new grow sites popping up like field mushrooms. Unlike field mushrooms, these sites tend to be capital-intensive operations with large light-deprivation greenhouses, extensive wooden fences, pungent odors, noisy cooling fans, and high worker traffic during harvest time. While cannabis provides a myriad of useful constituents and benefits, there are ways in which the gifts of the plant are being overshadowed by the detractions associated with its commercial cultivation.

While Jackson and Josephine counties account for less than ten percent of Oregon's total population, it's estimated that more than forty percent of the state's cannabis production occurs here.[1] The two adjacent counties are among the state's top six producers (joining Multnomah, Lane, Douglas, and Deschutes counties). This is a Green Rush, with newcomers and fortune seekers streaming into the area at a record pace. While southern Oregon has a long history of backwoods marijuana growing and the black-market economy associated with it, these newer recreational cannabis grows are regulated by the Oregon Liquor and Control Commission after Oregon voters successfully legalized recreational cannabis use in 2015 (medical cannabis was legalized in 2001).

Here are a few stats from the Oregon Liquor Control Commission and an early 2017 report by Beau Whitney of Whitney Economics:[2]

- Oregon has 685 licensed producers and another 800-plus applications pending.
- There are now over 12,500 cannabis jobs in Oregon.
- The average wage earned is $12.13 per hour.
- There are 917 cannabis businesses.
- Cannabis revenue in 2017 is $315 million (although an Oregon State Police report from 2017 estimates that the value of Oregon grown cannabis leaving the state to be $4.7-$9.4 billion – which means that only 3% to 5% of the cannabis grown in the state is being sold legally[3]).
- Jackson County has the second-highest number of medical marijuana cardholders of any county in the state, at 9,234, just behind Multnomah County at 10,890.
- The Department of Revenue estimated in 2016 that it should receive $13.46 million in taxes for the first quarter of 2017 if all stores file returns.

The explosion of jobs, agricultural businesses, and resultant revenue from a nascent agricultural sector is astounding and likely unprecedented. It would be interesting to look at figures for barley, grape, and grain production following the end of Prohibition in 1933. Chances are that farmers who were growing these crops before Prohibition started in 1920 found alternate markets for their grain or grapes, or rotated their crops to grow other things. Cannabis is unique in that it was prohibited in all fifty states in 1937 and then stopped being grown legally until the passage of Proposition 215 in California in 1996, followed by passage of similar laws in Oregon, Washington, and Alaska in 1998. It's unlikely that any farmer who grew cannabis before 1937 was still around, ready to resume growing it sixty years later. Many of the people jumping into cannabis production in recent years are new to agriculture, enticed by the high value from decades of prohibition and the subsequent black market economy. However, the inclusion of small-scale cannabis production has enabled many new and young farmers to establish biodiverse organic farms. Further, some existing regenerative farms have added cannabis to their rotation as a profitable crop, capitalizing on high prices still available for boutique organic dried flowers.

This has led to growing pains in the emergent industry, with former city folk moving to the country in hopes of striking it rich, and established farm and ranch communities struggling to understand and accommodate the newcomers. Rural folks are sometimes slow to accept change, and the rapid changes to some agrarian communities have led to friction with county land-use planners, increased traffic on rural roads, and farm labor shortages. Also, state regulations require fences, setbacks, security cameras, and so forth to comply with rules governing recreational cannabis production. The face of rural western Oregon has changed rapidly as a result of new cannabis grow sites. In addition, skyrocketing land prices in cannabis hot spots often prices out farmers wanting to grow crops other than cannabis.

Industrial hemp is regulated by the US Department of Agriculture and has less stringent requirements for production. However, by definition, cannabis produced as hemp must contain very low THC (the psychoactive constituent that most people associate with marijuana) and is being widely grown for use in high-CBD health products, from pills to lotions to juices, marketed to treat a host of health issues. The use of hemp will not produce any psychoactive effects.

What lies ahead in this burgeoning cannabis landscape? Peter Gendron, president of the Oregon Sungrown Growers Guild, which represents outdoor cannabis producers, said in mid-2017, "In the current market, a pound of marijuana fetches roughly $800 to $1,200 per pound for the grower. An experienced outdoor grower can produce a pound for $200 to $400, but an inexperienced grower may not generate any profit at all."[4] By the end of the season, a pound of legal marijuana could be had for as low as $500 per pound. The drop in prices resulted from massive overproduction. With so many people new to agriculture needing to invest heavily in cannabis-growing startups to cover the fencing, security cameras, reporting, and other expenses, and the price auguring into the ground as so many new players join the fray, many will probably wind up financially upside-down.[5] Nonetheless, savvy entrepreneurs with vertically integrated businesses and a solid market stand to profit well in the early stages before big ag joins the game and drives prices even lower. Oregon's new rules will probably cause medical marijuana production to "shrink and stabilize" over time, said Representative Carl Wilson, vice chairman of the Oregon Legislature's Joint Committee on Marijuana Regulation.[6] "In October 2017, Oregon growers harvested 2.5 million pounds of wet-weight cannabis—more than twice the previous year's crop." It's estimated that in 2017 "nearly one million pounds of usable flower remained unsold, about three times what consumers purchased in all of 2017."[7]

In other agricultural sectors, the production of raw commodities rarely if ever commands a high price. Some notable exceptions, like saffron pistils, do exist, but the trend is toward consolidating profit in the hands of the processors, distributors, and retailers. The relatively high prices being obtained for recreational cannabis and high-CBD hemp are likely temporary, and new pests and diseases are showing up as commercial production ramps up. Many growers are resorting to chemicals as the existing regulations for recreational cannabis largely preclude crop rotation and crop diversification, and there is little regulation of what agricultural chemicals are allowed in cannabis production. The regenerative cannabis movement is striving to encourage growing methods that foster healthy soils and biodiverse farming systems. Eventually cannabis will join all the other useful plants in the civilized human palette, and farmers will grow it in rotation and association with many other crops, and more people will be able to celebrate its beauty, resilience, and usefulness. My advice to any would-be new grower would be the same as for any crop: "Start small, harvest small mistakes, start big...."

NOTES

1. Associated Press, "Tiny Town Is Oregon's No. 1 Hub for Pot Growers," *CBS News,* August 3, 2011, cbsnews.com/news/tiny-town-is-oregons-no-1-hub-for-pot-growers/.
2. "Legal Marijuana a 'Powerful Force' in Oregon Economy," *Green Entrepreneur*, March 16, 2017, greenentrepreneur.com/article/290669.
3. Noelle Crombie, "Oregon Remains a Top Source for Black Market Pot, State Police Report Says," *The Oregonian*, May 2, 2017, oregonlive.com/marijuana/index.ssf/2017/03/oregon_a_top_source_for_black.html.
4. Mateusz Perkowski, "Oregon's Marijuana Industry Raises Land Use Tensions," *Capital Press*, August 3, 2017, capitalpress.com/Oregon/20170803/oregons-marijuana-industry-raises-land-use-tensions.
5. April Ehrlich, "Study Gauges Cannabis Farms' Impacts on Food Growers in Southern Oregon," Jefferson Public Radio, May 11, 2018, ijpr.org/post/study-gauges-cannabis-farms-impacts-food-growers-southern-oregon#stream/0.
6. Perkowski, "Oregon's Marijuana Industry."
7. Lily Raff McCaulou, "Oregon's Weed Glut: What Happens to Excess Pot?" *Rolling Stone*, July 19, 2018, rollingstone.com/culture/culture-features/oregons-weed-glut-what-happens-to-excess-pot-6.

[SINGALONG]

ALL THAT WEED

CASH JOHNSON

In February you planted seeds, by April you had females
Some folks you know grow lots of weed, so you asked them for details
You were in the ground by early May, at a friend's suggestion
By fall your plants were so damn tall that now it begs the question ...

What're you gonna do with all that weed, what're you gonna do with all that weed
Who woulda thunk just a thimble full of seeds, could ever have turned into all that weed

Now where you gonna dry all that weed, where you gonna dry all that weed
You could hang some in the dryin' shed, at least until it's full
You could hang some in the woodshed with some heaters don't ya know
You could hang some in the attic, or that bedroom down the hall
You could string some extra wires up and hang some on the walls

Well, what're you gonna do with all that weed, what're you gonna do with all that weed
Who woulda thunk, just a thimble full of seeds, could ever have turned into all that weed

Now who ya gonna get to clean up all that weed, who ya gonna get to clean up all that weed
You could hire the local trimmers, if nobody's hired 'em first
You could hire some drifters off the street but that idea's worse
You could get the new machine, let 'em do it all by hand
You still have to feed and pay 'em and keep track of every gram

So what're you gonna do with all that weed, what're you gonna do with all that weed
Who woulda thunk just a thimble full of seeds, could ever have turned into all that weed

Now how you gonna store all that weed, how you gonna store all that weed
Use some barrels in the bushes for those pounds already cleaned
You still got that monster pile you haven't processed yet it seems
Too damp and it gets moldy, too dry it turns to duff
And you're the one who's got to keep an eye on all that stuff

Well, what're you gonna do with all that weed, what're you gonna do with all that weed
Who woulda thunk just a thimble full of seeds, could ever have turned into all that weed

Now where you gonna sell all that weed, where you gonna sell all that weed
The local competition is a little bit intense
You could mail it out of state but that's a federal offense
You could hit up the dispensary, consignment's how they trade
And if the Feds come shut 'em down well, you ain't gettin' paid

What're you gonna do with all that weed, What're you gonna do with all that weed
Who woulda thunk just a thimble full of seeds, could ever have turned into all that weed

This song was written by Cash Johnson for the Camo Cowboys.
Recordings can be found on iTunes, CD Baby, and YouTube.

Nikki Lastreto | Swami in the Plants

HEY USDA, WE CAN GROW CANNABIS USING ORGANIC METHODS TOO

NIKKI LASTRETO AND SWAMI CHAITANYA

"Who grew this cannabis, and how was it grown?"

The smart cannabis connoisseur knows to ask that question before making a purchase. These facts can be more important than the actual strain of marijuana: a particular flower can get you extra high, sure, but if your weed is grown with chemicals, it can also make you very ill.[1]

To guarantee that whatever cannabis products you're imbibing are pure and clean, it's key that they've been officially tested and certified by one of the programs now available in the adult-use market. At a time when there are horrific reports of growers irresponsibly using chemical fertilizers and pesticides—including myclobutanil, a.k.a. Eagle 20 pesticide,[2] which becomes hydrogen cyanide when combusted in a smoking device of any kind—it's good advice to know where and how the cannabis you're consuming is cultivated. A year ago, NBC ran an investigative report alleging that pesticides were discovered in forty-one out of forty-five samples gathered from several Southern California dispensaries.[3] With news like that, it should come as no surprise that more and more consumers are starting to ask for organic ganja.

But wait—we can't use that word. The term *organic* is essentially owned by the US Food and Drug Administration (FDA) and the US Department of Agriculture (USDA). Since cannabis remains federally illegal, these institutions will not authorize standardized testing regulations for the legal cannabis industry. In other words, even if your flower is one hundred percent organic, it cannot be called "organic" when it comes to branding and packaging. The government's excuse is since cannabis is still a Schedule I drug, it is not officially a "crop" and cannot be certified.

"You must describe it as 'grown with organic methods,'" explains Chris Van Hook, founder of Clean Green (CG) Certified, the country's top certifier for cannabis cultivated using, well, organic methods. His company certifies cannabis in six states and has 225 clients—not just cultivators but processors, handlers, and retailers as well. In fact, his big break came in

Layton Hansen | digital drawings

2010, when Harborside Health Center contracted him to inspect and certify their operation. Subsequently, the dispensary began offering growers one hundred dollars more per pound if their product was Clean Green certified. Swami Select was already working with Harborside at the time, and this extra financial incentive was just another reason for us to become a part of the Clean Green program.

After growing up on a dairy farm in Pennsylvania, Van Hook got a degree in Environmental Studies at the University of California, Santa Barbara. After that, he became a marine biologist and abalone farmer, working for three decades in Crescent City on the northern California coast. He readily admits that smoking cannabis helped him get through the tedium, aches, and back pain when he was still in the abalone trade. His background enabled him to see that cannabis farmers face challenges similar to those of any other agricultural pursuit. When the abalone industry died out, Van Hook studied for his law degree online, and passed the bar on the first try.

The USDA started the National Organic Program in 2003, and Clean Green was accredited in 2004 to certify crops other than cannabis, or as Van Hook describes it, "from apples to zucchini, as well as all types of organic processed foods." When the USDA refused to inspect or certify cannabis because of its Schedule I status, Van Hook saw a need to protect the cannabis consumer. His company started certifying cannabis in 2004, and to this day Clean Green certification is still the closest thing available to an organic cannabis certification.

Van Hook has been doing an annual inspection on our farm in Mendocino since 2010. He's like a trusted family doctor who always makes us feel better and wiser after a visit. He spends upward of two hundred seventy days each year on the road in his Sprinter van, which doubles as a mobile law office and an RV—one equipped with a mountain bike, backpacking gear, and a surfboard. When he's at our ranch, Van Hook wanders around the garden, checking the leaves for bugs. "That's a good thing, seeing life on the leaves," he'll say, as he peers through his camera lens at some insect damage. If the leaves have no damage, it most likely means a pesticide was employed. Then he puts on sterile gloves, digs up soil samples from several places in the garden, puts them in a plastic bag, and sends them off to an independent government-certified testing lab for pesticides and nonorganic substances. Van Hook even inspects the lab tests, as he's somewhat critical of some of the testing companies that have sprung up for the cannabis industry. He's told us that many of these firms don't even have a PhD scientist in the office, although that's beginning to change as labs ramp up their staff qualifications.

We should clarify that there are different kinds of testing, as well as different times and sites for each individual test. Van Hook and the Clean Green staff insist on visiting the garden where the flowers are grown, looking at the plants on-site, sampling the growing medium (the soil) for testing, and verifying the water source—all in vivo. Every farmer submits an itemized list of all amendments, fertilizers, pesticides, compost, compost tea formulas, and anything else sprayed on the plant or added to the soil or irrigation water. Similarly, for certifying a cannabis processor or retailer, the site inspection is critical; operating procedures are reviewed and ingredients are itemized.

Cannabis labs that have opened in the last several years, like SC Labs, Steep Hill, and CW Analytical, all test the final flower product after harvest and processing, or, for products like edibles or concentrates, before and after manufacturing. This is the only kind of testing that the state requires, and these companies don't visit the farm or manufacturing site. They are testing for potency, pesticides, and pathogens, not for organic cultivation practices.

At first, these labs mostly tested for THC. This was done for marketing purposes, to put the highest price on the highest THC content. Soon, though, people found out about the beneficial medical properties of CBD, so it and other cannabinoids were added to the test results. Then terpene profiles became the rage, also for their value-added effect. The great benefit of all this detail was the rapid education of many of us on the mysteries of the ensemble effect, also known as the entourage effect, when all of these cannabinoids and terpenes combine to make inspirational and healing music together, which leads to a given cannabis strain's psychoactive and medical properties.

Prior to a Clean Green inspection visit at Swami Select's farm, we compile an exhaustive list of the products we are using. Generally, any product that is certified by the Organic Materials Review Institute is safe for our plants. Clean Green also examines our harvest drying areas for cleanliness, and the total fee is $2,500 per inspection. A few weeks later we get a packet in the mail with our Clean Green Certificate, along with a poster and a plaque for the garden. Best of all, Van Hook is always there for clarification. If a new product comes out that everyone is raving about, as happens every year, we can call him up and he will give his approval (or not) after appropriate research.

Clean Green also produces the California-Oregon Cannabis Exposition, featuring CG's growers, many of whom have won awards at major cannabis competitions,[4] demonstrating that the best cannabis is grown organically. All of us are proud to feature the Clean Green logo on our packaging because people can rely on it being a pure product. Van Hook has expanded over the years and now certifies distributors, manufacturers, and retailers as well as cultivators.

He's also begun Clean Green Real Property to help people with real estate transactions.

Integrated pest management (IPM) is another hot topic among growers these days. If you want a California cultivation license in the Golden State's new market, you need to submit a pest management plan. Furthermore, every state-sanctioned farm will be required to have a trained pest-control adviser, who instructs other workers on the farm in best practices for insect containment. The reality is that an infestation of bugs or pathogens threatening to attack the whole cannabis crop can tempt a fearful grower to use harmful pesticides. IPM is all about both preventative and emergency care through the use of beneficial insects and biological predators. As Richard Nathanson of Santa Cruz Integrated Pest Management advises, "Walk your crop every day and make notes of who is stressed, the water quality, is anyone root-bound, overfertilized, etc. If we do preventative measures, we don't need to spray pesticides."[5]

Not all farmers whose products test positive for pesticides are even aware of producing tainted bud—all the more reason for certification. At recent Emerald Cup and Golden Tarp cannabis competitions, a significant percentage of entries were rejected for pesticide content (especially Eagle 20) and for exceeding extremely strict microbial limits.[6] For the unwary cultivator, pests, pathogens, and pesticides, as well an nonorganic amendments, can enter a garden via store-bought soil, compost, clones, or even random visitors.

Since the USDA and FDA refuse to classify cannabis as a crop, and the state says it's not even a food, the result is that no pesticides (that is: none) are legal to use on cannabis, even if they are legal for other crops. Crops and products must test clean, or they will fail and be destroyed. In California, this is the responsibility of the distributor, who collects the material from the grower or manufacturer, sends a small sample to the lab, and, if it passes, then ships the product to the retailer—at the same time collecting the $9.25 per ounce cultivation tax from the farmer as well as the distribution fee. Some distributors are even differentiating themselves by working exclusively with organically minded farmers and cultivators.[7]

The word is starting to spread about the benefits of Clean Green cannabis. Having passed the CG inspection and testing, the farmer can be relatively confident they will pass the state-mandated tests, although CG certification will not exempt one from state-required testing. "Consumers who know about CG want it and are serious about it," said Blake Johansson of Pacific Wholesale Network, a distribution company in Los Angeles. "People are seeing the bifurcation now between a Clean Green product and not. People want it—clean is worth more."

Other companies that certify cannabis products are sprouting up too. Certified Kind, based primarily in Oregon, is an organization with an earth-friendly philosophy: "Kind to Life and Kind to Earth." They base their rules on the International Federation of Organic Agriculture Movements as well as USDA and international organic standards. In Washington state, the governor recently signed legislation for an organic certification program.[8] In Colorado, the Cannabis Certification Council has been working on legislation to create a state organic certification program. In the Emerald Triangle, other certification programs are in the works via several distributors and agricultural cooperatives. But these might not be independent third-party programs and would be more like a pretest before official testing for growers and manufacturers.

With all of these labs popping up, the question arises: who tests the testers? Once again, marijuana's federally prohibited status is the source of the problem. The result is that no federally certified lab will certify cannabis. Additionally, there are no enforceable nationwide standards or calibrations for the expensive testing equipment, leaving every lab to develop its own methods and protocols for testing cannabis. This lack of standards and government oversight means that testing for THC and CBD is notoriously unreliable. Another reason is that the lab has an incentive to inflate such numbers, because higher test scores for cannabinoids mean more profit for the grower, and greater likelihood they will return to that lab for more testing.

There are other problems with testing we

should mention. If the lab technician picks just three or four grams of bud out of a pound, is that truly representative of the whole pound? Farmers have told us they have taken three buds off the same stem and sent two samples to one lab with different strain names and the third to a different lab. All three test results came out different. Testing for mold and other pathogens is also hit-or-miss—more so with larger batches.

As it stands now, California is in the midst of finalizing its testing requirements. The state will inevitably mandate tests for pesticides, pathogens, microbes, and cannabinoid content.[9] Terpene profile testing remains optional. Predicted costs are hovering around $1,000 per batch of up to fifty pounds of product. The state is also establishing precise regulations for the testing labs themselves, requiring, for instance, PhDs in specific cannabis-related disciplines for the chief scientists.

When it all gets sorted out, regulated testing

and laboratory inspections will help guarantee safe medicine. And getting Clean Green or other "grown using organic methods" certification will enable both farmers and processors to send their products to the lab with confidence that they will pass. And the consumer can be sure they're purchasing the cleanest, best cannabis available. At a recent Clean Green meeting, gesturing in his sincere and paternal way, Van Hook simply stated, "I just want to take care of all my farmers." Turns out he's taking care of the average marijuana enthusiast too.

For more on Nikki and Swami, visit swamiselect.com.

This article was originally published on MERRY JANE *in April, 2018. It is reprinted here with permission from the publication's editors.*

NOTES

1. Madison Margolin, "The Curious Case of Cannabis Hyperemesis Syndrome," *Merry Jane*, July 27, 2017, merryjane.com/health/the-curious-case-of-cannnabis-hyperemesis-syndrome.
2. Frank Conrad and Cindy Blair, "Eagle 20 and Myclobutanil in the Context of Cannabis Cultivation and Consumption," *Colorado Green Lab*, May 14, 2015, coloradogreenlab.com/blog/eagle-20-and-myclobutanil-in-the-context-of-cannabis-cultivation-and-consumption.
3. Joel Grover and Matthew Glasser, "Pesticides and Pot: What's California Smoking?" NBC4, February 22, 2017, nbclosangeles.com/investigations/I-Team-Marijuana-Pot-Pesticide-California-414536763.html.
4. Nikki and Swami, "What It's Like to Judge the Best Cannabis in the World," *Merry Jane*, December 15, 2017, merryjane.com/culture/what-its-like-to-judge-the-best-cannabis-in-the-world-nikki-swami-the-emerald-cup.
5. Richard Nathanson: Integrated Pest Management 07/07/2017," *The Cannabis Connection*, July 26, 2017, thecannabisconnectionshow.com/index.php/2017/07/26/richard-nathanson-integrated-pest-management-07072017/.
6. Madison Margolin, "Regenerative Reefer: How Cannabis Farming Can Reduce Carbon Emissions and Save Water," *Merry Jane*, December 14, 2017, merryjane.com/news/regenerative-reefer-how-cannabis-farming-can-reduce-carbon-emissions-and-save-water; Nikki and Swami, "Emerald Triangle Growers Celebrate 'Light Deprivation' Cannabis at the Golden Tarp Awards," *Merry Jane*, November 29, 2017, merryjane.com/culture/dispatch-from-the-golden-tarp-awards-in-the-emerald-triangle.
7. Amanda Chicago-Lewis, "Inside the Weird and Wild Crusade for Clean Pot," *Rolling Stone*, March 1, 2018, rollingstone.com/culture/news/meet-crusaders-clean-pot-w517264.
8. Bernie Woodall, "State of Washington Law Paves Way for Organic Marijuana Market," Reuters, May 16, 2017, reuters.com/article/us-washington-marijuana/state-of-washington-law-paves-way-for-organic-marijuana-market-idUSKCN18D09M.
9. California expects to issue final testing requirements by December 31, 2018.

Layton Hansen | Cannabis Plant

CANNABIS AND THE DIVERSIFIED FARM

CASEY O'NEILL

At HappyDay Farms, we cultivate somewhere between thirty and forty annual crops during the course of the year. We run a small CSA and participate in two farmers markets during the late spring, summer, and fall. At times in the past we have run our CSA fifty weeks a year, but it proved to be too much work and we suffered from burnout.

Cannabis is a central part of our crop rotation, beginning with seed starting in hoop houses in March and finishing with an extended harvest period that runs from the beginning of September until the end of October. Processing of the dried, cured crop continues through the winter as flowers are taken off the branch and trimmed down to manicured buds ready for market.

Including cannabis in our crop rotation was a natural choice where we live in the northern part of Mendocino County, California. We have found that the plant operates very well in polyculture systems, providing a perfect stackable enterprise that meshes with our vegetable cultivation patterns.

We prep beds in early spring, getting them ready for planting, laying out irrigation and putting low hoop tunnels over them. In a sixty-foot row, we leave space for cannabis plants on a fifteen-foot center. Come fall, when the plants are large, they use up most of that space, but in early spring the plants are still tiny seedlings in a hoop house.

As the seedlings are growing in pots, we plant the beds to vegetable crops, leaving a space into which the cannabis is transplanted in late April or early May. Cannabis plants are either male or female, and you only want to plant the female ones to ensure that your final product is sinsemilla (without seed). While we are waiting for plants to demonstrate their sexual characteristics, the vegetable crops are booming under the protection of the low hoop tunnels.

In the spring we grow heading and leafy brassica, along with a wide variety of cooking greens and mesclun salad mixes. We also do things like green onions, radishes, turnips, carrots, and beets. Many of these crops are quick, and we're able to pull off a rotation of them between the time we prep beds and the time that the cannabis begins to take over the space in June.

As summer progresses from late June into July, we're pulling out the remains of the spring crops and planting smaller groups of hot crops. The cannabis takes up most of the space for the summer plantings, so we scatter in basil, okra, peppers, eggplants, and flowers. Some cannabis cultivars get much bigger than others, and the same is true for individual plants, so sometimes we have more space to fill with veggies and sometimes less.

Over the course of the summer, we undersow hot crops with buckwheat to serve as a green mulch cover crop. During that time, we seed trays of the same cool weather crops that we did in spring, so we can transplant them as the hot crops begin to come out. After they've established themselves, we undersow the cool weather crops with a cover crop blend of fava, vetch, peas, and oats. This way, after the cannabis is harvested, we're able to harvest a brassica crop and then see the cover crop come on uninterrupted. It doesn't always work this way, but that's what we strive for.

SOME PASTURES

A Cowboy's Almanac

SAM RYERSON

September, Weaning Pasture—Mescalero, New Mexico

We like to wean our calves early, in August or September. There's an old woven-wire trap someone built above the corrals that's just right for weaning calves, across the fence from their mothers. The trap is about a quarter mile long on each side of a square, with plenty of grass to keep five hundred calves for four days or so, and a little water trough in the corner by the gate. We never weaned them all at once. We split the calves off the cows in the corrals, up to about two hundred at a time, then vaccinated the calves, let the cows out, and moved the calves up the hill to their trap. They walked the fence and balled for the first day, with their mothers hanging along the fence, but by the second day, the older cows were ready to go back to grazing somewhere better, and the calves were hungry enough to put their heads down and start eating.

A weaning trap needs grass and water like any other pasture, and a strong fence to keep the cows and calves separate. Weaning the calves on green grass during the mild weather of late summer makes the time easier on them—easier than weaning through cold nights in late fall on old dry grass, or in dusty corrals feeding on hay. On a cow outfit, we spend all year preparing to wean, so we might as well take our time to make it as positive a parting as possible. The calves quit balling after a few days, and we drove them out to their bigger pasture to the east—a pasture many of them were born in. They walked quietly and slowly like little yearlings, with the dogs bringing up the drag.

Fall is a bittersweet time of year for ranchers and cowboys. We have plenty of good work, weaning, gathering, and shipping. It's work we do with neighbors and friends, like branding in the spring. Fall is usually when we get paid for the cattle we weigh and ship. I look forward to that, and to the work itself, but it's also a sad time of departure. Soon the cattle are gone, and the big crew and our friends too; we face the winter more or less alone.

Cow and Calf in Field, negative

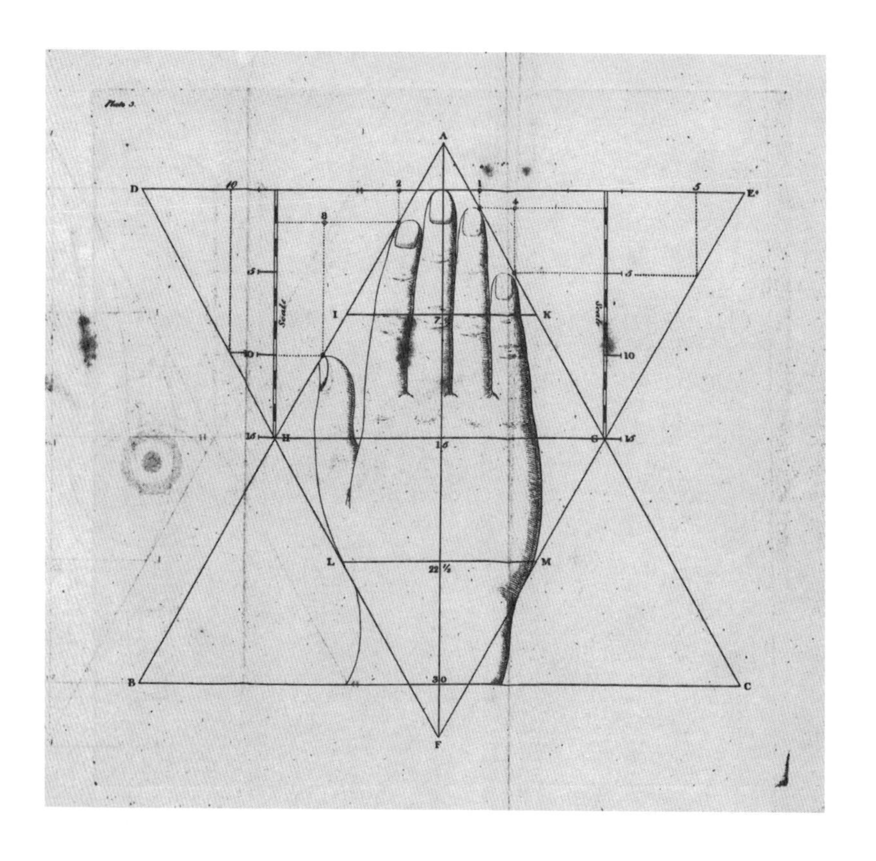
A
D
E
B
C
F
H
G
I
K
L
M
Scale
Scale
22 ½
30

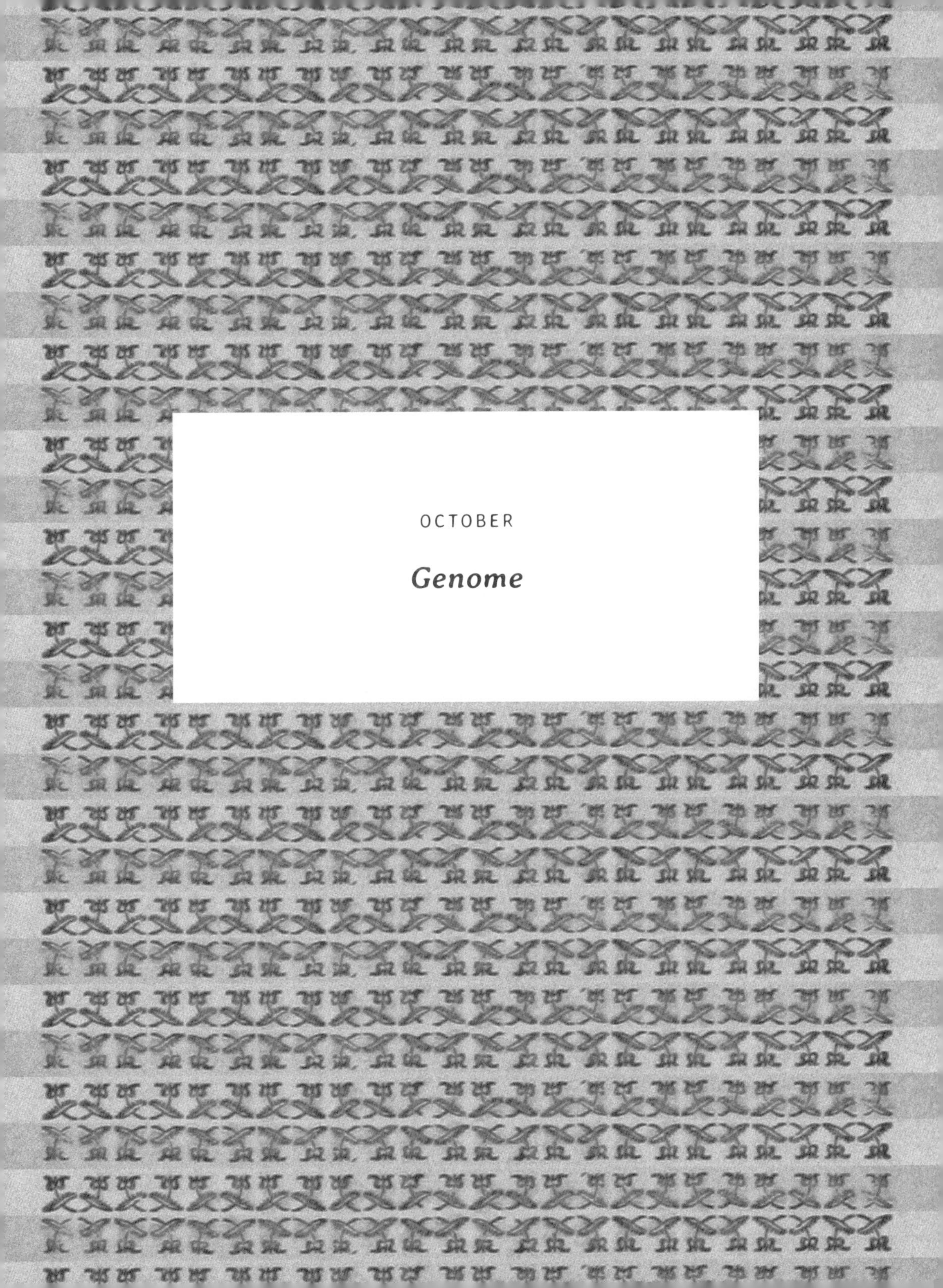

OCTOBER

Genome

TOGETHER, WE HAVE CHOICE

RENATA CHRISTEN

Cornell University student plant breeding symposium

Throughout my work, I've often heard the following narrative: that plant breeders are the arbiters of modern crop diversity. From past work as a farmer, I know that crop diversity is an expression of mass choice, the work of selection that has taken place for thousands of years by untold numbers of farmers. Presently, conditions are such that the genetic diversity inherent in farmer-selected seed is curated to fit a standard of what is in vogue regarding crops bred for the market. Genetics get whittled down through selection and interbreeding to fit specific niches, always ensuring yield as a constant. While plant breeders have demonstrated heroism—the Green Revolution did save purported millions from starvation—yield is not a one-size-fits-all cure for every social ill. There are many possible choices that get sidelined during this breeder curation process, and modern crop diversity is the result of farmers and gardeners stewarding traits that mattered to them in their personal lives. When breeders partner with, or work for, multinational companies, the influx of resources can fast-track certain kinds of scientific research. However, the profit motive inherent in such partnerships not only compromises the independence of research in breeding—if important qualities like nutrition or flavor encroach too heavily in time resources or offset more profitable traits like yield, they will not be prioritized—but also the opportunity for breeders to have freedom of choice in their work as well.

When we talk about the biggest seed companies (Bayer, DowDuPont, Monsanto, Syngenta, BASF), seed often constitutes a mere fraction of their product portfolio. The mammoth financial significance of industries like health care (Bayer, forty-eight percent) and performance plastics (Dow, thirty-eight percent) shape how science works for a corporate vision of our food supply. What we see in grocery stores is only a fraction of what could be possible; the ubiquity of subsidized corn in shelf-stable foods is a testament to such constraints. The larger the system, the more complex the process for average consumers to form linkages and advocate for core issues, like access to authentic choice in terms of, for example, diverse, nutritious food.

Taking another look at rhetoric, there's a disconnect between the narrative of "quality, improved seed" versus the "informal seed system" as used in development consortiums to distinguish between seeds shared among farmers and those delivered by these research institutes and company breeders. Such narratives render farmer knowledge illegitimate because scientific standardization creates norms about what qualifies as "good" and "bad"; such narratives also overlook the potential inherent in the informal seed system that operates outside the global market economy. Through corporate social networks, norms about "quality" get reinforced and calcified. "Quality" then becomes having a choice of different brands of potato chips, and less about the opportunity to grow and trial hundreds of different kale varieties.

Seed packet making in the Netherlands

Coming full circle, plant breeders select seeds to meet specific market demands, conducting research that is essentially a series of observational choices (phenotyping) about desirable plant characteristics that get codified as "science." Somehow, farmer knowledge is often not considered "scientific" even though it requires a lot of trial and error, experimentation, recorded observation, and time. Farmer knowledge has a lot to offer in terms of direct observation and interaction, and can help steward the ascendancy of localized community-driven choice taking precedence over global market directives. This is not an argument against science but against perpetuating the view that farmers are mere consumers of scientific research. The work of breeders like Michael Mazourek and Salvatore Ceccarelli is particularly noteworthy, as they integrate farmer knowledge into their breeding practices in a way that builds community and ensures that products aren't superimposed for consumption but cocreated.

We have the choice to invest in our own communities and the people we know. Buy seeds from people you know. Ask Rob Johnston who his suppliers are—Johnny's often sources from across the United States, and their seeds may not be adapted to your region. Over time, seed saved can become regionally adapted, ergo more resilient and shared. While it may be cheaper in the short term to buy all of your own seed, in the long run, it's a service to every community for its resident farmers and gardeners to save the seed from at least one crop per season. Seed saved is seed shared, and seed shared is freedom of choice.

 Seed Savers Exchange tomato tasting in Minneapolis | Photos by Renata Christensen

Josie Iselin | Spirula

Easy seed:
Corn *Zea mays*

Corn seed is another easy-to-save annual, but, though similar, growing **corn** for seed is a little different than growing it for fresh eating. The big question is… what type of **corn** do you want to grow? The varietal diversity of available **corn** seed is amazing. Your basic type choices are: dent, flour, flint, pop, sweet and parch. Flour types are dry **corns** with a powdery texture when ground and are good used for tortillas, cornbread, & cakes. Flint types are dry **corns** with a glassy coarse texture and are good for polenta, grits, hominy, and tamales. Dent types are half way between flint and flour. Popcorn pops and sweet **corn** is sweet, when you pick it in the *milk stage.* Parch types are dent **corns** that pop a little when dry roasted and turn into a kind of corn-nut. Even a dry **corn** variety can be eate in the milk stage. Although it may be not-so-sweet **corn**, it tastes great cooked up & smeared with butter and is more hearty, too. Mos sweet **corn** available today are *F1 Hybrids*, so if you grow **corn** and want to save seed, make sure it is an *open-pollinated* variety.

Corn is very much an out-breeding, wind-pollinated plant and optimum pollination and seed quality is achieved when it is planted i wide blocks of a 100+ plants. As the classic textbook out-breeder, **corn** is very susceptible to *inbreeding depression*: the seed saved will b weak and low quality if the population is not large enough. The tassels on the top of the plant are the male parts, and the cob contain the female parts. Each strand of corn silk is a pollen receptor for a single **corn** kernel. For a full cob, there needs to be enough pollen around at the right time to pollinate the several hundred kernels on each cob.

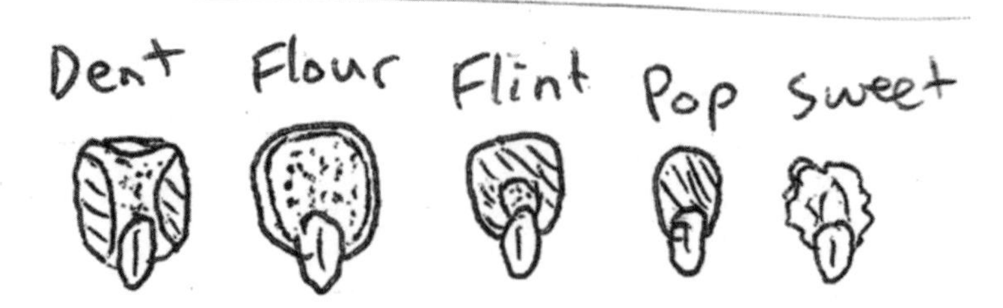

Corn pollen is very light and can travel very far, so isolate your corn from other varieties. This can be done with at least 2 miles distance (it is less important if you live in a wooded or urban environment with no giant fields of corn nearby). Or you can grow a very early variety and a late variety for "time isolation." Time isolation is also very useful when you live in a GMO growing region. Extra early **corn** varieties shed pollen earlier than the later tasseling GMO varieties.

Some people eat the first bigger ear on each plant and save seed from the smaller second ear. Others think this selects the lowest quality seed. **We recommend always saving the best ears for seed and the best of the best for *stock seed* for your own seed in the future.**

To save seed from **corn**, leave the ear on the plant until it dries down and the whole plant looks (and is) dead. Pick the cob, peel back the husk and let it dry for a few weeks. When it is crispy dry, thresh the kernels off of the cob with your hand or by rubbing two cobs together. If the germ peals off the kernels they need to dry a lot longer. If the seed shatters and doesn't smoosh when smashed with a hammer, it is dry enough to seal up in a moisture proof container. A dehydrator set below 95° F is very useful for drying corn kernels completely. If properly dried your **corn** seed will be viable for 5 or more years. Flint and flour **corn** types can survive for decades.

OATS

JULIA SHIPLEY

I didn't believe him—
so he stopped the car
and strode into stems
thin as spokes, up to his hips.
Doubtful it was cereal,
I waded in after him—this clumsy
man who'd accidentally whacked
a pillar with his bucket loader.

Now, deliberately, he kneeled
and split the grass's blade,
peeled away the sheath,
revealed the seeds layered in
like feathers, those nascent oats
packed in the grasses' ruff,
proved in his meaty hand.

Around us the whiskery tips
swayed, persuaded
when wind hit—
they wheeled like a flock;
they shook like a mane;
the whole grain field
thrashed—a secret school,
under whose command?
The animate land.

Liz Brindley | Seeds

LESSONS FROM THE CORN

KIRSTEN BONANZA

Years ago, when the kernels started to whisper to me their mysteries, I already loved to feel the soil between my fingers. My knowledge of planting had been passed down from my father and his father before him, and at the time was very science-based. I found, though, that there are some things only the plant itself can teach.

My relationship with corn sprouted from a deep, primal urge to make sure that our heirloom seeds kept growing and were not lost. As time went on, my service to seed keeping became a focused yearly endeavor on multiple plots of land, starting with my parents' backyard the first year, and then expanding to include a friend's home and my community garden. This creation was driven by a love of every part of the process of growing and saving seeds.

In the beginning, the corn seed came to me when I reached out to The Path, Grandmother Flordemayo's seed saving organization, to ask if I could help. My ability to become involved was guided and mentored by the connections shared with me to Dianna Henry, Jim Troudt, and Greg Shane—all legends in their own right. I was on my way; to what, I did not know.

The first lesson came before I held the seeds in my hand. It was the idea, which arrived during a vision in meditation, that the corn wanted to be grown in a spiral in such a way that one could walk through it like a labyrinth and sit in the center. I imagined the feel of the leaves and stalks moving around me as I walked the path and then sat in stillness that coming summer. Navajo Bird's Egg Corn—a lovely blue, white, cream, and white with blue speckles—was the first variety that I planted. It grew well that year in its labyrinthine shape in the garden where everything else grew in rows, as I'd been taught.

Every year a new variety speaks a shape, but always some combination of spirals, circles, or a handful of seeds clustered like the Pleiades. There was a point I thought I might be crazy, responding to corn asking to be planted in sacred geometric grids, but listening had brought me this far, and so I continued. The corn's next request was to share songs. I listened, periodically going out and singing my heart to the plants, treating them like an altar, filling my heart and the space with joy. That's how I became the woman in the fields singing to her corn regularly.

Here we are many harvests later. The corn has whispered all sorts of wisdom around how it wishes to be planted. Each year it asks for certain strains to be grown out. During this time I've listened beyond the wisdom of how to grow and learned personal life lessons from the relationship I've cultivated with the corn.

The most poignant is the deep understanding of how one must cultivate corn for seed keeping. While many plants can be near other varieties, corn must be kept at a distance in order to maintain purity of desired characteristics and variety. Believe me, it can be challenging to find a place to grow something that must have at least a mile between itself and another variety. This need for space exists because corn is the seed of a plant that is highly influenceable. GMO contamination, or cross-breeding, is common due to this mutability. The gift in this lesson is to see that we too are influenced not only by the nutrients we take in but also by the people and environments around us. While the plant does not have a choice where it is planted, we do have the ability to decide who and what we allow in.

I imagine over the years other kernels of wisdom will work their way into my awareness through my relationship growing this prolific seed. Like the blessings that arrive so fully with each harvest, I accept this knowledge as a gift and know that through my service and connection I invite myself to thrive.

Sarah Hughson | Substantia Nigra One

A STORY LIKE A SEED

GRANT JONES

In a packet, box,
That shimmers like a minnow
You lure a tongue
To lick or crunch,
To be swallowed
Carried for miles
To a dirt bed
Under trees where birds perch,
To be shat.

Or as letters with hooks
To be carried in your hair
Like gifts for your grandchildren.

Fragrant, you tempt
Gifts of pollen,
You baffle and dazzle me,
To lick to life
Arouse, to die
For a kiss, deeper,
To detour, until
My beak becomes
A scribe for love,
A wetted quill
To write a poem.

Coyote Springs Farm
Mouth of the Canyon of the Little Mosquito

 Opposite: Wall panel with garden urns

FARMERS GAIN PREMIUM PRICES FOR CERTIFIED MEAT AND EGGS

The American Poultry Association Will Certify Your Flock

CHRISTINE HEINRICHS

The National Restaurant Association placed Heritage Breed Meats at number thirteen on its list of Top 20 Food Trends for 2018. The American Poultry Association (APA) saw that coming. In 2014, the APA started a Flock Certification Program to allow poultry producers to sell their products as APA-certified poultry.

To issue the certification, APA-certified judges come to the farm to inspect flocks for their adherence to the APA-written *Standard of Perfection*, which lists all the breeds that have been described and officially recognized for exhibition at poultry shows. That's more than fifty different large fowl breeds in many varieties. The first *Standard* was published in 1874. Dates are given for every recognized breed. "The birds must have the general conformation of their breed," said former APA president Dave Anderson, who was instrumental in establishing the program. "The flock cannot have more than two percent significant disqualifications, such as roach back. The flock has to have less than fifteen percent minor variations from the *Standard*, such as the wrong eye color or side sprigs on the comb. Of course, the entire flock has to be healthy."

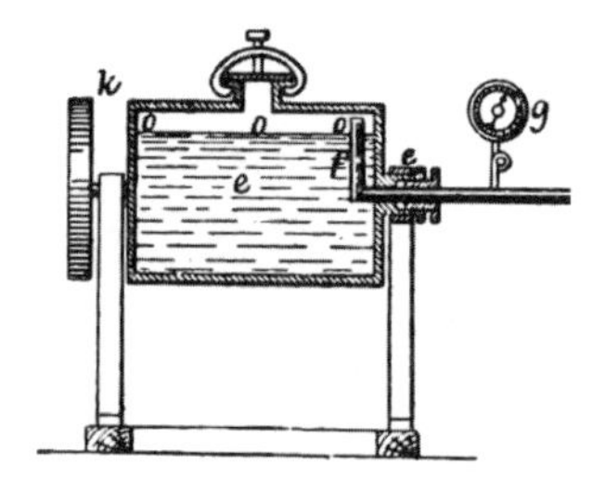

WHAT "STANDARD" MEANS

The original APA *Standard* was written to improve the quality, uniformity, and marketability of poultry flocks. Over the years, its emphasis changed to focus on poultry exhibitions. Utility became an afterthought, although the *Standard* still lists "Economic Qualities" in its breed descriptions. Standard is the operant word, meaning breeds that have been documented and officially recognized. Heritage, historic, traditional, antique, heirloom, and other words are descriptive, but their meanings vary slightly and can be stretched and distorted to cover anything. "Standard," for the APA, is a word with a defined meaning: if the breed wasn't in the *Standard* before 1950, it can't get the certification.

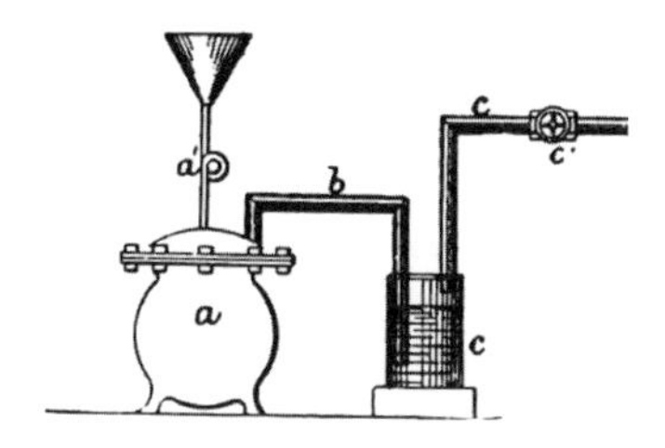

GETTING CERTIFIED

Judges can offer advice to help the producer improve his flock. They can help the farmer pick out the best birds for breeding. Their knowledge, and that of the Standard-breed producers whose flocks they inspect, will help USDA inspectors learn how to grade Standard-breed birds.

"They are not just coming to judge your birds," says Frank Reese of Good Shepherd Poultry Ranch in Kansas. "They are coming to help you. It should be a learning experience." His Barred Plymouth Rock, New Hampshire, Columbian Wyandotte, and White Cornish flocks have received certification. His Bronze, Bourbon Red, White Holland, and Narragansett turkey flocks are also certified.

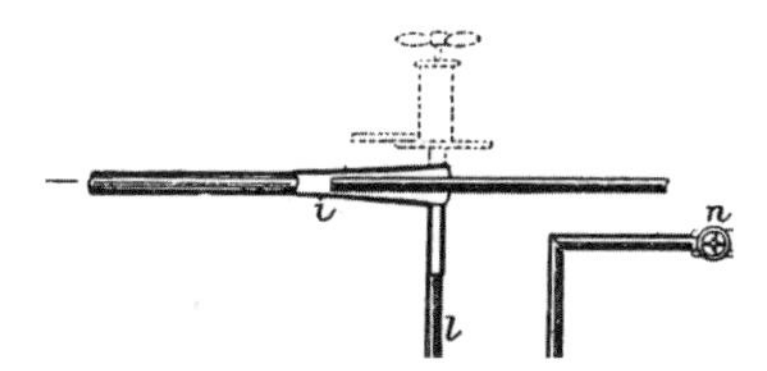

CERTIFIED PRODUCTS

Reese is the leading Standard-breed poultry producer in the country. He currently supplies Heritage Foods USA with five hundred chickens every three weeks for internet sales. He supplies the Ambassador Hotel's Siena Tuscan Steakhouse in Wichita with about fifty chickens a week, and another fifty to Chop Shop Meats butcher shop in Wichita. He and other producers who work with him also supply the steak house with 150 dozen eggs a week.

"Pretty much all chicken served, whether as fast food or in restaurants, has the same industrial genetics," says the steak house's executive chef, Josh Rathbun. "It doesn't matter what it's fed or what conditions it's raised in, they all have the same genetics."

Those fast-growing broilers—they mature and are processed at six to seven weeks old—don't develop the flavor that heritage breeds have. Heritage breeds aren't big enough to be processed until they are fourteen weeks old at the earliest. Many grow longer than that. At Good Shepherd, they live on pasture, developing muscle as they chase bugs and scratch for seeds. "For chefs, the real proof is the flavor," says Rathbun.

Not every chicken with a *Standard* name will make a good productive flock. Hatchery stock may have unacceptable defects. Birds bred for exhibition may have lost their productivity. Chickens are more than pretty feathers. "They need to have good muscle development, fertility, and egg production," says Reese. "This program should help people improve quality and production of these birds."

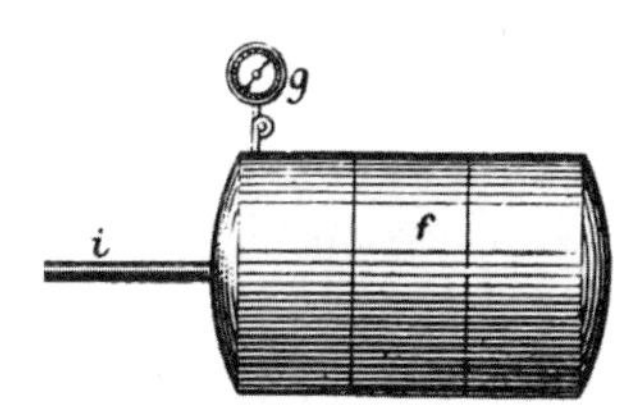

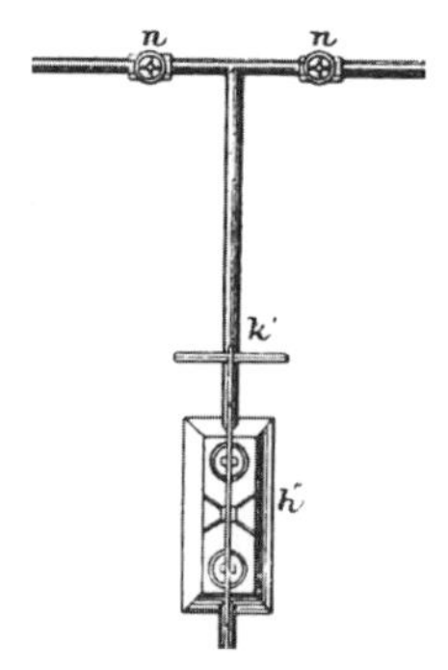

DEFINITION AND LABELING

Reese, in cooperation with the Livestock Conservancy and others, has developed a Heritage Breed definition that relies on the APA Standard. His label has been approved by the USDA and goes on each APA-certified bird packaged and sold.

Two other poultry producers have gotten their flocks certified: Greenfire Farms in Florida, for their Black Copper Marans and Light Sussex, and Old Time Farm in Pennsylvania, for Partridge Chantecler and Bronze Turkeys. Shelley Oswald of Old Time Farm in Pennsylvania got her birds certified to provide as much information as possible to her customers. She also raises grass-fed beef from Milking Devon cattle, and she had been frustrated by the confusion created by unregulated labeling. "Heritage is a breed claim," she says. "You need to have some certificate to make that claim. I want to let people know that they are getting what they think they are getting."

Organic is a legally defined term, but claims such as "free range" and "pasture raised" are not verified or certified. The idyllic notion of local farming conveys an image to the public that isn't necessarily backed by substance. "I got the certification because it was what I had to do to be authentic," Oswald says. She sells her chickens at the farmers market for $7 a pound, with more demand than she can supply. Telling the story is important to educate her customers and justify the higher price. "I explain that these chickens are a rare breed, raised for show and for production," she says. "I see the light go on in their eyes: 'This is what I think of when I think I'm buying a farm-raised chicken.'"

Explaining definitions clearly is a struggle, as industrial poultry companies try to muscle in on the market. Terms without regulated definitions are often used to mislead consumers. "We need to tell the public that chickens and turkeys must be historical, authentic, and pure-bred to carry that label," she says. "They have to be true representations of a historically accurate bird. Hybrids do not qualify. The USDA has labeling guidelines. You need to have documentation for it. That's why you need to have certification to verify that you have those birds."

Slow-growing heritage breeds that are active on range have more flavor but require different cooking methods to make them tender. Slow cooking at low heat works. Oswald has had success with Instant Pots, multiuse programmable cookers. Oswald, who sells her certified Bronze Turkey poults, also sells breeding stock, hatching eggs, and chicks, and she plans to scale up. She got a license to process her poultry on-farm in 2017 and says she finds the public unprepared for some products, such as schmaltz, but eager to learn.

Greenfire Farms is targeting a different market with its Black Copper Marans and Light Sussex. Paul Bradshaw is selling certified day-old chicks to hobbyists. He went to France in 2017 to bring back exhibition-quality Black Copper

Marans. Marans lay dark-brown eggs. He's getting a premium price for them, $49 a chick. Hatcheries selling BC Marans chicks that do not have that certification sell for $10 to $12. So far, he's the only one selling Marans chicks from a certified flock. "I'd recommend it to anyone who sells chicks," he said. "It's worth doing. The backyard hobbyist who is only going to have six birds is willing to pay the price." As Oswald has experienced, Bradshaw can sell as much as he can produce.

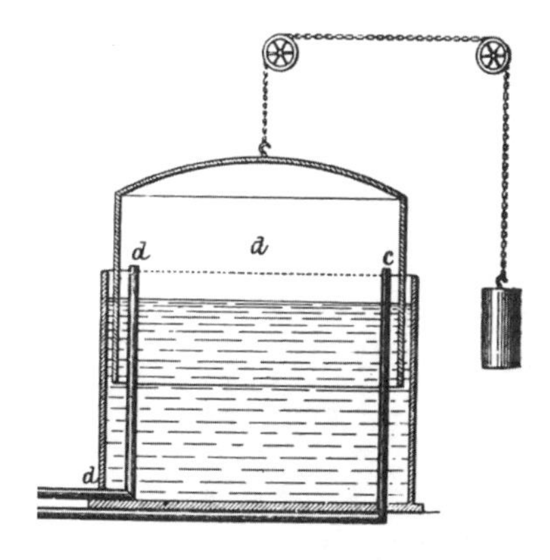

BRINGING CERTIFICATION BACK

In the past, the APA inspected flocks, but it abandoned that responsibility fifty years ago. Commercial poultry farms overwhelmed smaller Standard-breed flocks after World War II. The chicken meat business turned to genetically similar, industrially developed chickens, which are unable to mate and reproduce naturally. They grow to market size in six to seven weeks. If allowed to grow to maturity, they are hardly able even to walk. Their underdeveloped immune systems can't protect them against even ordinary diseases.

Modern hybrids with flashy names such as Freedom Ranger and Golden Nugget have been developed to take advantage of the market for chickens that are raised in better conditions. They may be raised on pasture and fed an organic diet, but their genetics doom them. They may have unseen internal abnormalities such as cardiac and skeletal problems. As hybrids, they do not reproduce chicks with similar qualities. "Chickens have several serious welfare problems that come from bad genetics and can be fixed only with good genetics," animal welfare advisor Temple Grandin wrote in her book, *Animals Make Us Human*. Those are the factors that affect the flavor too. "Pretty much every fast-food outlet and every restaurant is serving the same bird," said chef Rathbun. "It doesn't matter what it's fed, what condition it's raised in. It's still the same genetics as the Tyson bird."

Standard breeds have recognizable identities and documented histories. Reviving the inspection program in the 21st century will help Standard-breed producers justify the higher prices their products deserve.

The Sower

GOOD INTENTIONS

DOUGLASS DECANDIA

Be patient
with the seed
inside,

Awaiting its
spring to grow.

Love
but
do not hover.

For good intentions
mean little

If in effect
they block
the light.

SOME PASTURES

A Cowboy's Almanac

SAM RYERSON

October, Shipping Pastures—Negrito, New Mexico

I'd been part of some good big crews before, where we got along and the work went well, but one year everything just came together. I knew it would be my last year on that place since I had decided to give my notice, so maybe I cherished the days more. We spent a week gathering steers into the holding pastures above the corrals. They were good generous shipping pastures. The two bigger pastures were each more than a full section (a square mile) of grama grass, headed out and purple in the fall, with some open stands of ponderosa pine around the edges. A shipping pasture should be big enough to keep the cattle happy, so they stay full before they get weighed across the scales. We were careful to rest the holding pastures during the growing season so they had good grass in the fall.

By the time we were ready to start sorting, we had about one thousand steers in the north holding pasture. We had two brands to sort between. When we finished the steers we would start gathering the cows and calves, sort them from the yearling heifers, test them all, and ship them too. We sorted the cattle horseback, down in a little grassy draw near the south fence. Most of the crew held the herd in a loose bunch. Tuff and I took turns and rode in and sorted out the cuts, the steers of the brand we needed to ship first. The other riders turned back the rest and held them in the herd. Don and Jeannie held the cuts, another herd of which started to shape up fifty yards or so to the west. It was generally quiet and smooth. We knew each other and how to work together without saying much, and the cattle were used to us riding through them. When we were done sorting, just before dark, we took the cuts over the ridge and down the hill to the smaller trap on the creek below the corrals. We shipped steers the next two mornings. They gained nearly three hundred pounds per head over the summer, to weigh about nine hundred pounds. It had been an exceptionally good year.

Cattle, well, shadoof

BREEDING THE HEIRLOOMS OF TOMORROW

Cultivating Resilience to Climate Change

DON TIPPING

Svalbard Global Seed Vault

Plant domestication is an ongoing process, a relationship rather than an endpoint. When I first began seed saving nearly twenty-five years ago, I perceived our relationship with domesticated plants as one akin to enslavement and servitude. The more time I work closely with saving seeds and making selections across a wide diversity of plant families, the more I see how we are working toward one another's best interests. Through understanding what traits enable a plant to adapt more proficiently to its environment, we can tease out the best aspects of a given variety.

When I take a big step back and look at the long view of the history of agriculture, I believe that seed saving is likely the defining element of agriculture, if not civilization itself. Consider that before humanity was harvesting, saving, and replanting seeds, we were hunter-gatherers. In this way we may equate seed saving with agriculture. Ironically, agriculture has become so specialized that it is the rare farm that actually saves any seeds. Most farms, even most organic farms, buy all their seed every year. Seed saving has become a task relegated to the experts. This is unfortunate, because along with the maintenance of soil fertility through on-farm nutrient cycling, saving your own seed is the most important feedback loop to fine-tune the whole farm organism to the climate, pests, diseases, stresses, and consumer preferences. If more farmers could witness what is possible through thoughtful on-farm selection, they might be inclined to abandon the expensive fancy Dutch hybrids that fill the glossy pages of seed catalogs in favor of growing and selecting their own seeds.

It's estimated that before the Green Revolution, India had over thirty thousand varieties of rice. Afterward, it had eight. Where did all of that agri-biodiversity go? And why were there so many to begin with? The likely explanation is that farmers in every microclimate saved their own seed, allowing plants to constantly adapt to subtle differences in soil, climate, and other stresses. Put most simply, a plant that performs the best out of a population will make more seeds, thereby producing more seeds than a plant that does poorly. Over time, simply saving seeds leads to adaptation to the environment within which the plant is growing.

INTENTIONAL PLANT BREEDING STARTS WITH VARIETY TRIALS

From my perspective, any time that you are saving seeds from anything less than a whole population of plants, you are engaging in plant breeding. Obviously, applying some intention to the process can improve it greatly. The process of breeding first involves identifying the traits that you want to select for. To do that, you must have a comprehensive knowledge of what is possible for a given species or type. This starts with variety trials—growing numerous individuals of a given species in order to thoroughly understand what is possible within the breadth of that species' diversity. This helps you to identify the endpoints of a spectrum for each trait and all points in between.

We have a lettuce variety trial planted right now at our home farm in southern Oregon, with twenty plants each of seventy-five varieties that we steward for our family farm organic seed company, Siskiyou Seeds. This will reveal a lot of information about the lettuces that we grow when we can observe how they stack up against each other in the field. We will evaluate and score them for appearance, vigor, pest and disease resistance, flavor, bolt resistance, and an overall grade. What may happen as a result is that we realize that some varieties are inferior to another in a given category, say Red Romaines, and we may wind up dropping a variety. Another outcome may be that we will note that we don't have a good variety within a given category, such as French Batavians, which are noted for their excellent performance in hot weather, and we may need to source new strains and trial them again to see if we want to add a new one. Our results may also highlight plant-breeding opportunities for us to breed for better bolt resistance by crossing similar strains, for instance one with good bolt resistance, and another with our preferred disease resistance and leaf shape. The possibilities become staggering once you try and get a handle on all the plant biodiversity out there, but performing regular trials is a crucial component of any plant-breeding program.

A number of years ago, we planted a good sized kale trial with one hundred plants each of twenty-two varieties of kale that represented three basic types: Siberian/Russian (*Brassica napus*), Lacinato/Black Tuscan (*Brassica oleracea*), and Curly/Vates (*Brassica oleracea*). We planted them in the late summer and observed their performance in the fall, winter, and spring. We had Red Russian Kale from five sources: (1) our strain that we had saved for over a decade, (2) one from Southern California, (3) one from Colorado, (4) one from western Washington, and (5) another from a colleague two hundred fifty miles north of us. We noted that the kale from Washington truly disliked our hot early fall weather and suffered. The kale from Southern California had never experienced cold like we had that winter (ten degrees Fahrenheit for a week in January) and died. The kale from Colorado did well with the heat and the cold. And the kale from a bit north did great, as expected.

This taught us that bioregional adaptation is an important consideration, but even more important is adaptation to climate patterns. So while we may actually be in the same general bioregion of the Pacific Northwest as western Washington, we have observed that the climate of the high desert of Colorado is more functionally similar to that of the Klamath Siskiyou region that we are in. Once again, understanding these patterns is only possible through repeated variety observation trials.

TRADITIONAL PLANT BREEDING

Before the advent of modern hybrid plant breeding, we had many skilled traditional plant breeders across the world. In the United States, many of these were affiliated with our land grant universities, helping to select and breed for varieties that in turn helped farmers in their area to thrive. This was a wonderful feedback loop that worked fairly well to serve the needs of small-scale family farmers. With the corporate takeover of plant breeding through proprietary hybrid plant breeding, much knowledge about traditional plant breeding was lost, consolidated in the hands of profit-driven companies.

I am grateful to have had the opportunity to study with John Navazio, PhD in plant breeding, who was able to learn from some of the greats of a bygone era like Dr. Henry Munger from Cornell University. Navazio is currently employed at Johnny's Selected Seeds, where he breeds open pollinated vegetables. Thanks to the influence of him and others like him, there exists a new vanguard of amateur plant breeders, many of whom now have their own small seed companies (Wild Garden Seeds, Adaptive Seeds, and Uprising Seeds, to name a few) to showcase their varieties and provide an economic vehicle to support this slow and sometimes laborious work.

Before one embarks upon a plant-breeding project, a thorough understanding of the species that we aim to improve is vital. In the words of John Navazio, "We should strive to become a samurai warrior for the species we want to work with." This entails an exhaustive study of its botanical relatives, currently available varieties through variety trials, and detailing of the important traits and agronomic considerations that we should focus our attention on. I believe that this is vital work, so for the past five years we have been offering a five-day Seed Academy training at our farm, which is a deep dive in a whole systems approach to seed saving, plant improvement, breeding, seed cleaning, and seed stewardship, with guest appearances from luminaries in this movement.

WHY BREED NEW VARIETIES?

Paging through the *Seed Savers Exchange Yearbook* is an astounding glimpse into the tremendous diversity of open pollinated varieties that are currently commercially available. Nonetheless, this diversity represents less than ten percent of what was available fifty years prior. It is estimated that we have lost up to ninety percent of the commercially available varieties since World War II. Why would we need so many different varieties? There are so many different unique microclimates in this country; add to this our widely varying culinary palates and regional preferences, and you can begin to understand that there are limitless ideas as to what constitutes the perfect variety.

If, after performing thorough variety trials, you come to the conclusion that there is still something lacking—what to do? Consider that what was once suitable within a region or among a community of growers may no longer be ideal. Our ideals must align with the dual moving targets of climate change and shifting consumer preferences. Now it gets complicated. The way I see it, seed saving is the doorway into cultivating a deeper relationship of seed stewardship. At Seven Seeds Farm, we have been working with many of the varieties that we grow for two decades or longer. In addition, we inherited a seed collection that includes varieties that have been grown in the area since 1978. So we have strains of vegetables that have forty years of adaptation to the growing conditions of our region. Baby steps from an indigenous agriculture perspective, but also a good start.

Through evaluating the collection that we steward here on our farm and make available through our retail seed company, Siskiyou Seeds, we are firmly convinced that seeds "imprint" upon their locale. I believe that we have two paths that we can choose as permaculture horticulturalists: one option is to adapt to climate change through working with adapting our seeds—from the inside out—taking a cue from how water flows around obstacles. The other option is the path that is being taken by most organic farmers and conventional agriculturists: to adapt to change by manipulating the environment—from the outside in—with row covers, herbicides, insecticides, high tunnels, and prima donna hybrid genetics that do great when everything is optimized perfectly. I've heard stories from John Navazio of Henry Munger, at Cornell University, who would grow thousands of cucumber starts and then taste the cotyledons of each, selecting for the nonbitter-tasting ones that he had observed were less favored by cucumber beetles, thereby breeding for cucumber beetle resistance.

The more I ponder the question of what is the way forward, considering the enormity and potential severity of the impact of anthropogenic climate change on the metastability of human civilization, the more I am directed to conclude that the thoughtful breeding of our annual food crops, domesticated animals, and perennial plants represents our best chance for thriving and maintaining resilient agrarian communities. I do not place my stock in techno-fixes. I believe in biology. Can I get an Amen?

A VARIETY IMPROVEMENT CASE STUDY

From the kale trial described above, we determined that there was room in the available open pollinated seed offerings for a workhorse curly green Vates kale. This conclusion was underscored by the shortage of Winterbor F1 seed that was occurring at that time—a variety that organic growers were relying upon to fill boxes for kale-hungry customers. We let the seven varieties of Vates kale that constituted both open pollinated and hybrid F1 types to intermate, or cross-pollinate. That was in 2010, and the total population was about seven hundred plants of seven varieties. That winter we experienced a week of sustained lows at about ten degrees Fahrenheit. Nevertheless, about twenty-five percent of the plants survived and managed to make flowers and seeds, freely cross-pollinating among one another. What we had created was a diverse gene pool of potentiality, also known as a *grex*, which is Latin for "herd." Another term used for this is "synthetic cross," or "syn" for short. It was anything but uniform.

We harvested seed in summer 2011 and replanted right away, overwintering and then flowering and intermating again in spring 2012. Once again we harvested seed and replanted in fall 2012, allowing the cycle to complete one more time with no selection pressure other than the climate. In fall 2013 we planted a large population of about two thousand plants and overwintered. In the early spring of 2014 we identified our favorite hundred plants based on the following traits: deep green color, vigor, deeply curly leaf shape, upright stature, and disease resistance. We tagged our favorite mothers with surveyor's tape and allowed them to flower and set seed. Once seed pods had dried down, we harvested the seed from our hundred favorite "mothers" into one hundred separate bags. The other nineteen hundred plants were allowed to flower and make seed, so these hundred mothers had the potential of two thousand different "fathers." Kale is a monoecious perfect flowering plant, which has both male and female flowers on each plant.

This project is known as a half-sibling progeny row breeding program. From each of these hundred different bags of seed, we planted fifty plants, giving us a total population of five thousand plants that we transplanted in the early fall of 2014 in small blocks that were labeled with stakes numbered one to one hundred. This represented a sampling of the progeny of our hundred mothers, meaning that they all had the same mother, but that their fathers were unknown, hence the term half-sibling. As the season progressed, we evaluated the plants for the traits listed above, and then we "rogued" or eliminated about thirty percent of the mothers that we'd determined unsuitable. From the remaining seventy percent, we removed individuals that failed to meet our breeding criteria, such as not having a dark-green leaf color, or not frilly enough, or more susceptible to aphid damage.

This resulted in a total population of about two thousand plants that went into the winter of 2014–2015. In summer 2015, we harvested the resulting seed from this vastly improved population. This became the new Alive Vates Grex Kale, and we sold its seed widely. In fall 2015, in order to evaluate if we had made any genuine improvements, we planted another trial, including the new seed along with a wide diversity of kale from our stock and strains from other growers. This allowed us to compare our efforts against other commercially available strains.

Fortunately, we learned that we had made great strides in improving selection for the desired traits; however, we realized that it might be worthwhile to breed more height into the Alive Kale population, so we selected the tallest and best specimens from the other Vates types in our trial and allowed them to overwinter with the fifty plants of Alive Vates in our trial. In the spring of 2017, these 150 plants all flowered and intermated. They are in the field now and seedpods are maturing. We will harvest this seed in July and then replant it alongside our original Alive Vates to evaluate our progress.

I share this case study to highlight what a long process plant breeding becomes, especially with a biennial like kale. But do not be dismayed: through employing traditional plant breeding techniques such as the half-sibling progeny rows and then the back-crossing to tall Vates Kale, we can greatly accelerate the process over simple mass selection, or simply saving seed of the best plants in a population. The parent lines of hybrids, even GMOs, are created using similar techniques. Not everything can be done in the lab. Sexual reproduction in the field is absolutely necessary.

WHERE TO START?

I encourage anyone reading this not to become overwhelmed by the seeming complexity of plant breeding. Start small and choose a crop that you have affinity with, and begin on the journey of a deeper relationship with plants. In a way I see that selection, adaptation, and breeding allow us access to the fine-tuning control knobs of the genetic potential of a species. In doing so we can begin to steer a variety toward specific goals, fostering greater adaptation to the stresses that variety will encounter in our systems. We also become witness to the great diversity of interactions that take place between soil, insects, environment, our diets, and our creativity that converge in a given variety. See plants as your allies, and they will guide you to learn how we may be of service to them and all life in the process.

Ray Ray Mitrano

WORDS FOR FARMING AND FARMERS

JOHN D. GALUSKA

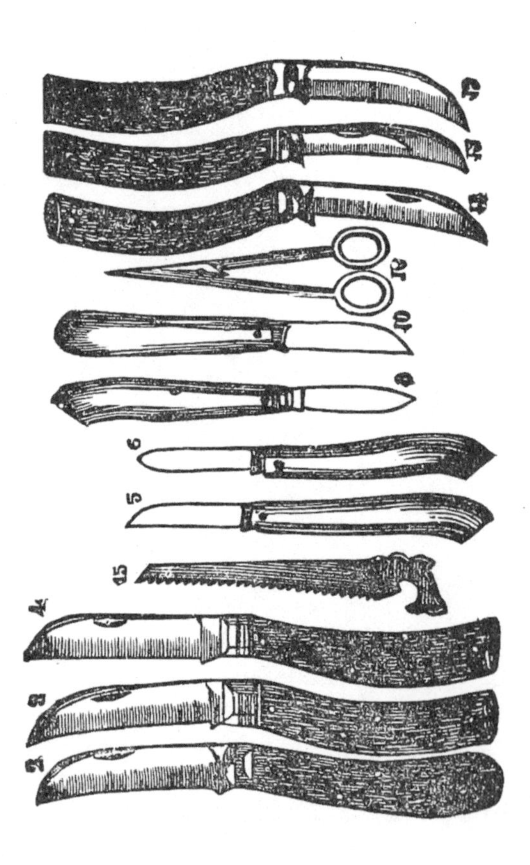

A

aeroponics
agricology
agriculturally-supported
community (ASC)
agrihood
agritourism
agroecologist
agroforestry
agronomy
allotment gardens
animal husbandry
apiary
aquaponics
arable farming
arboriculture

B

backyard farmer/gardener
bee keeping
biodynamic farming
biointensive farm/garden
bog farming
botanical garden

C

calorie farming
campus farm/garden
carbon farming
cash crop farming
chinampa (Mexico)
civic agriculture
climate-smart agriculture
collective farming
commercial farming
commercial grower
commons farming/gardening
communal farming/gardening
community-supported
agriculture (CSA)
companion planting

composter
container farm/garden
continually productive urban landscapes (CPULs)
conuco garden
conventional farming
corporate agriculture
croft (Scotland)
curbside gardening
custodian farmer

D

dairy farmer
demonstration farm/garden
demonstration landscape
desert farming/gardening
dooryard garden
dryland farming

E

easement gardens
ecological agriculture
edible estates
edible garden
edible hotel/resort
edible landscaping
emergency agriculture
estate farm/garden
experimental agriculture
experimental farm/garden
export agriculture

F

farm to table landscape
farminary (farm seminary)
farmscapes
fisherman
floating farm
flower farmer
food forest farming
food pantry garden
foodscaping
forager
free-range farming
French intensive
fringe farmer
front yard garden

G

garden homes
gardener
gardenscapes
gentleman farmer
geoponics
gleaner/gleaning
GMO agriculture
grain farmer
grass farmer
Greenhorn
greenhouse farming
guerrilla gardener

H

hay farmer
haylage farmer
herb garden
high-brix farming
hillside farmer
hobby grower
home gardening
homesteader (rural or urban)
horse-powered farming
horticulture
hospital garden
hugelkultur
hydroponic farmer
hydroponics

I

indigenous agriculture/farming
indoor farming
industrial agriculture
insect farming
institutional farming
intensive agriculture
intensive gardening
intercropping

K

kibbutz farming/gardening
kitchen garden

L

landrace grower
large-scale farmer
learning farm/garden
livestock farmer
lunatic farmer

M

mariculture
market gardener
medicinal garden/farm
micro-farm/farmer
migrant farmer
mixed farming
modern farming
moon farming
multifamily landscapes
mushroom grower

N

neighborhood farm
noncommercial grower
nonorganic farmer
nontraditional farming
no-till farming

O

orchardist
organic agriculture
organic farmer

P

paddy cultivation
paradise gardening
park & farm
pastoral farming
peasant farmer
peri-urban farmer
permaculture garden/farm
plasticulture (five-billion dollar industry)
potager garden
prison garden
professional grower
provision grounds ("slave" garden)
pulse farmer

R

rabi farming
radical farming/gardening
rain gardens
raised bed gardening
rancher/ranching
refugee gardens
regenerative agriculture
research farm/garden
restaurant garden
revolutionary gardening
riparian farming
rooftop farm/garden
rotational farming
rural farmer

S

school garden
seed library
seed saver
seedsman
seed-to-table farm/garden
sharecropping
sheep herder
side yard garden
silage farmer
smallholding farmer
small-plot-intensive (SPIN) farmer
small-scale farmer
square foot gardening
streetscapes
subsistence agriculture
subsistence farmer/gardener
sustainable farmer/gardener

T

temporary gardens
terrace farming
traditional farming
tree pit
truck farmer
turf (sphagnum) farming

U

unconventional agriculture
underground farms
urban farmer/gardener
urban grower
urban homesteader
urban horticulture
urban livestock

V

vegitecture
vermiculture
vertical forest
vertical gardening/farming
victory gardens
village food factory
virtual farm
viticulture

W

wellness gardens
wetland farming
wildcrafting
wildlife gardening
window gardening
winter farming
worm farmer

X

xeriscaping

Y

yardshare garden/farm
young agrarian

Z

Z-farming (zero acreage)
Zymology (zymurgy)

Approximately eight years ago, I started compiling a short list to document the range of words and terms commonly used for different kinds of farming and gardening. The list now stands at over two hundred terms. This is a work in progress. It is my hope to provide some sense of the diverse range of farming, gardening, and other types of growing activities being practiced in the United States and abroad. In addition, I hope the list inspires readers to share feedback and to provide suggestions for additional words and terms to be added to the list.

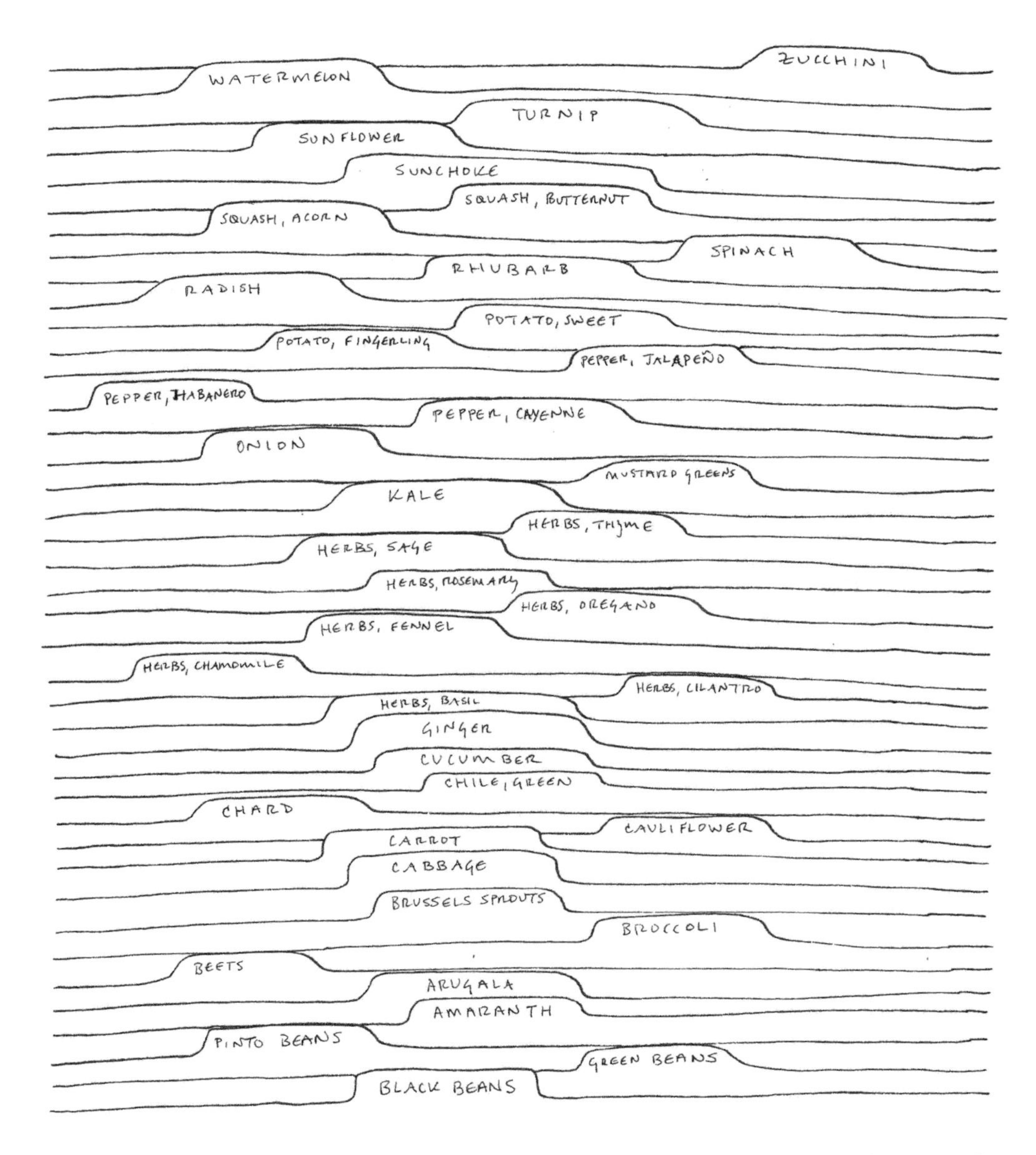

Liz Brindley | Filing Seeds

Graffiti during the occupation of Alcatraz, 1969

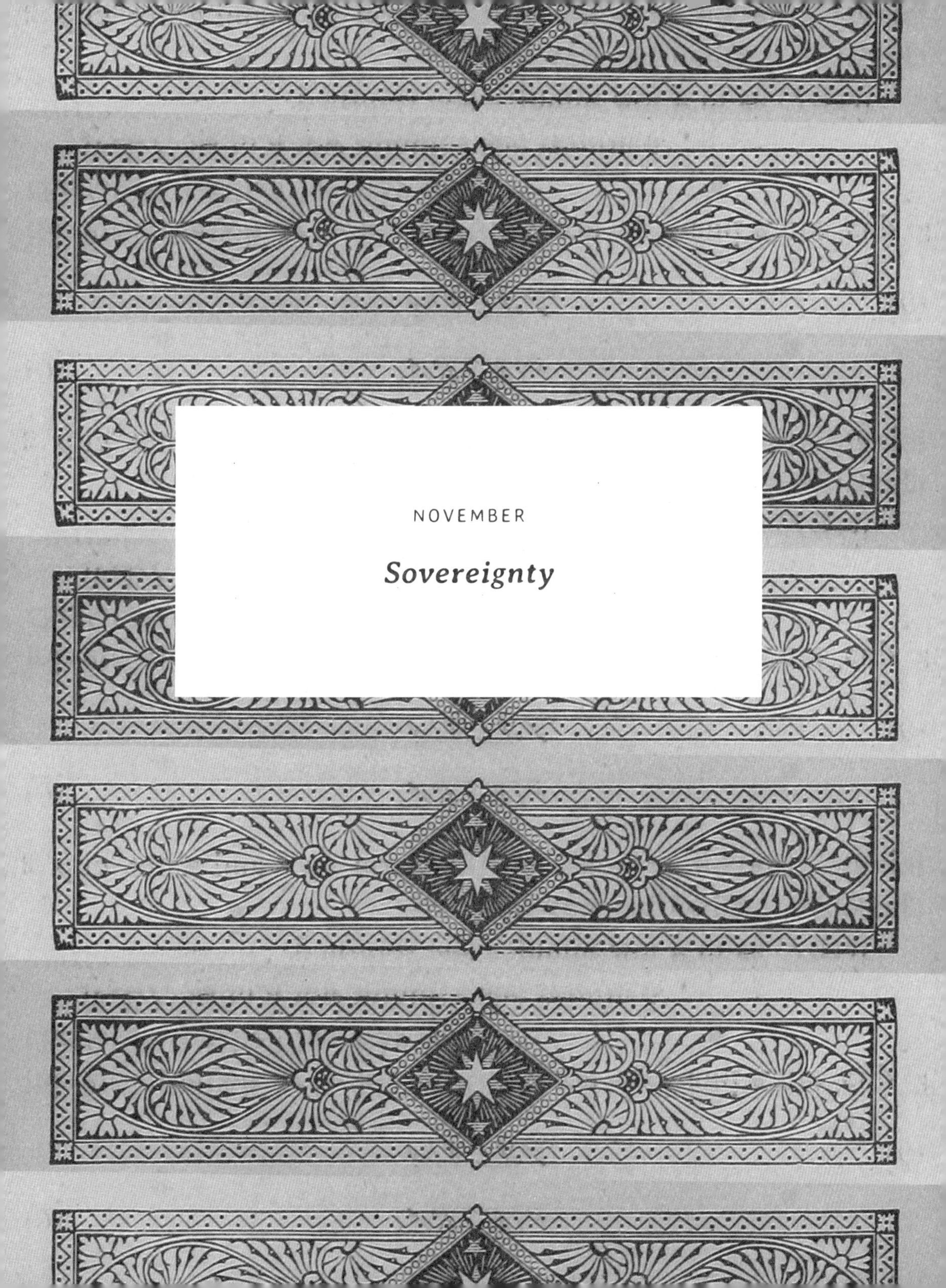

NOVEMBER

Sovereignty

OUR LIVING RELATIVES

Maintaining Resilience and Seed Diversity in Native American Communities

ROWEN WHITE AND ELIZABETH HOOVER

Seeds are a vibrant and vital foundation for food sovereignty, and are the basis for a sustainable, healthy agriculture. As seed keepers, we understand that seeds are our precious collective inheritance and it is our responsibility to care for the seeds as part of our responsibility to feed and nourish ourselves and future generations. Many of the important food crop varieties in this world come from the western hemisphere—many kinds of corn, beans, squash, potatoes, peppers, tomatoes, and cacao, just to name a few. Foods taken from the western hemisphere to Europe, and then to other parts of the globe, changed how the world eats. Each Indigenous community developed their own distinct varieties of these crops. These landraces, what we now refer to as heritage seeds or heirloom seeds, were adapted to a community's climate as well as their cultural preferences for taste and appearance. Each seed was shaped by the hands of generations of seed keepers who decided at the end of a season what would be saved to be planted the following year. These seeds nourished families and communities, filling stomachs and connecting gardeners and eaters to previous generations. But many of the Indigenous communities who developed these varieties no longer have access to them.

Because of a variety of factors related to the colonization of Indian country, many of these seeds were lost to communities. In many cases, food supplies were targeted by colonial armies, like in 1779 when General Sullivan and his troops followed George Washington's order to "lay waste all the settlements," and thus burned acres of cornfields and millions of bushels of stored Haudenosaunee corn. In the 19th century, many tribes were forcibly relocated from their homelands to territories with very different soils and climates—tribes were forced from places like North Carolina, Florida, and Nebraska to Indian Territory in Oklahoma. When they arrived, many found that the seeds they might have been able to bring with them wouldn't grow. The Bureau of Indian Affairs pushed for the elimination of traditional farming by replacing dryland farming in some areas with canal irrigation, ranching, and agribusiness enterprises—in the process replacing the genetic diversity present in many traditional seeds with a small number of introduced crops or hybrid seeds. In more recent years, climate change, which has led to unpredictable weather, has made it difficult to grow out seeds adapted to a particular climate. Over the past century, many people have come to rely on seed catalogs and stores to purchase commercial seeds that are designed for the same performance and taste regardless of where they are planted.

Seeds are a reflection of the people, and are the foundation of any resilient food system. Our indigenous ancestors were seed savers by necessity, and faithfully passed down a beautiful and diverse collection of varieties of corn, beans, squash, and other food crops to subsequent generations to ensure a bountiful connection from the land we descend from. In addition to these varieties becoming bioregionally adapted to each community's climate, many of these seeds carry vibrant cultural memory and stories that make them a critical link in the restoration of our cultural heritage. As Native communities rise up from the era of colonization and displacement, we look to our traditional foodways to bring balance and alignment in our endeavors to cultural restoration.

Out of a collective desire and need for solidarity in this seed sovereignty work, indigenous people have come together in an intertribal seed network called the Indigenous Seed Keepers Network (ISKN). The mission of the ISKN is to nourish and assist the growing Seed Sovereignty movement across Turtle Island (North America). As a national network, we leverage resources and cultivate solidarity and communication within the matrix of regional grassroots tribal seed sovereignty projects. We accomplish this mission by providing educational resources, mentorship training, and outreach and advocacy support on seed policy issues, and organizing national and regional events and convenings to connect many communities who are engaging in this vital work. We aim to create a collaborative framework and declaration for ethical seed stewardship and Indigenous seed guidelines for tribal communities to guide them as they protect their seeds from patenting and bio-piracy. We support the creation of solutions-oriented programs for adaptive resilient seed systems within tribal communities to enhance the creative capacity to continue to evolve as the face of our Mother Earth changes.

Across Turtle Island, there is a growing intergenerational movement of Indigenous people proud to carry the message of the grand rematriation of seeds and foods back into our Indigenous communities. Some of these seeds have been missing from our communities for centuries, carried on long journeys in smoky buckskin pouches, on the necks of peoples who were forced to relocate from the land of their births, their ancestral grounds. Generations later, these seeds are now coming back home to communities of origin, from the vaults of public institutions, seed banks, universities, seed keeper collections, and some lying on dusty pantry shelves of foresighted elders, seeds patiently sleeping and dreaming. Seeds waiting for loving hands to patiently place them into welcoming soil once more so that they can continue to fulfill their original agreement to help feed the people.

In the seed movement, we have begun to use the word "rematriation," instead of the more patriarchal repatriation, as it relates to bringing these seeds home again. In many communities, including our Mohawk tradition, the responsibility of caring for the seeds over the generations is ultimately within the women's realm. Both men and women farm and plant

Havasupai corn fields, Arizona, early 1900s

seeds, but their care and stewardship are part of the women's bundle of responsibility. So the word rematriation reflects the restoration of the feminine seeds back into the communities of origin. The Indigenous concept of rematriation can also encompass the reclaiming of ancestral remains, spirituality, culture, knowledge, and resources. It simply means back to Mother Earth, a return to our origins, to life and cocreation, honoring the life-giving force of the Divine Feminine.

Relationships and partnerships are being developed to ensure that communities who have been absent of their seeds are reunited with them, and that in many cases those seeds are then reunited with their ancestral soil. This process often entails policy documents and negotiations, but also spiritual and emotional work—developing new ceremonies and protocols to welcome home these relatives, as well as acknowledging the elders who have worked for so long to make this movement possible.

One example of a successful seed rematriation project that reunited Native growers with seeds took place in Minnesota, between the Minnesota Museum of Science and the Dream

Three boys of the Red Lake Band of Chippewa in a corn field, Minnesota, circa 1938

of Wild Health project, among other affiliated projects. In the 1930s and 1940s, a dentist and amateur anthropologist from Minneapolis named Wesley Hiller traveled across the Upper Midwest, the Northern Plains, and the Southwest, collecting Native American seeds. In 1977 his estate donated his collection to the Minnesota Science Museum. Through the Upper Midwest Seed Keepers Alliance, and the work of Miami museum curator Scott Shoemaker, the museum agreed to allow seeds from this collection to be grown out and rematriated to Native communities. Dream of Wild Health, run by director Diane Wilson (Dakota), is a program that gives urban Native youth the opportunity to learn about farming and healthy eating. They have been growing out some of these seeds for several years now, and this year are pleased to be growing out enough Dakota yellow flour corn, a variety that had all but disappeared from the Dakota community but had lain waiting for them in the museum's collections, for the community to consume.

This past year, ISKN has initiated a relationship with one of the largest public access heritage seed banks in the world, Seed Savers

Exchange (SSE). SSE safeguards over thirty thousand diverse and unique heritage seed varieties, all of which have cultural significance in North America. Working cooperatively with SSE, we identified hundreds, potentially thousands, of varieties within their seed vault collection that have origins within tribal communities within North America. Over the winter of 2018, we came into an agreement between the Native American Food Sovereignty Alliance (NAFSA) and SSE to collaborate on a seed rematriation project, where we would sponsor the rematriation process of an initial twenty-five native varieties of corn, beans, and squash back to tribal communities. As a part of this process, we are working with a focus group of Native farmers and seed keepers, as well as other stakeholders from similar seed rematriation projects working with regional seed banks, museums, and universities, to develop a shared action framework and best practices guideline document for other communities wishing to engage in the seed rematriation process. ISKN is part of an advisory group for seed rematriations that are happening with the University of Michigan, Minnesota Museum of Science, and the University of Wisconsin– Madison. This year we are rematriating seeds back to communities of the following tribes: Mohawk, Seneca and other Haudenosaunee, Odawa, Ojibwe, Taos Pueblo, Cherokee, Arikara, Mandan, Hidatsa, Stockbridge Munsee, Pawnee, Kickapoo, and Narragansett.

Working collaboratively with SSE has been very inspirational and mutually beneficial. As a nonnative organization, they have had incredible respect and willingness to allow Native leadership in this process, and are doing internal work on educating staff as well as adopting this process to inform future community engagement. They are setting an important precedent for other seed banks and nonprofit organizations to work respectfully and collaboratively with Native communities in the process of reclamation of traditional seeds. It has been a very insightful process for all involved. SSE has been willing to share information about their comprehensive seed banking methods and database; protocols and methods for seed history and cultural memory documentation; and critical methods for seed storage with seed keepers from Native communities who wish to establish seed banks or libraries from their own communities. ISKN and our circle of Native seed keepers have helped inform this organization on how we view these seeds as part of our cosmology, and offer a wider perspective outside the Western scientific view of seeds, as well as inform them of how to move forward in culturally appropriate ways to honor not only the seeds but the communities of origin as well. There has been a cultivation of mutual respect and benefit, as well as building trust to ensure that our beloved and sacred traditional seeds are returned to our communities with honor and dignity. This is healing work that spans many diverse cultural communities.

Not only does seed rematriation focus on bringing seeds back to communities of origin, it also is looking at restoring connection to ancestral farmland from which tribal communities have been displaced. Haudenosaunee (Iroquois) territory historically stretched across the eastern side what is now New York State and southern Ontario and Quebec, until the nations that make up this confederacy were pushed north by settlers. Akwesasne Mohawks were displaced from the fertile farm soils of southern New York into the swampy and rocky banks of the St. Lawrence River, which was traditionally used for hunting and fishing. This has made the stewardship of traditional seed varieties difficult, as there is only marginal farmland to grow on. Loss of land base, in addition to the forces of colonization and acculturation, have resulted in the erosion of the health and vitality of dozens of heritage seed varieties in the Northeast.

In a beautiful collaboration between the Hudson Valley Farm Hub and Seed Shed and the Akwesasne Mohawk community, the Native Seed Sanctuary was established to return these heirloom seed varieties back to ancestral planting grounds. Leveraging resources from a diverse array of people, including Mohawk elders and young adults, migrant farmworkers, inner-city youth, and new Farm Hub organic farmers, a beautiful field of heirloom Haudenosaunee seeds was planted in 2016 in Hurley, New York, in the spirit of working together in solidarity, cross-culturally, to ensure the health and well-being of ancestral seed varieties. This work of reconciliation has been deeply healing for all who have been involved. The Native Seed Sanctuary is in its third season of planting, and each year the garden grows heritage seeds that are returned to the Mohawk people for planting and nourishment.

The corn planted in the first two seasons was Mohawk Red Bread corn, which is a variety that was nearly lost—it came down to one cob that Mohawk elder and seed keeper Steve McComber began to grow out again, over twenty years ago. This collaborative farm project was able to send two thousand pounds of Mohawk Red Bread corn back to Akwesasne in 2017 to feed the youth taking part in the Ohero:kon Rites of Passage ceremonies.

Seed rematriation is deep and multilayered, encompassing spiritual, emotional, practical, scientific, and legal and political realms. As we carry these sacred bundles of our seed relatives home to their mother communities, we reawaken time-honored relationships once again. When we come together to cultivate the earth and sing our seed songs and prayers on behalf of future generations, we embody the great generosity and benevolence of our own beloved Mother Earth. Knowing that our hearts beat in promise to carry this bundle of seed gifts to share with the world, these seeds help us to hydrate the stories that make up the constellation of who we are. May we continue to be there for these seeds as we turn our prayers into actions for this new day.

Each one of us reading this has the ancient and cellular memory of being a seed keeper; the communities we descend from are resilient and tenacious survivors. Just like our seeds, we have overcome so much adversity. In honor of the grand lineage of seed keepers who passed down seeds through the generations, we make renewed commitment to steward these seeds for those yet to come.

SIDEWALKS, SQUARE NAILS, AND OPEN FIELDS

Farming with History in Southwest Colorado

ELICIA WHITTLESEY

On this six-thousand-acre property known as the Old Fort, there's an endless list of projects. Buildings need renovating. Fences need repairing. Irrigation needs replacing. I try to bite back the suggestions that prioritize beauty over function. But when my boss mentions a project replacing old sidewalks around some of the buildings, I can't help proposing that we also take out the sidewalk that parallels the west side of the Education Garden. There's a dirt road there that makes for steadier walking than the old sidewalk heaved up by young cottonwoods.

We can't take it up, my boss says. It's considered historical, and not a safety hazard, unlike the sidewalks going up to the buildings we use every day.

Ah. I should have known.

All of our farming and working at this place, the Old Fort, is informed by its history. We encounter—or avoid—the past through its physical remnants, the layers of lead paint on the walls or asbestos in the floors of the buildings, the square nails and metal remnants unearthed by the shovelful, the sidewalks squaring off in a field, the old cellars where red foxes raise kits and watch us. Historical artifacts dictate the look of the place and its uses. We can't take down old signs, so they contradict the new ones. We tiptoe, forge ahead, ask questions, and for forgiveness.

There's an invisible history too. This land was inhabited by the Utes and Navajos for hundreds of years, and then overtaken and claimed as a US Army fort in the 1880s. And before the land and buildings became a high school turned college turned research station turned multipurpose site with sustainable farming, it was an Indian boarding school. From about 1892 to 1910, the school housed and trained Ute, Navajo, and Apache children in general education and vocational skills. In a photograph from the time, the children and staff are dressed formally and posed on the front steps of a now-vanished building. The boys wear little suits with bright brass buttons, holding their caps at their sides; the girls wear dark, modest dresses, waists and collars outlined in light ribbon. Of course, no one smiled for pictures in that period, but these children's indistinct sober expressions speak to me of homesickness.

"Haunted!" A friend teased when I got this

Parade Feature, by sixteen Indian girls from the Government School at Chilocco, O. T. (These girls composed the Mandolin Orchestra that played in the booth of the Santa Fe Indian Feature.)

Native American children at forced government boarding schools job and mentioned the site's history. But I think there is nothing eerie about this history, which is different from the time the office door was locked from the inside, or the way my skin crawls going into one of the old houses. Haunted is too light a word to describe the loss, disconnection, sadness, fear, and cruelty of the federal policy of removing Native children from their homes and cultures to promote "integration." The intention behind these policies and actions—to erase language, culture, and identity—renders our Halloween sense of haunted completely inadequate. This past feels too solemn and worldly to haunt our present in a palpable way, leaving me uncertain exactly how to invite it in.

One legacy of the boarding school is that Native students from all over the country can receive a tuition waiver to attend Fort Lewis College, the college that began at this site and moved to nearby Durango, taking the name with it. Native students make up about a third of the student body at this small public liberal arts college. Some of these students reconnect with farming in our programs. I teach what I know, which is small-scale vegetable farming with influences from the French biointensive method. Yet sometimes this newfound farming knowledge resonates with students' memories of their grandparents growing corn and squash.

It's possible that I'm the wrong person to be farming on this particular ground, to be growing a few token varieties of Navajo squash alongside romaine lettuce and green curly kale. When my aunt shared the results from her 23andMe genetic test, the "big news" was that in addition to our known German, French, and British ancestry, my family is Finnish too. My whiteness spells out complicity in colonialism.

How long does it take to be the right person for a piece of land? Various strands of my relations have been in North America for over four hundred years. Could we all be better stewards of the land if we gave ourselves time to wonder about the students at this place who walked the sidewalks and worked the fields, rather than being paralyzed by historical awareness and appropriate grief? Can we let ourselves be haunted gently, and in all seriousness, by the small and large tragedies of a place? At the Old Fort, my approach is tentative, inquiring, uncomfortable. I vow to be open about the past. I want to let this invisible history insert itself into how we farm here and now.

Though the Old Fort has a particular concentration of sadness, we all face the invitation to think of the people who came before us on our farmland, and more than likely they're not our ancestors. How can we all find a way to farm respectfully and effectively on land that is only temporarily in our care?

ON THE CAPTURE AND IMPRISONMENT OF CRAZY SNAKE, JANUARY, 1900

ALEXANDER POSEY

Down with him! chain him! bind him fast!
    Slam to the iron door and turn the key!
The one true Creek, perhaps the last
    To dare declare, "You have wronged me!"
Defiant, stoical, silent,
    Suffers imprisonment!
Such coarse black hair! such eagle eye!
    Such stately mien!—how arrow-straight!
Such will! such courage to defy
    The powerful makers of his fate!
A traitor, outlaw,—what you will,
    He is the noble red man still.
Condemn him and his kind to shame!
    I bow to him, exalt his name!

From The Poems of Alexander Lawrence Posey (Crane & Co., 1910).

NOTES

Chitto "Crazy Snake" Harjo was a Muscogee (Creek) leader who served in the Union army during the Civil War and defended the sovereignty of the Creek Nation. From 1898 to his disappearance in 1909, he resisted the federal government's dissolution of the Creek Nation and fought the redistribution of communally held tribal lands.

AN INDIGENOUS PERSPECTIVE ON MOVEMENT BUILDING

Native Sovereignty through Industrial Hemp

MARC GRIGNON

The Native Sovereignty movement in North America can be traced back to the 1680 Pueblo Revolt. A coalition of pueblos throughout what is now New Mexico banded together to drive the Spanish conquistadors out. From 1680 to 1692, the Pueblos of New Mexico held their sovereignty, until the Spanish returned and reoccupied. This revolt is a true testament of the resilience that drives Native Sovereignty movements.

Flash forward to the 19th century; the Native Sovereignty movement rekindled the spirit of the Pueblo Revolt to fight to protect the natural world for all first nations across North America. Tecumseh, a Shawnee from present-day Ohio, guided by his younger brother, Tenskwatawa, led a pan-Indian movement east of the Mississippi River from 1808 to 1813. Together, the brothers worked to uphold tribal sovereignty in the wake of the American and British fight over the ancestral homelands of Indigenous people. Tecumseh's struggle was short-lived, but he would inspire the next generation of the movement's torchbearers to build coalitions.

In the mid-20th century, the federal Indian policy era of Termination, a policy devoted to stripping tribal nations of their treaty rights, led many natives throughout the United States to rise and reignite the fire. "Terminated" tribes, including the Menominee Tribe of Wisconsin, sued the federal government, seeking to restore their treaty rights and federal recognition. The American Indian Movement organized in 1968 and the Indians of All Tribes in 1969. These two groups intuitively channeled the spirit of the Pueblo Revolt when, in 1971, they formed a coalition that took over Alcatraz Island just off San Francisco. The Native Sovereignty Movement was reborn. The political acts of Indigenous activism that followed the Alcatraz takeover shaped later Native Sovereignty activism, shifting the focus to community development, and the era of self-determination within federal Indian policy was born.

Self-determination is the ability to choose your own way of life and structure of government. The Indians of All Tribes and the American Indian Movement entreated tribal nations throughout the United States to shape their tribal governments after their traditional roots.

Opposite: Velvet panel with flowering plants

The spirit of the Native Sovereignty Movement carried on as tribes in the United States began to fight back with the pen instead of the sword, taking legislative action and voicing resistance to federal policies. What came out of this time was an understanding that native and nonnative alliances are the most powerful grassroots forces we can wield as people.

The year of 1998 proved to be a significant test for the policy of self-determination. The Oglala Lakota Tribe passed a tribal law to grow industrial hemp to generate economic development on their impoverished reservation. Upholding their terms of the Fort Laramie Treaty of 1868, the Oglala Lakota Tribe passed the hemp law to drum up support among their tribal members to become farmers. An act of Native Sovereignty through the flick of the pen brought tribal member and farmer Alex White Plume into the fold of a four-century-old struggle.

At the turn of the 21st century, under the Fort Laramie Treaty of 1868, Alex White Plume grew industrial hemp to fulfill the obligations under treaty rights that the Oglala Lakota become farmers instead of nomadic hunters. As a hemp farmer, White Plume saw the benefits of cultivating hemp for the people and the planet. He knew hemp could support a thriving economy on Pine Ridge Reservation. Unfortunately, the United States did not share the same vision. In 2000, federal agents descended onto the Pine Ridge Reservation and took away White Plume's hemp crop, stating that his actions were in violation of the Controlled Substances Act. This disregard for the 1998 tribal law would continue in 2001 and 2002, when the US government ultimately banned White Plume from growing hemp.

John Trudell, a core member of the Indians of All Tribes and the American Indian Movement, saw this infringement of Native Sovereignty on Pine Ridge Reservation and fought for the tribe's self-determination. After witnessing the wrongful seizure of White Plume's hemp crop, Trudell sought a way to channel his frustration into action. For ten years, he brainstormed, consulted friends, and began to lay the foundation for an organization to educate the 21st-century public on the benefits of industrial hemp.

Hempstead Project Heart was founded by John Trudell and Willie Nelson in 2012 to raise awareness of the benefits of growing industrial hemp for the people and the planet. Trudell envisioned that industrial hemp could bring strong economies back to tribal, rural, and urban communities throughout the United States. Hempstead Project Heart started in California, educating residents on the many opportunities that could be created with industrial hemp. Through fashion shows, concerts, education booths, and public-speaking engagements, Trudell brought industrial hemp forward as a solution to some of the challenges we face in the 21st century.

Two years later, the Agricultural Act of 2014 opened the door for industrial hemp cultivation and research in the United States. American farmers could now grow industrial hemp if they worked alongside their state Department of Agriculture or an institution of higher education. In 2015, the Menominee Indian Tribe of Wisconsin passed a tribal law that directed their tribal members and the College of Menominee Nation, an institution of higher education, to grow industrial hemp. The tribe hired consultants and a team of tribal members to grow and research industrial hemp under the supervision of the College of Menominee Nation on a three-acre plot on the Menominee Reservation.

During the summer of 2015, the Menominee Tribe and the US government deliberated the legality of growing and researching industrial hemp on tribal lands. The Menominee argued that the tribe is sovereign, and that as a sovereign nation, they have the right to pass their own laws and could act as a state to oversee the industrial hemp research plot. The US Department of Justice argued that the Menominee Nation is a domestic dependent nation within the United States and does not hold the same power as states' rights. During these discussions between nations, a letter was sent to Hempstead Project Heart to ask for help in restoring the Menominee's sovereign right to cultivate industrial hemp.

Trudell received the letter and began preparations to educate and build national awareness around the benefits of industrial hemp. On October 23, 2015, the US government, assisted by county sheriff's departments, descended on the Menominee Reservation and uprooted over thirty thousand industrial hemp plants. When Trudell found out about the raid, he jumped into action to begin the formation of a legal defense fund for the tribe to utilize while fighting their case in court. However, not too long after, Trudell's health started to decline and his mobility wavered.

Still fueled by the fire of past movements, Trudell set into motion plans to carry the Native Sovereignty movement through the 21st century using industrial hemp as a tool. He advised the caretakers of Hempstead Project Heart to act for the people and the planet rather than ego, fame, or power—the crutches that had weakened the Indians of All Tribes and the American Indian Movement. His last wish was to see an alliance of native and nonnative people working together to bring thriving hemp economies back to tribal, urban, and rural communities throughout America. He predicted Wisconsin would be the linchpin in securing the future of an environmentally friendly industry fueled by industrial hemp cultivation and sent Hempstead Project Heart there to build awareness of the benefits of industrial hemp for the people and the planet.

Hempstead Project Heart sprang into action in Wisconsin. Guided by Trudell's strategies, members of Hempstead Project Heart built a coalition of native and nonnative peoples and organizations to turn Wisconsin back into an industrial hemp state. (Wisconsin once led the nation in hemp production, producing seventy-five percent of hemp used during World War II). Trudell's approach proved worthwhile. On November 30, 2017, Wisconsin governor Scott Walker signed into law SB 119, which makes industrial hemp legal to grow. Since Wisconsin legalized hemp cultivation, scores of other states are discussing or have passed similar legislation.

In Congress, a new piece of legislation is being debated in the halls of the Senate and the House of Representatives. The Hemp Farming Act of 2018 treats industrial hemp as an agricultural commodity instead of a controlled substance, which will give the US Department of Agriculture authority over industrial hemp state programs. This act will open the door for federal research dollars to fund projects and provide crop insurance, and it carves out a space for Native Sovereignty, where tribes across America can create their own laws for industrial hemp cultivation.

The Hemp Farming Act of 2018 provides a tremendous opportunity among native and nonnative farmers to grow a crop that has been illegal for the past eighty years in America. The untapped potential industrial hemp has for tribal, urban, and rural communities is limitless. Advance manufacturing opportunities in 3-D printing, home building and construction, supercapacitors, automotive parts, textiles, and a new line of superfoods as well as dietary supplements are just a few of the many products that can be made from industrial hemp. The 21st century is poised to be an era of creativity and innovation, and now is the time to #bringbackhemp to America's tribal, urban, and rural communities. Hemp can help regenerate America's economy and strengthen native sovereignty.

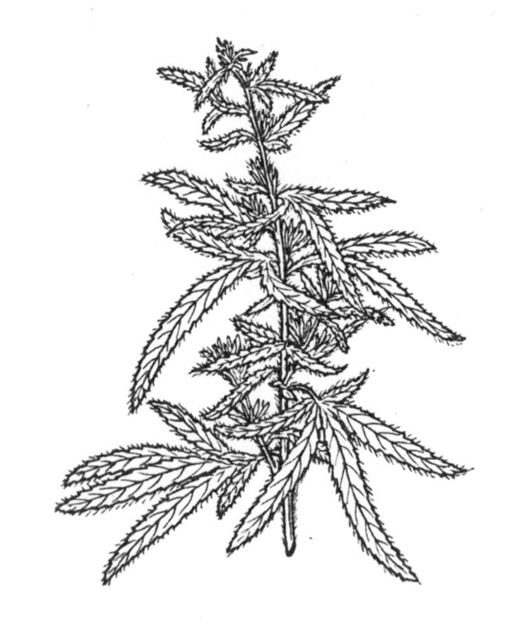

Gathering wild rice, Minnesota, circa 1897

A TOOLBOX FOR INDIGENOUS LAND REPATRIATION

ADAM HUGGINS

Groundplan of Great Bear Mound Group, Indian Effigy Mounds National Monument, Wisconsin

If you've worked with the earth, if you've coaxed the soil and formed relationships with its inhabitants and beneficiaries, you have probably thought about the past stewards of the land underneath your boots (or bare feet, as the case may be). Who were they, and what challenges did they face? Is their influence still felt in the landscape? How did they approach the stewardship of the land that you live on?

Despite centuries of oppression, Indigenous peoples across North America still hold many of the answers to these questions within their cultures, languages, and oral histories. In recent decades, working alongside archaeologists, anthropologists, and ethnobotanists, Indigenous peoples have uncovered and documented rich traditions of land stewardship that extend back thousands of years. These traditions demonstrate an unparalleled knowledge of natural ecosystems and a deeply-rooted conservation ethic that have much to offer farmers and land managers of all ethnic backgrounds.

Today, Indigenous communities in the United States and Canada are mobilizing in a cultural, spiritual, and political resurgence. Standing Rock and #IdleNoMore are high-profile manifestations of this resurgence, which provides all of us with a critical opportunity: now, more than ever, those of us who do not claim Indigenous ancestry need to work together with Indigenous peoples to heal the land, heal each other, and find better ways to feed and provide for ourselves.

(If I've lost you at this point, now might be a good time to pick up a book by Robin Wall Kimmerer, Kat Anderson, Nancy Turner, Leanne Betasamosake Simpson, or Gary Paul Nabhan, then join us later.)

Obviously, this is a long road. A bitter history of genocide, forced migrations, residential schools, economic isolation, and prejudice has left most tribes in states of hardship and disadvantage; many remain landless and legally unrecognized to this day. While Indigenous knowledge and wisdom are essential guides for land managers, the application of this knowledge out of context can be detrimental: not only can it constitute yet another appropriation of Indigenous identity and intellectual property, but it can also backfire if the knowledge is applied outside the cultural context that made it functional.

In order to contribute their knowledge effectively and equitably, Indigenous peoples need access to land, roles of equal partnership in public and private land management organizations, and funds to educate and employ younger generations in the stewardship of these lands. Unfortunately, given the track records of federal, state, and provincial governments, the current turmoil, and the gridlock that has gripped them, it will be up to individuals, organizations, and local governments to acknowledge and meet these needs.

Here are some examples of ways that this is already happening. They are offered in the hope that they will inspire and inform further action by those who recognize the critical importance of these efforts.

PUT YOUR LAND IN AN INDIGENOUS LAND TRUST

When Gavin Raders, cofounder and executive director of Planting Justice, arrived at Standing Rock with his family, something changed for all of them. "We were called while we were there to go home and recognize that all land is sacred," he told me one morning in February 2017, sitting in a cozy greenhouse in East Oakland during a heavy but welcome winter downpour. In 2015, Planting Justice had purchased the land underneath our feet—sandwiched between I-880, the train tracks, and San Leandro Creek, near the historical site of the Jalquin Ohlone village—to house an organic fruit tree nursery. Then, in January of 2017, Planting Justice gifted a quarter acre of this two-acre property to Sogorea Te', an Indigenous women-led land trust working to acquire land for Ohlone people living in the East Bay Area of California.

Sitting next to Gavin was Johnella LaRose, cofounder with Corrina Gould of the Sogorea Te' Land Trust. "When we lit the fire back there for the folks that had gone to Standing Rock … that was the first sacred fire on free Ohlone land in two hundred years," she said, gesturing toward the corner of the nursery where the ceremonial circle was held. Planting Justice and Sogorea Te' have formed a remarkable synergistic relationship on the property, with stewards from Sogorea Te' helping with nursery work while simultaneously working to establish native plants and a sacred arbor on the Trust's first acquisition. Despite the sometimes bleak social and ecological conditions in this part of Oakland, the feeling of opportunity and possibility at the Planting Justice nursery is palpable. When she and Corrina started the land trust, Johnella told me, they set out to save the land, "but the land really saved us."

A quarter acre is admittedly a modest beginning, yet the significance of this land to the Ohlone—who faced such extreme persecution that they remained in hiding for much of the 20th century—can't be overstated. Johnella, Corrina, and Gavin see this as an important first step. "This is hopefully just the first of many landscapes throughout the Bay Area that go back into Indigenous stewardship, and we hope that Planting Justice can serve as an inspiration for other landowners," Gavin told me. Ultimately, after they've cleared all the debt from the property, Planting Justice intends to transfer ownership of the full two acres into the Sogorea Te' Land Trust. He feels confident that Sogorea Te' will continue to support the nursery's presence on the land. Johnella summed it up: "We have to be really creative … because all of our lives depend on that creativity and that goodness."

CULTURAL CONSERVATION EASEMENTS

About forty miles south of the Planting Justice nursery, the iconic Mount Umunhum presides over the southern end of the vast shimmering expanse of Silicon Valley. As the site of a former Air Force station designed to monitor US airspace during the Cold War, the peak was closed to the public for years. When I was a teenager growing up in the Valley, we used to look up at the box-shaped radio tower that sits atop the summit and wonder what was going on up there. We imagined all sorts of outlandish experiments (the likes of which, ironically, were probably taking place in one of the hundreds of warehouses populating the valley floor).

Anyone can now visit the summit as the result of what may be a far more remarkable experiment: a decade-long collaboration between the Midpeninsula Regional Open Space District (Midpen for short) and the Amah Mutsun Ohlone tribe. Midpen, a taxpayer-funded entity created by ballot initiative in 1972 to acquire and protect open space in the Bay Area, acquired Mount Umunhum from the federal government in 1986, but it wasn't until 2009, following an infusion of new federal funding, that they embarked on restoring the peak. It was then that they contacted Valentin Lopez, chairman of the Amah Mutsun tribal band. "The more we worked with [the Amah Mutsun], the more we realized we share such strong values about land, about stewardship, about conservation … it just seemed like the right fit," says Ana Ruiz, Midpen's general manager.

For Lopez and the Amah Mutsun, the journey back to Mount Umunhum—which means "place of the hummingbird" in the Ohlone languages—has been a long time coming. "Mount Umunhum has always been a place of prayer for our people, our ancestors, and it needed to be restored as a place for prayer," Lopez told me. "Our creation story takes place on Mount Umunhum … the Creator very specifically picked our people to live within our traditional tribal territory, and [the] Creator gave us a responsibility to take care of Mother Earth and all living things. That's a sacred covenant that our tribe has with the Creator." After years spent out of the public eye, the Amah Mutsun are reclaiming that covenant, forming partnerships with the University of California at Berkeley and Santa Cruz, Pinnacles National Monument, and now Midpen, as well as leveraging their resources to help put Amah Mutsun youth out on the land as stewards.

The Amah Mutsun created their own land trust, which is slowly acquiring easements and covenants on public lands throughout their traditional territory—which runs south from Mount Umunhum to encompass the fertile Salinas Valley—through partnerships like the one with Midpen. On December 13, 2017, Midpen's board of directors unanimously approved the creation of a cultural conservation easement over thirty-six acres at the summit of Mount Umunhum, to be held by the Amah Mutsun Land Trust. The agreement prohibits commercial development and permits public access and recreation, but allows the Amah Mutsun to restore, conserve, and steward the natural and cultural resources of the site, including the construction of a Tribal Garden and a Ceremonial Space. In return, the Amah Mutsun will provide interpretation and education to site visitors, as well as contribute their wisdom to the management and restoration of the unique mountaintop ecology. Lopez hopes that the site will serve as a "place for prayer for all people, regardless of their background or religious beliefs."

Conservation easements, long employed by land trusts to protect the ecological features of the landscape, can be placed on any piece of land by the landowner in partnership with a land trust. These agreements are tailored to each unique situation, allowing the landowner to retain title and necessary rights for themselves while at the same time welcoming input, stewardship, monitoring, and access on behalf of the land trust. The Amah Mutsun, whose understanding of their obligations to the Creator does not involve the ownership of private property, have been very successful in using conservation easements to fulfill these obligations. For local governments or organizations with unused, degraded, or ecologically sensitive lands, these easements present a unique opportunity for partnership with Indigenous peoples.

PAY YOUR RENT

Those of us without access to large pieces of land—city dwellers, urban farmers, scientists, entrepreneurs—can still contribute to the restoration of Indigenous knowledge and management of local lands. Recently, parallel initiatives were launched in the Bay Area and Seattle for urban residents to voluntarily "pay rent" to the Indigenous groups on whose occupied territories they live. Real Rent Duwamish allows Seattleites to give a small percentage of their annual income, rent, or mortgage monthly to the Duwamish tribe. Similarly, the Shuumi Land Tax allows East Bay residents to support the work of the Sogorea Te' Land Trust through a voluntary "tax" on the land they occupy. Your personal *shuumi*, which means "gift" in the Ohlone languages, can be calculated based on how long you've lived in the area, whether you rent or own, and how many bedrooms your home has.

These initiatives are simply ways to reframe the idea of making a monthly donation, but they have struck a chord with Seattle and Bay Area residents. If local tribes in your region do not offer the option of voluntary "rent," you can, of course, simply send a monthly donation. If you have more energy, you can float the idea yourself. Both of these projects were created and are maintained by settler solidarity groups, working in conjunction with tribes to best meet their needs. For Sogorea Te', the Shuumi Land Tax supports the land stewards who are currently working to restore and develop that first parcel at the Planting Justice nursery.

OPEN-ENDED COLLABORATION

I've been working for a couple of years now on Galiano Island, part of an archipelago in British Columbia. Galiano was and remains the unceded territory of Hul'qumi'num-speaking peoples, but today the island is primarily populated by people of European descent. In 2012, a local organization called the Galiano Conservancy Association (GCA) bought a large piece of partly logged land in order to restore it and develop a learning center for conservation, sustainable agriculture, and land stewardship. When I first had the opportunity to work for the conservancy in the summer of 2016, they were already well on the way to realizing this vision.

In the two years since, we've worked to design and establish a native plant food forest in one of the logged areas. This effort—to restore a large, degraded site using primarily edible, medicinal, and useful native species in mimicry of a natural ecosystem—is to our knowledge the first of its kind in our bioregion. Or is it? The more we learn from the work of ethnobotanists, archaeologists, and anthropologists, the more it appears that these sorts of vegetation management systems may not be so novel after all. Still, to us, it feels like uncharted territory, as there are few other contemporary projects that we can reference with a similar range of considerations and objectives.

On a cloudy morning in October 2017, the staff of the GCA and a group of elders from the Penelakut tribe—one of several Hul'qumi'num-speaking groups that traditionally resided on Galiano Island—sat across from one another in order to discuss Penelakut involvement in the project and on the site in general. The talks were wide-ranging, personal, emotional, and encouraging. Afterward, as we walked the site together, listening to the elders' thoughts and soaking in their knowledge about the plants, it seemed to me that we had formed an important relationship that would transform and give renewed life to the site. For me, it was like holding the seedling of an unfamiliar plant: it felt fragile and uncertain, but full of potential for growth and discovery.

REACH OUT

The legacy of colonization and displacement has had far-reaching effects on Indigenous peoples, and as a result, not all tribes have the resources or ability to participate in stewardship projects or land management on their traditional territories. The concept of a "tribe" as a political, legal, and social unit is fraught to begin with, and depending on where you live, tribes may own land or be landless; they may be recognized as sovereign nations or lack state recognition. Wherever you live, you can be assured that the ancestors of local Indigenous people walked and stewarded the land under your feet, and their descendants hold valuable knowledge, wisdom, and insight pertaining to it. We are all the current beneficiaries of the land these ancestors stewarded, the soils they protected, and the plants they formed relationships with. Those of us with comparatively shallow time-depth in North America, we owe it to ourselves, to Indigenous peoples, and to the earth to reach out and support Indigenous resurgence.

This article is adapted from several episodes of the podcast Future Ecologies. *To hear the full story: futureecologies.net.*

MORE INFORMATION

Planting Justice: plantingjustice.org
Sogorea Te' Land Trust: sogoreate-landtrust.com
Amah Mutsun Tribal Band: amahmutsun.org/land-trust
Midpeninsula Regional Open Space District: openspace.org
Real Rent Duwamish: realrentduwamish.org
Galiano Conservancy Association: galianoconservancy.ca

Native woman with Suak River Stump, circa 1945

SOME PASTURES

A Cowboy's Almanac

SAM RYERSON

November, Pajarita Pasture—Mescalero, New Mexico

On the mountain ranches we always miss some cattle when we gather the big pastures. They're called the remnants. The rougher timber pastures, with live water in springs and creeks, hide the cattle where they can live year-round in the brush. Left alone too long, cows like that become feral and raise unbranded calves that grow up wild, and they call them mavericks. Pajarita is a solitary peak in the middle of such a pasture, rising gradually from the northeast plateau of the Sacramento Mountains and dropping steeply to the east above its old caldera. On top is an abandoned fire lookout tower. Down in the canyons are five perennial springs that freshen late in the summer, after the rains. The fence circles the mountain so that the pasture includes it all—ten thousand acres or so. It's good country for mavericks.

We were still missing a few remnants of the summer. I like the work of gathering the last remnants. It's a kind of hunting: looking for cow tracks and following them; finding the cattle in the brush and holding them with dogs; calming them or sometimes roping them and bringing them home. I like big pastures—or country without pastures or fences at all—where you locate your cows by the lay of the land itself. In early November we were short just a handful, and it was time to pregnancy-test the cows and ship the culls. We sorted the opens and the dry cows that had lost calves and about one hundred fifty of the little weaned heifer calves to ship. Jim and Hana and I gathered them up from the traps and drove them to the pens at headquarters, ten miles south, around the other side of Pajarita, to load on trucks the next morning.

We loaded the trucks and Jim took off to go home. It was Election Day. Hana and I drove back in the pickup. I said, "It might be a big day for the country, but you'd never know it out here." We thought there might be a few cows left in Pajarita. We drove up the road to the top of the mountain. Maybe we'd find some cow sign along the way. You can see the whole ranch from the top, and all the way down to the Guadalupe Rim, right into Texas. We sat down out of the wind in the lee of the old lookout's cabin and ate lunch in the afternoon sun, and went home when it set behind Whitetail Summit. We heard the election news the next morning—that it had been a shocking day for the country. It didn't seem to matter much, except to be another day beginning, another sunrise across the hills. We all went down to a conference in the city. Sure enough, the city people were talking about the election, but most of the farmers and ranchers were just glad to see each other and didn't mention the news.

Mountain pasture

ON THE PUEBLO FOOD EXPERIENCE

ROXANNE SWENTZELL

Sometimes, when we are limited by circumstances, our creative humanness rises to the occasion. This seems true with me. I was born into a time and place that held many conflicting forces, so it was easy to get swept away by the mainstream society and hard to figure out how to hold onto my unique identity and honor my existence.

As a Native American woman, born between worlds and in a time of great environmental changes, I found myself inspired to learn new old ways—ways of not only surviving but also creating a meaningful life. In the 1980s, I was privileged to be introduced to permaculture. I remember a conversation with Bill Mollison about how he came up with the name *permaculture*. He spoke about it being two words combined: *permanent* and *culture*—permaculture—the permanence of culture. I had no idea at the time that my journey into permaculture would lead me backward, or rather, deeper into what culture is.

Like most permaculturists, we were interested in creating a sustainable homestead based on the models of Nature. Joel Glanzburg, my husband at the time, and I cofounded Flowering Tree Permaculture Institute and spent the next ten years creating a homestead that could feed a family of four and an array of different farm animals. We grew a "food forest" on one-eighth of an acre of high desert in northern New Mexico. We taught classes on farming, animal husbandry, food storage, pottery, you name it—anything that had to do with hands-on survival skills. The more we learned, the more there was to learn. Just like our small yard: the more we added, the more complicated the system became, so what was once a very small piece of dry wasteland became whole ecosystems that multiplied with each new species of plant, animal, insect, bird, fish, worm, or fungus. That one-eighth of an acre became a deep, rich forest of unbelievable diversity and life within its many layers. I feel overwhelmed and small within it.

Watching it evolve over thirty years, I've learned some things about "culture" that I apply directly to myself. Not everything prospered in the environment of the yard. The microclimates, species, and planned design methods were not always the answer to success. The situation was far more complicated than any plan our human brains could create. I didn't know at the time that I was watching a family evolve based on belonging to a place.

What did not "belong" died or struggled endlessly to remain. I watched this orchestra struggle from making just noise into playing something that sounded like music. I realized that Nature is doing this dance always. If you leave it alone long enough, it will balance into a most harmonious culture of life. People forget that they do the same. We think we can go anywhere and do anything our minds think up. We forget that we are part of the family of life on earth and that some places and conditions fit together better than others. I define *culture* as that which belongs to a place and a way of life because it has been formed by, and it has formed, what surrounds it over long periods of time. A culture takes time to form. Even cultures in a petri dish take time to grow.

Our seed bank was a constant reminder of where we were. Only certain plants—mostly our pueblo crops—thrived in this high desert landscape. As a Native person myself, I was already interested in the preservation of Native crops and lifeways, but I hadn't put it all together quite yet. I was still in the mind-set of being a seed saver or a permaculturist or a farmer or a teacher. I wasn't seeing myself in the context of culture. Yes, I thought of myself as a Native person partaking in community activities once in a while, but I wasn't seeing how I was part of the ecosystem of a larger cultural belonging that involved my local surroundings and my DNA. We are a species that thrives or doesn't because of being in the right place or the wrong place.

I became obsessed with how it all fit together. I had been saving our traditional crops for years; I stored most of them away just to save the seeds from extinction. I bought organic products from natural food stores. I struggled with weight and health issues and never seemed to find real help—just excuses and more Band-Aid remedies. My son, Porter, is a historian, and we talk a lot about our common interests. One year he researched what exactly our ancestors ate before the introduction of European foods. He decided to eat that way for a few months to see if you could still do it in modern times. Yes, you can.

But something happened that we didn't expect. Putting the environment and us together created a profound thing. Some unexpected "fit" started a domino effect of health, not just in our physical bodies but also in our spiritual lives and lifestyle. I realized that we, the people of this landscape, needed to find our right place again in the system. I once read that it takes twenty generations in the same location for a species (humans included) to genetically adapt to that environment. The Pueblo Food Experience was the result of this experiment.

It is a project designed to help us remember where we fit in the permanence of culture. A group of Native people (whose ancestors have lived in northern New Mexico for more than twenty generations) agreed to eat only their original foods for three months. They took blood tests before and after the project. The results proved that we were doing the right thing. Everyone got healthier, and the diet seemed to spur inspiration to change our lives for the better. The project was self-empowering. As a permaculturist, I love how this approach could save us in a time when we might have to find new ways to survive without our beloved grocery stores and fast food. We just might have to learn to feed ourselves again and to once again make things with our hands.

If our economy dried up tomorrow, if the stores closed, what could you make to eat with the resources you have now? This thought has always led me back to my ancestors, who were the true permaculturists of this arid Southwest. They figured out how to live sustainably in this environment for thousands of years. I look to them for answers. We got lazy and depressed and took all the easy handouts and forgot who we were and where we came from. There is a scary price for easy food—fast food is killing us, and I don't just mean McDonald's and Dairy Queen. I mean all the packaged foods in stores and all the unknown processed ingredients. There has been no cornmeal thrown for the lives taken. There have been no harvest dances for gratitude for the rain on our fields. There have been no prayers of thanks along the food's journey to our tables. Our connection to food is part of who we are as Native people, and when we disconnect from it, we are lost and easily tricked into buying whatever corporate greed dictates. The result is more than bad health; it is blindness and starvation of the soul.

The Pueblo Food Experience helped pave the path back to ourselves. It has been a long walk back to Mother Earth, but each piece of the puzzle we figure out is one more piece of ourselves that we have taken back from those who would wish us gone.

It is empowering to connect with any part of our surroundings. It is empowering to make something to eat from scratch from what is available in our environment. Growing food ourselves is also empowering. Walk instead of drive once in a while. Fix something yourself and see how it works or how you could make it better. We ourselves truly are the ones we have been waiting for! *The Pueblo Food Experience Cookbook* was one of the outcomes of our journey. We wish to share this journey with others who want to really know how we fit and dance within the mystery of life, time, and place—within "the story of us."

Introduction to The Pueblo Food Experience Cookbook *by Roxanne Swentzell reprinted in its entirety from* The Pueblo Food Experience Cookbook: Whole Food of Our Ancestors, *edited by Roxanne Swentzell and Patricia M. Perea (Santa Fe: Museum of New Mexico Press, 2016). This book can be ordered online at mnmpress.org or by calling (800) 621-2736.*

Karen Cantor | Rose Simpson, Marian Naranjo, and Roxanne Swentzell dancing on salt flats

Return: Native American Women Reclaim Foodways for Health & Spirit *features Swentzell (Santa Clara Pueblo) and five other Native American women exploring the connection between ancient foodways, food sovereignty, and health. Along with Swentzell, Kibbe Conti (Oglala Sioux), Jennifer Gauthier (Menominee), Desiree Bergeron Jackson (Tlingit), Andrea John (Seneca Nation), and Valerie Segrest (Muckleshoot) share their experiences reviving traditions within their communities.*

THANK YOU EARTH

LINCOLN GEIGER

Thank you earth so soft and strong,
Thank you meadow filled with song.
Thank you mountain, forest and stream,
By you we rest and find our dream.
Thank you creatures wild and tame,
Your trust we love and hope to gain.
Thank you for your milk and fleece
And for your meat that we may eat.
Thank you root and leaf and seed
We'll not forget your wondrous deed.
You hold the earth you catch the rain
You fill the world with air again.
Thank you wind for bringing rain,
Please help our friends who are in pain.
For us who thirst and cry from hunger
Please bring hope, life and wonder.
Thank you moon for guidance and grace,
For heart-bent flowers with dew drop lace.
Thank you sun as day begins,
For golden light by angel wings.
With thankful hearts and open hands,
We ask to share your loving lands.

ALIGNMENT AT PARC-Y-MARW ("THE FIELD OF THE DEAD"), NEAR FISHGUARD.*

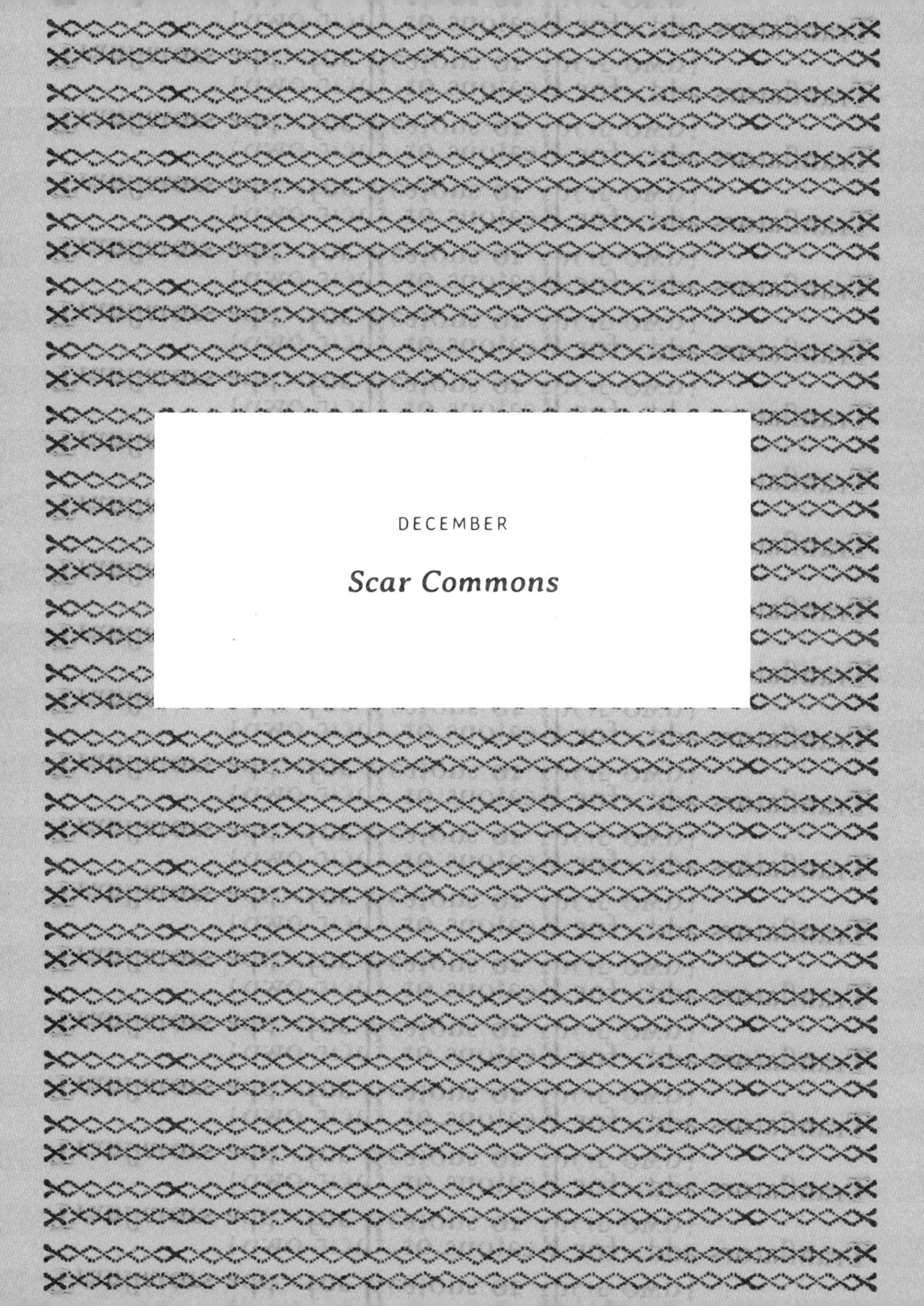

DECEMBER

Scar Commons

THE CHANGE

ALLISON ADELLE HEDGE COKE

Thirteen years ago,   before bulk barns   and
fifth gear diesel tractors,   we rode royal blue tractors with
toolboxes big enough   to hold a six pack on ice.
In the one hundred fifteen degree   summer
heat   with air   so thick with moisture
you drink as you breathe.
Before the year dusters sprayed
malathion over our clustered bodies,   perspiring
while we primed bottom lugs,
those ground level leaves of tobacco,
and it clung to us with black tar so sticky we rolled
eight-inch balls off our arms at night   and
Cloroxed our clothes for hours and hours.
Before we were poisoned   and
the hospital thought we had been burned in fires,
at least to the third degree,
when the raw, oozing hives that
covered ninety-eight percent of our bodies
from the sprays ordered by the FDA
and spread by   landowners,
before anyone had seen
automated machines that top and prime.
While we topped the lavender
blooms of many tiny flowers
gathered into one,   gorgeous.
By grasping hold below the petals
with our bare, calloused hands
and twisting downward, quick, hard,
only one time,   snapped them off.

Before edgers and herbicides took
what they call weeds,
when we walked for days
through thirty acres and
chopped them out with hoes.
Hoes, made long before from wood and steel
and sometimes (even longer ago)
from wood and deer scapula.
Before the bulk primers came
and we primed all the leaves by hand,
stooped over at the waist for the
lower ones and through the season
gradually rising higher until we stood
and worked simultaneously,
as married to the fields as we were to each other,
carrying up to fifty pounds of fresh
leaves under each arm and sewing them onto
sticks four feet long on a looper
under the shade of a tin-roofed barn, made of shingle,
and poking it up through the rafters inside
to be caught by a hanger who
poked it up higher in the rafters to another
who held a higher position
and so they filled the barn.
And the leaves hung down
like butterfly wings, though
sometimes the color of
luna moths, or Carolina parakeets, when just
an hour ago they had been
laid upon the old wooden
cart trailers pulled behind
the orange Allis-Chalmers tractor
with huge, round fenders and only
a screwdriver and salt in the toolbox,
picked by primers so hot
we would race through the rows
to reach the twenty-five gallon
jugs of water placed throughout

the field to encourage and in attempt to
satisfy our insatiable thirsts
from drinking air which poured
through our pores without breaking
through to our need for more
water in the sun.
Sun we imagined to disappear
yet respected for growing all things on earth
when quenched with rains called forth
by our song and drumming.
Leaves, which weeks later, would be
taken down and the strings pulled
like string on top of a large dog-food bag
and sheeted up into burlap sheets
that bundled over a hundred pounds
when we smashed down with our feet,
but gently smashing,
then thrown up high to
a catcher on a big clapboard trailer
pulled behind two-ton trucks and
taken to market in Fuquay-Varina
and sold to Philip Morris and
Winston-Salem for around a buck a pound.
Leaves cured to a bright leaf,
a golden yellow with the strongest
aroma of tobacco barn curing
and hand grown quality
before the encroachment of
big business in the Reagan era
and the slow murder of method
from a hundred years before.
When the loons cried out in
laughter by the springs and
the bass popped the surface on
the pond, early on, next to
the fields, before that time
when it was unfashionable to
transplant each individual baby plant,

the infant tobacco we nurtured, to
transplant those seedlings to each hill
in the field, the space for that particular plant
and we watched as they would grow.
Before all of this new age, new way,
I was a sharecropper in Willow Springs, North Carolina,
as were you and we were proud to be Tsa la gi
wishing for winter so we could make camp
at Qualla Boundary and Oconaluftee
would be free of tourists and filled with snow
and those of us who held out forever
and had no CIBs would be home again
with our people, while the BIA forgot to watch.
When we still remembered before even the Europeans,
working now shoulder to shoulder with descendants
of their slaves they brought from Africa
when they sold our ancestors as slaves in the Middle East,
that then the tobacco was sacred to all of us and we
prayed whenever we smoked and
did not smoke for pleasure and
I was content and free.
Then they came and changed things
and you left me for a fancy white girl
and I waited on the land
until you brought her back
in that brand new white Trans Am,
purchased from our crop, you gave her
and left her waiting in a motel.
The nearest one was forty miles away,
but near enough for you
and for her and I knew though
I never spoke a word to you
about it, I knew and I kept it to
myself to this day and time and
I never let on
until I left on our anniversary.
I drove the pick-up
down the dirt path by the empty fields

and rented a shack for eighty dollars,
the one with cardboard windows
and a Gillespie house floor design,
with torn and faded floral paper on walls
and linoleum so thin over rotted board
that the floor gave if you weighed over
a hundred pounds.   I did not.
And with no running water of any kind,   or bathroom.
The one at hilltop,   where I could
see out across all the fields
and hunt for meat when I wanted
and find peace.
I heard you remarried
and went into automated farming
and kept up with America.
I watched all of you   from the hill
and I waited for the lavender blooms
to return   and when it was spring
even the blooms   had turned white.
I rolled up my bedroll,   remembering before,
when the fields were like waves on a green ocean,
and turned away,   away from the change
and corruption   of big business on small farms
of traditional agricultural people,   and sharecroppers.
Away,   so that I could always   hold this concise image
of before   that   time   and it
floods   my memory.

"The Change" is reprinted by permission from Off-Season City Pipe *(Coffee House Press, 2005).*

Women

SOME PASTURES

A Cowboy's Almanac

SAM RYERSON

December, Wagon Mound—New Mexico

This is who we are; these are my friends and partners. We are cowpeople, horsemen, and horsewomen. We didn't all start out this way. We are young and old and middle-aged. We are from Cambridge, Massachusetts; Wikieup, Arizona; Chugwater, Wyoming; and many other places. Our lives choose us; we choose our lives. Fate does not rule us, but faith. We are missing fathers and sons and daughters and wives. We hold each other together, and remain. Our horses buck us off, step on us, kick us in the face, and carry us forward again. We are half deaf and half blind. We eat a lot of dirt and cow dust. We learn to get better every time. We keep hurting and healing. We have a community of common scars, pastures full of scars.

We are, all of us, marked, like the land, with scars. We have old broken bones and pinched nerves, which no doctor diagnosed, but there they are to remind us on cold mornings. We tell stories about them, and the broncs or bulls that caused them, and we laugh, and improve the stories in the telling. But we have worse wounds—deeper, older—that don't heal, that leave empty places inside, that we don't talk about. We are wounded, and sometimes weary, like our horses after long days trotting home. Yet we keep trotting out early in the morning just as glad to go again as the old dogs at home are to see us when we return. When there are dogs around, and children, baby calves, and young horses to ride, we have reasons to smile. We all get bucked off sooner or later. We work on the land to heal ourselves, and in our work we mark ourselves with scars. When we work on the land to heal the land, we also scar the land.

Pastures are markers on the land where we work. Pastures are marked and scarred by fences, by the trails our cattle crease and follow coming to water, by the troughs and ponds we build to hold water and the pipelines to carry it, by the roads we drive on, the corrals we work in, the houses we sleep in. The deeper wounds to the land are the falling water tables, eroded gullies, bare ground, encroaching brush, invasive monocultures, the loss of fire and fuel and bison and wolves.

We work on the big rangelands because we love the land and the animals and one another, and we find unconditional joy in the work. The big country lends itself to sadness, beauty, quiet, and powerful community. The work on the land draws us in and wraps us in quiet joy, in days we cherish—and then it tears us apart, sometimes violently. We come within inches of death or dismemberment: when we dally our long ropes around our saddle horns to catch cows at a flat-out run, when we drive home at night on long dirt roads after drinking a lot, when we keep guns everywhere with us. We witness birth and death on a daily basis. The way we work lends itself to depression and desperation, addiction,

divorce, and death. Our work out here draws the people—the passionate, intense, powerful, creative, solitary people—who are often prone to such patterns. When we work together, we are better.

When we ride or work cattle on the land, we begin to be free of the old patterns that afflict us. We feel of our horses and through them and our cows, who see before them and remember behind them just what we show them. We move across the land, leaving not scars but hoofprints, and manure here and there, and the positive effects of animal impact. When we adopt old transhuman patterns across the range, we work without the scars of fences, roads, corrals, and houses. We ride to keep the cattle located across the ridges and canyons, following a grazing plan we have designed to give them the best forage and to heal the land. We ride to watch for sick and injured cattle, and give them medicine when we find them. We build improvements that support the range as an ecosystem, as a whole. We find communion, working alone with animals on the land.

This winter I'm living in a tent on the edge of the plains. On this dormant grass, I feed my cattle salt and protein from a wagon. I go horseback everywhere I go across this ranch of mesas and canyons that I lease with my partners in northeastern New Mexico. It's one big pasture and a little holding trap, with some temporary electric fence to hold my horses. This winter has been warm and dry. I'm worried for the springtime, but hopeful because I think there is still moisture deep in the soil. My days are full of work I enjoy. My back aches if I sit for very long, so I don't sit for very long, but I stand when I drive my wagon. I stand and sing to my team and my voice drives them forward. I sing for the days when I didn't, for the sadness that remains, for the hope that wells deep, that persists in my community, my friends and partners, all of us neighbors.

Michelle J. Felicetti | Prak Soy's Garden

PRAK SOY

MICHELLE J. FELICETTI

Prak Soy came from Cambodia; a survivor of genocide.

She crocheted bicycle seat covers with pom-poms
and shrouded coat hangers in colorful yarn;
wore a patchwork bucket hat and spoke English, stranding each word a sentence.

Next to her apartment where a building was demolished and buried, she gardened basil, leafy greens, eggplant, lemongrass, tropical-looking flowers, and beans;
waving to us from her porch, beckoning, Eat, eat.

Her living room was the scent of drying herbs, collected seeds, and rice steam.
Prak Soy preserved lemons in jars;
Cambodian music videos played on the television screen,
pictures of her family hung on the wall;
she sipped cognac during a get together once around a hot pot,
shuffled in flip-flops a few sizes too small, her heels touched the ground.

TEATIME

KATHLEEN SHANNON

1. I hadn't seen Rachel in a while. She told me she'd be harvesting stinging nettles soon. I didn't know a nettle from anything else. Sting, I knew. My favorite swimming hole too.

When we met there, we walked right past the old dam: too early for swimming. We looked out across a narrow portion of the Sheepscot River and headed down a winding path.

"You'll probably get stung. But, they're young yet. If you snap the stems toward the base, it might not be too bad."

I knelt to the ground, fingers poised for my first wild harvest. "Where are they?" I asked.

She gestured around us, smiling. "Everywhere."

Within an hour, we each held a sack of greenery as we stood in the shallow river—the water just past our itchy ankles. Nettles' neighbor kindly soothed its sting.

2. Holding the bouquet above my soaking wet head, I exclaimed triumphantly, "It's still dry!" I'd flipped my kayak trying to climb back in after harvesting St. John's wort from the sides of the railroad bridge.

Weeks later, I showered myself with dusty dried leaves as I removed them from their high perch—a screen bridging the gap between two tall bookcases. I thought it ironic that a plant so well known for promoting happiness grew best in soils most void of life. I kept the thought to myself, though, as I sat beside my partner, stripping leaves and flowers into jars. It was the elephant in the room: I'd saved the flowers on that sunny day but, somehow, my head was still underwater.

3. I used my harvest knife to cut red clover by the handful from the ends of the vegetable beds. I was thinking over a conversation I'd had earlier with the head farmer. I'd been late to work, I explained, because I'd suffered a panic attack. "I'm going to therapy, though," I admitted in a low voice.

"Good for you," she said, sincerely.

I looked up, surprised.

"I've been seeing the same therapist for ten years," she said. "It's been amazing."

Now, exhausted and sweaty, I considered tasting the sweet end of the clover petals—the way I used to as a kid. Instead, I capped my harvest jar and headed home. My ill stomach didn't have much taste for culinary adventure these days.

4. It was my birthday weekend—twenty-five. I spent a portion of it in the elderberry patch, bending the tall stalks down from over my head and combing the small berries from their heavy stems into my bucket until my hands were a happy shade of magenta. My kitchen turned that color too as I mashed the berries and mixed them with spices: the first steps toward wine.

I got a little tipsy decanting the potion, and thought about how I was building a liquid time capsule. Who would be drinking this next year? It'd be me, surely, but who would I be then? Still depressed? Anxious? Still facing relentless intestinal pains? Still living with my partner in a deeply unhealthy relationship?

I hid the three bottles away in the closet, to be pulled out a year later: a message in a bottle for my future self, having crossed uncharted waters.

5. We could hear the tide rolling in as we harvested rose petals from the wild bushes along the coast. The petals were surprisingly easy to pick. With gentle fingers, they simply fell from the flowers. The scent lingered on my hands and wafted from my harvest bag.

"This kind of makes me feel like a goddess," I said to Emily, dreamily. She agreed.

As it turns out, an abundance of rose petals is a fabulous problem to have. Our solution: lemon-rose cake. Just in time, we cut two huge slices and ate them on the back porch in blissful silence, watching the full moon glide up and over the horizon.

"Imagine you're lying at the bottom of a river," my therapist said, "and you're watching the water move above you."

The image left me feeling claustrophobic.

Instead, I closed my eyes and pictured the Sheepscot—as far upriver as my mind's eye could recall. As far up as I'd ever been.

There, I let go. I bobbed along with the gentle current and smiled at all the familiar faces and places: the nettle bed, the railroad bridge, the veggie fields bordered by clover, the elderberry patch, and, just as I was sent out to sea, I nodded to the rosebushes dotting the rocky coast.

I opened my eyes and took a sip of my corresponding tea. It was only a five-minute meditation, but I'd traveled rivers.

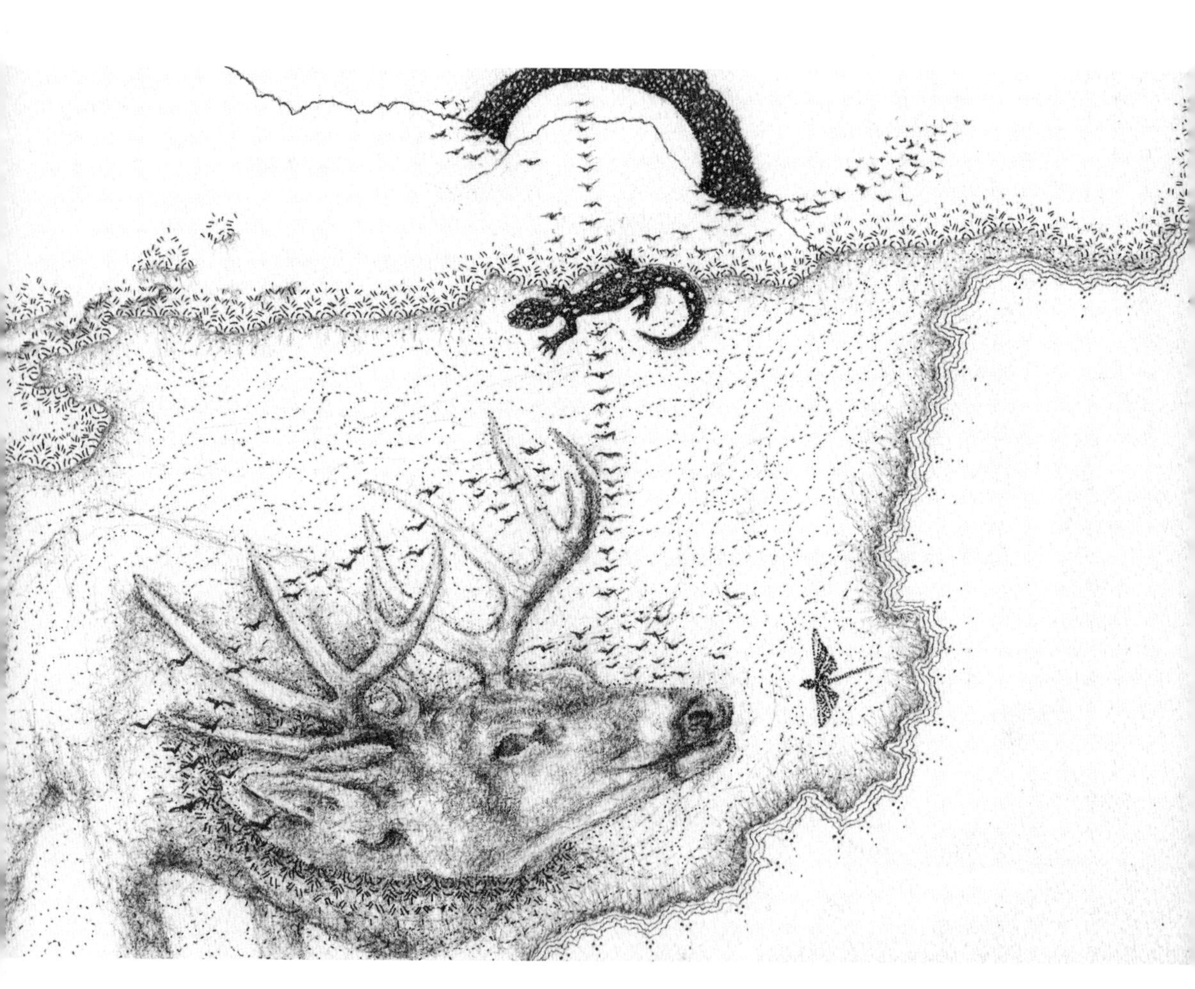

Maureen Walrath | Cooling our Wounds: frozen deer in the Rochester Creek on the day of the longest night

EXPLAIN: WHAT WE DID WITH THE HONEY

JULIA SHIPLEY

The day we learned you were gone, Howie says he knocked, nothing answered.
He'd shoveled the slush-lined steps to the garden, through our marriage arbor, all the way out to the hive—your wedding gift to us—were any still inside? They'd stung him all summer: his face, his ankle, his arm, but never me, though sometimes they veered for my curtain of hair. Come fall we dared not break their combs for fear they'd starve; in late winter's thrall, I worried a famished bear might bash the frames and slash the sweet from it.

Then your daughter called and said your heart had quit; you were halfway up the stairs, but they restarted you. And then a clot shut part of you down; but before you got used to the stroke's ongoingness—your speech thickened to the viscosity of honey—another stroke broke you for good.
So they unhitched your electricity and permitted egress; but I hear your real voice—steady, sandpapered bass—and picture your tight mind zooming into the night, while daughter and wife stand by the closed house of you, surrounded by flowers.

Now Howie pokes a stick in the entrance to clear winter's harvest: bee carcasses. On returning, he says, So many. His dowel pierced more and more deceased. If they're all gone, I interrupt, distressed—just tell me.
No, he replies, I stuck the dowel in again and they roared!
Go put your ear against and listen, he said, they're alive.

Originally published on Riverteeth: A Journal of Nonficton Narrative. *Reprinted with permission.*

ON AIR: AN INTRODUCTION TO TRANSDISCIPLINARY SOUNDSCAPE ECOLOGY

RACHEL WEAVER

We are enveloped in the sensory experience of environmental impacts. The chaotic and incoherent elements of soundscapes become encounters that create space for active concentration. Deep listening leads to a maieutic experience, where latent ideas may be brought into clear consciousness. Soundscape ecology provides spatial context for the reservoir of expressions in surrounding environments.

Soundscape ecology is made up of geophony (non-biological audio coming from different types of habitats, whether marine or terrestrial); biophony (the collective sound that vocalizing animals create in each environment); and anthrophony (the human sounds/noise that occur within a soundscape). In the words of soundscape ecologist Bernie Krause, "Every soundscape that springs from a wild habitat generates its own unique signature, one that contains incredible amounts of information."[1]

The wind sets the mood in north Texas. Feel the ambient noise and the heaviness in the troubled air. Guided by air, we attempt to cohabitate within our environmental soundscapes. Listen to the songs of the cottonwood, a tree that can be classified through the rustling of its leaves in the wind. Oak trees sing while the wind dances through their branches. The calls of red-tailed hawks resonate through the air, and I feel them sense me in the garden below. I try to catch sight of them as they soar above the prairie, but the sun catches my eye and I can only catch a glimpse. "Understanding is not only a matter of a conceptual or linguistic identification," writes artist Aliza Shvarts, "but also a sensate one, wherein the viewer not only knows, but feels—or perhaps more specifically, knows precisely by feeling."[2]

When we tune into the soundscape ecology of our restorative ecosystems, we begin to comprehend the impacts of industrialization and urbanization on the biological and non-biological inhabitants of our shared habitats. Patience, active listening, and dedication to the land reveal the signatures of the native beings and natural entities around us. In her seminal "Cyborg Manifesto," Donna Haraway observes, "We are living through a movement from an organic, industrial society to a polymorphous, information system—from all work to all play, a deadly game. Simultaneously material and ideological."[3] As we enter this deadly game,

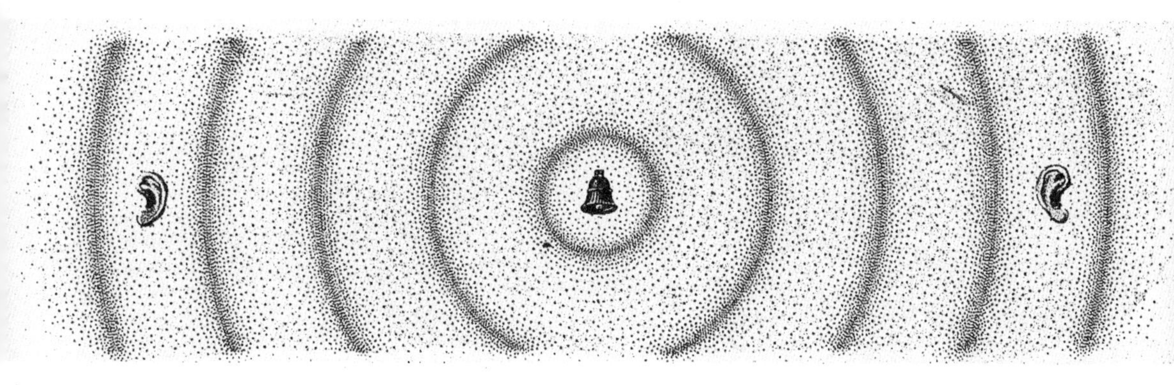

we should be the coddiwomple—we travel purposefully toward an as-yet-unknown destination. To travel with purpose is to engage our senses with our surroundings, and to contemplate our interconnectedness. Despite a fraught horizon, we aspire towards an agrarian utopia with intention and awareness.

Our performance of radical acts shifts in the polymorphous information system of our society. "We cannot go back ideologically or materially," says Haraway, to thinking in terms of essential biological properties and determined sexual reproduction. Rather, we formulate understandings in terms of frequencies, rates of flow, and we realize that organic sex role determination is only "one kind of reproductive strategy among many."[4] Transdisciplinary environmental education bends boundaries and mends consciousness. Queer ecology alters our conceptions of the normative environment, and we comprehend: there will always be more uncertain knowledge than certain knowledge about the environment. In soundscape ecology and active listening, queer ecology can amplify our perception of the whole environment of sounds, sharpening our perception of eccentric aspects within particular sounds. Queer ecology acknowledges that ambient soundscapes and raw sensations have no boundaries, and the heaviness, the low-end sounds, the loudness, the exceptional endurance of the sleepless planet.[5] Embracing the unknown is the trade skill of the agrarian and environmentalist. Acknowledging the mysteries and disarray of our soundscapes, and how they are changing, requires active listening situated in place.

"The level of awareness of soundscape brought about by deep listening," taught composer Pauline Oliveros, "can lead to the possibility of shaping the chaotic sounds of technology and of urban environments and machinery."[6] Wind whistles through a monarch butterfly's broken wing, which tells a generational story of unbroken determination. Airplanes and sirens soar with constant hum from flyways and highways in the distance—boundaryless in air. Imagine the loudness of the cotyledon bursting through the seed epidermis and soil, clawing its way to fresh air and light. Seed saving is our voice, conserving its heritage is storytelling. The loss of our seed biodiversity represents the silencing of our *vox populi*. Corporate consolidation of seeds and chemical impacts of industrial agriculture are gags and muzzles on our diverse soundscapes.

The signature voices that express our communal storytelling require shared breath, "and as we hear them, we're endowed with a sense of place and the true story of the world we live in."[7] We must work to align community heartbeats with ecological pulsations. At KUZU 92.9 LPFM, a community-based low power radio station in Denton, Texas, we realize public airwaves are public space.[8] We are community supported airwaves, encouraging a community to share vested interest in terrestrial radio. KUZU creates a sense of place through aural media platforms, fostering experimentation and cultivating awareness.

Transdisciplinary environmental education and queer ecology are subversive and crucial in the age of anthropogenic climate change and global environmental disruptions. The radical acts needed to protect restorative ecosystems are found in the minutiae of our agrarian work and labors of love. Soundscape ecology and maieutic practice through deep listening are immersive and vital. We perceive connections between our natural and urban environments, aesthetics and ecosystems, and living beings and non-living entities.

Rachel Weaver is a board member, studio volunteer, and show producer at KUZU radio.

NOTES

1. Krause, Bernie. "The Voice of the Natural World." Filmed June 2013. TED video, 14:48. ted.com/talks/bernie_krause_the_voice_of_the_natural_world.
2. Shvarts, Aliza. "Troubled Air: The Drone and Doom of Reproduction in SunnO)))'s Metal Maieutic." *Women & Performance: A Journal of Feminist Theory*, Volume 24, Issue 2–3 (2014): 203–219.
3. Haraway, Donna. "A Cyborg Manifesto: Science, Technology, and Socialist-Feminism in the Late Twentieth Century." *The Cybercultures Reader,* edited by David Bell and Barbara M. Kennedy, 291–324. London: Routledge, 2000.
4. Haraway. "A Cyborg Manifesto: Science, Technology, and Socialist-Feminism in the Late Twentieth Century."
5. The Institute of Queer Ecology: queerecology.org.
6. Oliveros, Pauline. "The Difference Between Hearing and Listening." Filmed October 2015. TEDx Indianapolis video, 11:37. youtube.com/watch?v=_QHfOuRrJB8.
7. Krause. "The Voice of the Natural World."
8. KUZU 92.9 LPFM Denton, TX Community Supported Radio: kuzu.fm.

PART OF THE WEAVE

GRANT JONES

I started feeling younger when
I took up residence in
the mouth of this canyon.
I became less hurried here,
relaxed, quieter, more awake.

When I sat still and rotated my head
like I was cranking a director's chair,
tiny movements caught my attention
tugged at the tapestry that unfurled
before me, like when I waded
the Sauk and baby cutthroat
would dart behind the boulders
or rotating fins of big browns
fanned sand in eddies up the Flathead:
my skin tightened its weave.

It's like that now—a pair
of red-shafted northern flickers flick
from one vertical face of crumbling gneiss
to the next as they belay
to peck off the ants,
or a ground squirrel dashes
between balsam roots
or sometimes it's just a patch
of leaves that flash in a swirl
of warm air, or something else
I can't name like the shadows
that twitch in the antelope bitterbrush.

The tapestry out here is mine.
I am part of its weave.

Watch Over Us Hill
Mouth of the Canyon of the Little Mosquito

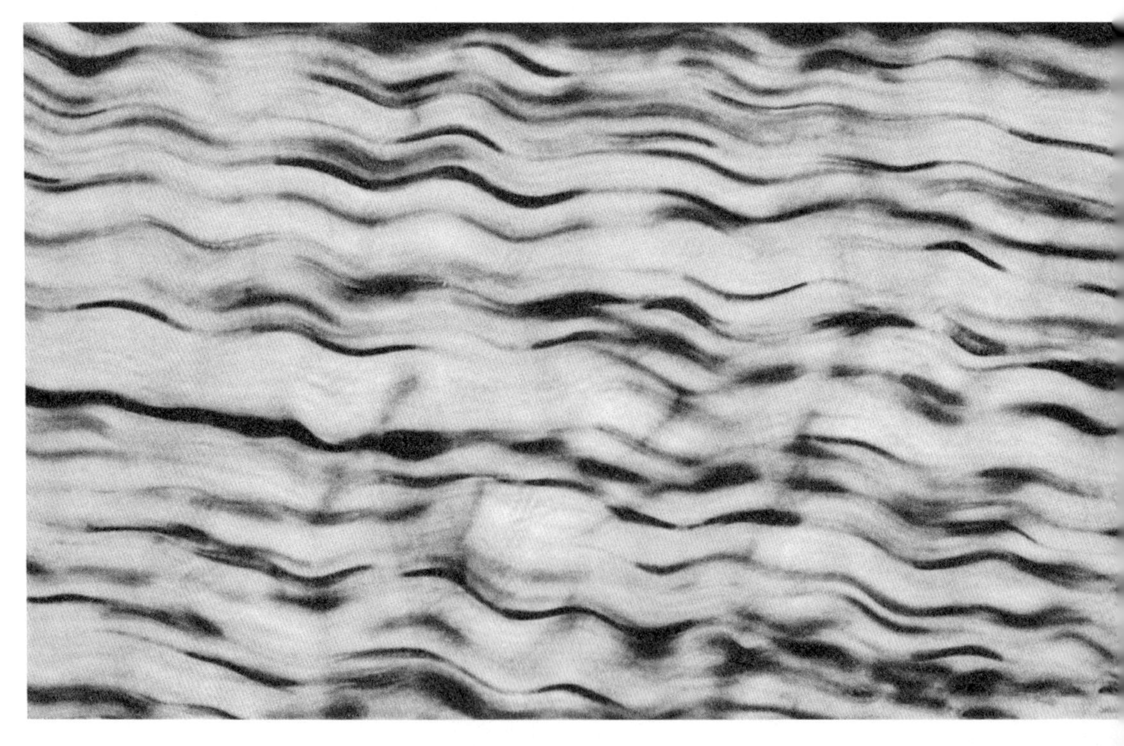

Connective tissue, close-up

FARM COMMONS

IAN MCSWEENEY

We abuse the land because we see it as a commodity belonging to us. When we see land as a community to which we belong, we may begin to use it with love and respect.
—ALDO LEOPOLD

False beliefs in manifest destiny and boundless land provide false justification to land being commodified and extracted from. Theft, separation, violence, discrimination, enslavement, and acculturation are used to separate the people from the land, and to facilitate privatization and abuse of the land.

We, as humans, residents of America, farmers, growers, parents, and children, must create community with the land. We must let go of private commodified ownership of land. We must give access to those who have been excluded, removed, or taken from the land—whether for their skin color, gender, beliefs, or place of birth.

Farmland is a scarce and threatened resource in a significant ownership transition. The next two decades will see four hundred million acres of farmland transfer from its present owners to the next. These acres include farms that have provided nutrient-dense food, supported health, invested in soil, aligned with ecology of place, engaged in the broader foodshed, and bolstered local agrarian economy. These acres include farms that have created and continue to foster communities that love and respect the land, that belong to the land. These farms and their legacy, future ownership, stewardship, and use will have significant impacts on the future of land as a community to which we belong.

We must—and can—create community ownership to cooperatively own and steward farmland. Community ownership can provide secure land tenure for the next generation of farmers. Community ownership is connected to the land—the land to which we belong.

Ian McSweeney is director of Agrarian Trust, which is creating local community Farm Commons to bring about and foster the transition of healthy farm communities. Follow the work of Agrarian Trust at agrariantrust.org to learn more.

SAVE YARD

AZURÉ KEAHI

At the mercy of the giant spoon,
here:
scoops of chocolate earth,
berms for perennial-filled bellies,
spit stones from the past.
Buried
living rooms are garden beds.
Doll houses — chicken coops.
Beer bottles (decades old) — terrariums.
Drained snow globes — insect motels.
All the same,
we find homes in detritus,
life in waste.

Colm Harrigan | Turf stack
Photo by Deirdre O'Mahony
Part of T.U.R.F. (Transitional Understandings of Rural Futures)
Limerick City Gallery of Art 2013

SLOW TIME

BINYAMIN KLEMPNER

The guy I work for has some space available for an orchard. He wants to see his land being used productively. He wants to see trees. Beauty and life taking root on his property. We have an agreement: If I plant the trees and lay the irrigation, the fruit is mine to harvest and sell. I get an income and he receives personal satisfaction. A good deal. I'm looking into planting Peruvian cactus apples and jackfruit. High value fruits I can sell directly to exotic fruit markets and through Etsy. If I plant one hundred of each, it will be a solid four or five years until I get my first harvest. But that's to be expected. For the next five years I'll have a nice boost in income to look forward to, arriving just when I'll need it most.

Time takes time. Within the scope of five years, as I harvest my first jackfruits and Peruvian cactus apples, there will have been many fast-tracked, high-tech startups in nearby Tel Aviv and around the globe that have come and gone and perhaps come back and gone out several times. Technology will have reached new, and for some, perhaps many, exciting new plateaus. Self-driving technology will be more of a commercial reality. The world of the city will look different, and, in all probability, be lived in differently. Yet the field will look much the same as a field of one hundred years ago and of a field one hundred years before that.

I will use flower pots to plant my seeds. The seedlings will grow at the pace at which they grow. I will use farm-grown worm castings (we are, after all, a worm farm). While the worm casting and the compost tea may enhance the vigor and growth of the plants, there is only so much one can do to speed a process that takes the time it takes. The seedlings will be transplanted in the field. They will grow. Slowly. Imperceptibly. In the meantime months and years will go by. My children will grow. My one-year-old son will be five or six. Talking. Playing. My fifteen-year-old will be nineteen or twenty. She might be interested in marriage by then. Meanwhile, I'll be well into my middle age. Then, we will remember the trees. The years that have gone by, the world that has changed around us, just as much as we have changed within the world; and now, our first harvest. Then our second, our third. The jackfruit trees will continue to produce for another sixty to seventy years, while the Peruvian cactuses may bear apples for well over one hundred years, allowing the fruit to be tasted and appreciated by generations that I and my children will never live to see. With roots buried in the ground, these trees become time capsules of a sort. Great-great-great grandchildren may

RIDING THROUGH THE TREE TRUNK.

well look at these trees and ponder with an awed reverence, "This orchard was planted by my ancestor well over one hundred years ago." With that, they can begin the narrative of their ancestry, of which they, as much as the trees and the orchard, are an ongoing part.

This area has many olive trees. Some nearing two thousand years old. Silent chroniclers of history. Gnarly branches that self-fertilize with their own fallen fruit. Oil from these olives, so sublime, the flavor and texture, inspires heavenly gratitude. Perhaps only a tree that grows slowly is capable of producing such awesome culinary complexity. With the passage of time, when my orchard has been forgotten by humans, the olive trees will remember.

DEEP IN THE QUIET WOOD

JAMES WELDON JOHNSON

Are you bowed down in heart?
Do you but hear the clashing discords and the din of life?
Then come away, come to the peaceful wood,
Here bathe your soul in silence. Listen! Now,
From out the palpitating solitude
Do you not catch, yet faint, elusive strains?
They are above, around, within you, everywhere.
Silently listen! Clear, and still more clear, they come.
They bubble up in rippling notes, and swell in singing tones.
Now let your soul run the whole gamut of the wondrous scale
Until, responsive to the tonic chord,
It touches the diapason of God's grand cathedral organ,
Filling earth for you with heavenly peace
And holy harmonies.

AN INVITATION TO SEVERAL PLACES THAT ONCE WERE SAVANNA

KEEFE KEELEY

Come raise tree crops and shepherd flocks. Flash mob the cattle and pigs and goats into the Tartarian honeysuckle—we need real bulldozers and brush hogs and weed eaters to knock it back in favor of bluestem and trefoil and shagbark hickory seedlings. Help with a chainsaw if you must. Keep a hedgerow at the toe of the slope. Train Great Pyrenees to guard your herds. Come steward a Turtle Island dehesa.

Come keep in mind the millennia of people who have called this place home. Remember the many Menominee, Ho-Chunk, Dakota, Ioway (Báxoje), Ojibwe, Potawatomi, Odawa, Sauk and Fox (Mesquakie), Mescouten, Miami, Kickapoo, and other peoples who were forced to leave. Honor and learn from those who have returned to land that their families were removed from. Avoid appropriation as you adapt traditional ways to contemporary challenges.

Come learn from the aging hippie back-to-the-land farmers here. They came, in their turn, to learn from the aging Norwegian farmers who had worked with Aldo Leopold and Hugh Hammond Bennett to meet the climate challenge of their generation—rampant erosion and the Dust Bowl. The Norwegians experimented and helped their neighbors to learn contour plowing and strip cropping, to plant hillside conifers, and to install check dams. The back-to-the-landers went on to pioneer organics, community-supported agriculture, and management-intensive rotational grazing.

Come follow in their footsteps, go farther with perennial staple crops, silvopasture, native super fruits, multistory agroforestry, and all sorts of eco-mimicry. Meet the climate challenge of our time. Draw carbon down into tree trunks, grass roots, and uninterrupted mycelium. Turn buckthorn into biochar. Respongify the fine loess soils, restore floodplain hydrology, and recharge the karst aquifers. Haul valley-harvested hay bales, lush of the anthropogenic alluvium, back uphill to replenish the depleted ridge-edge fields. Come slow down the runoff from these crusted and compacted soils; these catastrophic thousand-year floods are killing us.

Come make catch-and-release brook trout the most valuable crop on your farm. Play well with the otter whose trail runs through the oxbow marsh across the creek. Her family is still recovering from John Jacob Astor's commodification of their hides. Make room for monarch butterflies, and also Karner blues. Sure, save the world with soil carbon, but also save

space for the xeric clovers up on the goat prairie.

Come play with fire. Try a patchwork of decadal fire return intervals to maintain successional stage habitats and oak regeneration. Whip-poor-wills prefer the brushy and grassy stages. Raise guinea fowl to keep down the ticks. Bless thy opossum who hoovers up hundreds o' ticks off his hide each day.

Come to Sinsinawa. There's an empty dairy barn and missing teeth in the old orchard. Come fill in the smile and ask the Dominican sisters for support. Come to the learning farms, scattered throughout the biome. Come to the Farley Center and Silverwood. To Liberty Prairie and Angelic Organics. To Main Street Project and the Dairy Grazing Apprenticeship. To Michael Fields and Wellspring. To Troy Community Gardens and the Agroforestry Academy. To Dreamtime Village and Dancing Waters. Come pick apples at Turkey Ridge.

Come to the widows on the land, and to the absentee mid-career inheritors. Come to them with a vision beyond monocultures for the land. Sell them on your vision—they want to believe in it, so make it into a business plan. Work out a long-term lease, or a custom land contract. Plan a future bequest into a community land trust. Aim for symbiosis. Beware of feudalism.

Come develop CSAs that realign economy and ecology. Run against the grain of the tyranny of convenience. Get your members to the farm to pick up their food.

Come build equity in emerging cooperative companies. Find an equitable and helpful way to use blockchain. Offer carbon bonds to crowdfund your local Nutella-inspired start-up. Spread your risk and generate cash flow with laying hens and broilers under the nut tree canopies.

Come establish farms that provide more than food, fiber, and fuel. Be a farm that provides paths out of the virtual idiotic world and into interdependent reality. Be a farm where souls connect with the soil, where spirits connect with sunshine. Practice jubilee especially with the disenfranchised and dispossessed. Dismantle sexism and racism and classism and colonialism in any way you can.

Come build an agrarian economy with the resilience of savannas that once were here, with farms made to weather the whim and tumult of the modern world, and with roots as deep as you can sink them.

IMAGE CREDITS

This is a noncommercial, nonprofit publication. To reprint original work, request permission of the contributor.

4 Public domain image. Unknown, Elizabeth Barrett Browning, *The Poems of Elizabeth Barrett Browning. Collected Poems* (1893), p. 7, flickr.com/photos/britishlibrary
5 Public domain image. Jean L. Watson, *Round the Grange Farm: or, Good Old Times,* (1872), p. 13, flickr.com/photos/britishlibrary
13 Public domain image. Unknown*, Català: Almanac per a 1928 de la revista taurina d'editorial carceller* (1928), commons.wikimedia.org
14 Public domain image. Unknown, Paul Villars, translated by Henry Frith, *England, Scotland and Ireland. A Picturesque Survey of the United Kingdom and Its Institutions* (1887), p. 512, flickr.com/photos/britishlibrary
15 Public domain image. Unknown, James Vincent Elsden, *Applied Geology … With Numerous Illustrations* (1898), p. 167, flickr.com/photos/britishlibrary
16 Original work by Josie Iselin.
27 Public domain image. Unknown, William John Loftie, *Orient Line Guide, etc., Fifth Edition, Rewritten* (1894), p. 267, flickr.com/photos/britishlibrary
28, 29 *2019 Moon Calendar*, Public domain clip art.
30 Public domain image. Unknown, William Hughes, *A Class-Book of Modern Geography … New Edition, Revised and Largely Rewritten (by Albert Hill)* (1898), p. 28, flickr.com/photos/britishlibrary
31 Public domain image. Yan' Dargent, Emmanuel Liais, L'Espace céleste et la nature tropicale, description physique de l'univers (1866), p. 199, flickr.com/photos/britishlibrary
32 Public domain image. Unknown, William Chambers, *Chambers's Alternative Geography Readers. Standard IV.(-VII.)* (1898), p. 29, flickr.com/photos/britishlibrary
34 Original work by Nikki Mokrzycki.
37 Public domain image. Unknown, William Latham Bevan, *The Student's Manual of Modern Geography Mathematical, Physical, and Descriptive* (1869), p. 579, flickr.com/photos/britishlibrary
39 Public domain image. NAL poster.
40 Public domain image. NAL poster.
41 Original work by Vincent Sheridan.
43 Public domain image. Unknown, Gordon Stables, *Aileen Aroon: a Memoir of a Dog. With Other Tales of Faithful Friends and Favourites* (1884), p. 259, flickr.com/photos/britishlibrary
45 Public domain image. Unknown, *London (illustrated). A Complete Guide to the Leading Hotels, Places of Amusement* (1872), p. 251, flickr.com/photos/britishlibrary
46 Public domain image. Eric Polk, *Untitled*, 2006, commons.wikimedia.org
49 Public domain image. Unknown, *The Pasture in Summer: The Drinking Trough*, 1867, Library of Congress, commons.wikimedia.org
50, 51, 53, 55, 56, 57 Public domain image. Unknown, Robert Brown MA, PhD, *Our Earth and Its Story: A Popular Treatise on Physical Geography* (1893), p. 29, flickr.com/photos/britishlibrary
58 Public domain image. Unknown, Henry Seebohm, *Siberia in Asia: A Visit to the Valley of the Yenesay in East Siberia* (1882), p. 215, flickr.com/photos/britishlibrary
61 Frank R. Rupert, Reproduced from Lane, "Florida's Geological History," p. 22, 19.
63 Public domain image. Unknown, John Rutty, *Observations on Mr. Archer's Statistical Survey of the County of Dublin* (1802), p. 357, flickr.com/photos/britishlibrary
65 Public domain image. Unknown, David Low, *On the Domesticated Animals of the British Islands* (1845), p. 110, flickr.com/photos/britishlibrary
66 Original work by Áine Mullan.
67 Public domain image. Unknown, David Page, *Introductory Text-book of Geology* (1861), p. 151, flickr.com/photos/britishlibrary
68 Public domain image. Reederei Wessels/Fotoflite, *Untitled*, 2011, commons.wikimedia.org
72 Original work by Sophie Wanzer.
73 Original work by Rachel Alexandrou.
76, 77, 78, 79, 90, 91 Original work by Ginny McClure.
83 Public domain image. H.S. Bender Photographs, HM4-083 Box 1 Folder 3 Photo 117. Mennonite Church USA Archives—Goshen. Goshen, Indiana, commons.wikimedia.org
85 Public domain image. Lebreton, Guillaume Louis Figuier, *La terre et les mers* (1864), p. 213, flickr.com/photos/britishlibrary
86 Public domain image. Jaroslav A. Polák, *Barbed Wire Study*, 2014, commons.wikimedia.org
88 Public domain image. Unknown, David Low, *On the Domesticated Animals of the British Islands* (1845), p. 127, flickr.com/photos/britishlibrary
92 Public domain image. Unknown, *Encyclopædia Britannica, 11th ed., Vol. 13* (1911), p. 448, commons.wikimedia.org
93 Original work by Charlotte X.C. Sullivan and Alayna Rasile Digrindakis.
95 Public domain image. Unknown, *Ginseng*, Archives of Pearson Scott Foresman, commons.wikimedia.org
97 Public domain image. Martin Proll, *A View across the Desert Landscape of Big Bend National Park, Texas*, 2008, commons.wikimedia.org
98 Original work by Rachel Alexandrou.

IMAGE CREDITS

101 Public domain image. Source unknown.
103, 104, 105 Original work by Kirk Crippens and Gretchen LeMaistre.
107 Original work by Rachel Alexandrou.
109 Original work by Christie Green.
112 Original work by Olivia Green.
115 Public domain image. Unknown, Gerrit Engelberts, *Ons Vaderland* (1856), p. 322, flickr.com/photos/britishlibrary
116 Public domain image. Unknown, Harry Ellington Brook, *The Land of Sunshine* (1893), p. 134, flickr.com/photos/britishlibrary
118 Public domain image. Unknown, Mary Elizabeth Dodge, *When Life Is Young* (1894), p. 58, flickr.com/photos/britishlibrary
121 Original work by Gabriella Marks.
125 Public domain image. Unknown, André Laurie, *The Conquest of the Moon* (1889), p. 322, flickr.com/photos/britishlibrary
126 Public domain image. Unknown, J. Leyland, *Adventures in the Far Interior of South Africa* (1866), p. 332, flickr.com/photos/britishlibrary
129 Public domain image. Unknown, John Arthur Phillips, *Elements of Metallurgy* (1891), p. 767, flickr.com/photos/britishlibrary
130 Public domain image. Johney Bearrup, *The Farmer's Advocate* (1905).
132 Public domain image. Charles Roscoe Savage, *Train Running on Dale Creek Iron Viaduct, Wyoming*, 1876–1885, commons.wikimedia.org
135 Original work by Frank Utpatel.
137 Public domain image. Unknown, *The Half Hour Library of Travel* (1896), p. 187, flickr.com/photos/britishlibrary
139 Original works by Savannah Volkoff.
140 Public domain image. Boussole Folle, *Empreinte de spore Psilocybe semilanceata*, 2015, commons.wikimedia.org
142 Public domain image. Unknown, Edward Douglas Fawcett, *Hartmann the Anarchist* (1893), p. 203, flickr.com/photos/britishlibrary
145 Public domain image. Daido Bunka, *The Character for "Heart/Mind" as an Enso (image 1 of 2)*, 18th century, Los Angeles County Museum of Art, commons.wikimedia.org
146 Original work by Briana Olson.
148 Public domain image. Unknown, Joseph Beete Jukes, *The Student's Manual of Geology* (1857), p. 208, flickr.com/photos/britishlibrary
151 Public domain image. Adam Nuttall, *Fire Engine at Work* (1760), reprinted from George Lyman Kittredge, *The Old Farmer and His Almanack* (1920), p. 146.
153 Public domain image. Mennonite Church USA Archives, *Project work—building Fence for Pasture Land*, commons.wikimedia.org
154 Public domain image. A. Y. Bingham, Annie Brassey, *A Voyage in the "Sunbeam"* (1878), p. 114, flickr.com/photos/britishlibrary
156 Public domain image. Hila Shaked, *Untitled*, 2010, commons.wikimedia.org
159 Public domain image. Jim Peaco, *Firestorm Rages across Mirror Plateau*, 1988, Yellowstone Digital Slide File, commons.wikimedia.org
161 Original work by Rachel Alexandrou.
163 Public domain image. E. Marillier, Forbes Edward Winslow, *The Children's Fairy History of England* (1889), p. 188, flickr.com/photos/britishlibrary
166 Public domain image. Unknown, Charles Darwin, *What Mr. Darwin Saw in His Voyage Round the World in the Ship "Beagle"* (1879), p. 173, flickr.com/photos/britishlibrary
168 Public domain image. Anonymous, *Untitled*, 2010, commons.wikimedia.org
171, 174, 175 Public domain image. Unknown, Robert Jameson, *System of Mineralogy, Comprehending Oryctognosie, Geognosie, Mineralogical Chemistry, Mineralogical Geography, and Œconomical Mineralogy. Vols. 1–3* (1804), p. 689, flickr.com/photos/britishlibrary
177 Original work by Ursula MacFarlane.
180 Public domain image. Unknown, Charles C. Chapman, *History of La Porte County, Indiana* (1880), p. 600, flickr.com/photos/britishlibrary
181 Original work by Vincent Sheridan.
184 Public domain image. Marisol Grandon/Department for International Development, *Untitled*, 2011, commons.wikimedia.org
189 Public domain image. Frederic G. Cooper, *Food—Don't Waste It*, 1917, Library of Congress, commons.wikimedia.org
190 Public domain image. Unknown, Henry Lemonnier, *Nouveau cours d'instruction primaire* (1886), p. 11, flickr.com/photos/britishlibrary
192 Public domain image. Unknown, *Cuneiform Tablet: Assumption of Debt by Guarantor, Archive of Bel-remanni*, circa 509 B.C. Metropolitan Museum of Art, Purchase, 1886 commons.wikimedia.org
194 Public domain image. Unknown, *The Half Hour Library of Travel* (1896), p. 343, flickr.com/photos/britishlibrary
199 Public domain image. E. Fitzpatrick, Philip Bennett Power, *Cured by an Incurable* (1888), p. 49, flickr.com/photos/britishlibrary
200, 201, 202, 203 Public domain image. Unknown, *Crowds Listening to the President's Speech*, 1907, Library of Congress, commons.wikimedia.org
205 Original work by Frank Utpatel.

IMAGE CREDITS

209 Public domain image. Dorothea Lange, *Mexicans bound for the Imperial Valley to harvest peas. Near Bakersfield, California*, 1936, United States Farm Security Administration, commons.wikimedia.org
209 Public domain image. Charles O'Rear, *In lettuce fields along the Colorado River, Mexican farm worker carries boxes to field pickers*, 1972, US National Archives and Records Administration, commons.wikimedia.org
209 Public domain image. Bob Nichols, *Migrant workers harvest lettuce at Lakeside Organic Gardens in Watsonville, CA*, 2013, commons.wikimedia.org
209 Public domain image. William Allen White, *The Farmer's Advocate*, (1906).
212, 213 Public domain image. L. T. Breeling, "The Price of Being a Radical," *The Farmers Union Herald, (1930).*
216 Public domain image. Nils Otto Gustaf Nordenskjöld, *Från Eldslandet* (1898), p. 210, flickr.com/photos/britishlibrary
217, 221 Original works by Jenny Zhao.
224 Public domain image. Allart van Everdingen, *The Lion Takes the Advice of the Other Animals for Renard's Punishment*, 1650-1675, Metropolitan Museum of Art, The Elisha Whittelsey Collection, The Elisha Whittelsey Fund, 1962, metmuseum.org
226 Public domain image. Unknown, Ambroise Tardieu, *Grand Dictionnaire historique* (1894), p. 173, flickr.com/photos/britishlibrary
227 Public domain image. Unknown, Martin Farquhar Tupper, *Proverbial Philosophy in Four Series* (1881), p. 137, flickr.com/photos/britishlibrary
228 Public domain image. Unknown, George Thomas Bettany, *The World's Inhabitants* (1889), p. 324, flickr.com/photos/britishlibrary
230 Public domain image. Unknown, *Fort Larned,* U.S. Geological Survey.
238 Anna Mielke, *Wealth*, kitchenterroir.com/2014/10/08/fermentation-fest/
241 Public domain image. Unknown, Unknown, flickr.com/photos/britishlibrary
244 Original work by Neil Thapar.
246 "Minority Principal Operators, 2007 and 2012," USDA NASS, 2012 Census of Agriculture.
248 Public domain image. John Vachon, *Black refugees evicted from sharecropping, now on the roadside. Parkin, Arkansas*, 1936, US Library of Congress, commons.wikimedia.org
253 Public domain image. Unknown, *Black female sharecropper picking cotton*, 1939, Farm Security Administration, commons.wikimedia.org
256, 257 Original work by Lucia Gaia.
258 Public domain image. Paul Conklin, *Members of the black-owned Bluffton oyster cooperative crabbing is the off-season occupation for these men*, 1973, National Archives and Records Administration, commons.wikimedia.org
260, 261, 262, 263, 264 Public domain image. Unknown, Jean-Jacques Rousseau, *Œuvres complètes de Jean Jacques Rousseau* (1788), p. 168, flickr.com/photos/britishlibrary
267 Public domain image. Unknown, *An American having struck a bear but not killed him, escapes into a tree,* 1810, US Library of Congress, commons.wikimedia.org
268, 269 Original work by Danjo Paluska.
271 Public domain image. Unknown, H. Billet, *Beaujolais-Forez-Dombes* (1899), p. 51, flickr.com/photos/britishlibrary
272 Public domain image. Aleksandr Kun, *Places of Social Gathering. Preparation of Hemp Beer*, 1865-1872, US Library of Congress, commons.wikimedia.org
274, 277 Original works by Casey O'Neill.
278 Public domain image. Unknown, *Gold Miners, c1850. Prospectors posing at their sluice box*, The Granger Collection, commons.wikimedia.org
281 Public domain image. Unknown, ad for the American film *Fool's Gold*, *Moving Picture World*, p. 1105, 1919, commons.wikimedia.org
284 Original work by Nikki Lastreto.
285, 286, 287, 288, 289, 290, 291 Original works by Layton Hansen.
292, 293 Original work by Casey O'Neill.
295 Public domain image. Tudor Washington Collins, *Cow and calf in field*, 1940, Auckland War Memorial Museum, commons.wikimedia.org
296 Public domain image. Unknown, Williams Seaman Stevens, *Homographia* (1806), p. 95, flickr.com/photos/britishlibrary
298, 299, 300 Original works by Renata Christensen.
301 Original work by Josie Iselin.
302, 303 The Seed Ambassadors Project, *A Guide to SeedSaving, SeedStewardship & Seed Sovereignty, 4th edition*, 2010, p. 14-15.
305 Original work by Liz Brindley.
307 Original work by Sarah Hughson.
309 Public domain image. Unknown, *Wall panel with garden urns*, late 18th century, Metropolitan Museum of Art, Rogers Fund, 1980, metmuseum.org
310, 311, 312, 313 Public domain image. Unknown, Florence O'Driscoll, *Notes on the Treatment of Gold Ores* (1889), p. 142, flickr.com/photos/britishlibrary
314 Public domain image. After Sir John Everett Millais, *The Sower (The Parables of Our Lord and Saviour Jesus Christ)* 1864, Metropolitan Museum of Art, Rogers Fund, 1921, metmuseum.org

IMAGE CREDITS

317 Public domain image. Unknown, *Untitled*, 1938, commons.wikimedia.org
318, 319 Unknown, *Svalbard Global Seed Vault*, the Crop Trust, croptrust.org
325 Original work by Ray Ray Mitrano.
326 Public domain image. Unknown, *Annual catalogue of Price & Reed*, 1894, US Department of Agriculture, National Agricultural Library, commons.wikimedia.org
329 Original work by Liz Brindley.
330 Public domain image. Unknown, *Graffiti during the occupation of Alcatraz by Red Power activists*, 1969, National Park Service, commons.wikimedia.org
334 Unknown, *Havasupai Corn Fields*, Mathers Museum of World Cultures, Wanamaker Collection, flickr.com/photos/mathers_museum
335 Public domain image. Unknown, *3 boys of the Red Lake Band of Chippewa in a corn field, Red Lake Minnesota ca1938*, National Archives, flickr.com/photos
339 Public domain image. Unknown, *Untitled*, archive.org
339 Public domain image. Unknown, *Parade Feature, by sixteen Indian girls from the Government School at Chilocco, O.T.*, flickr.com/photos
340 Public domain image. Unknown, Richard Ashe King, *Passion's Slave* (1889), p. 315, flickr.com/photos/britishlibrary
343 Public domain image. Unknown, *Velvet Panel with Flowering Plants*, first half 17th century, Metropolitan Museum of Art, Rogers Fund, 1912, metmuseum.org
345 Public domain image. Title screen of the film *Hemp for Victory*, 1942, U.S. Government, commons.wikimedia.org
346 Public domain image. Pearson Scott Foresman, *Untitled*, commons.wikimedia.org
347 Public domain image. Unknown, *Gathering wild rice, Minnesota, circa 1897,* archive.org
348 Public domain image. Unknown, *Effigy Mounds National Monument, Aerial View of Great Bear Mound Group*, National Park Service, commons.wikimedia.org
353 Public domain image. Unknown, *Sauk River Stump, Mt Baker NF, WA*, 1945, U.S. Forest Service, flickr.com/photos/forestservicenw
355 Public domain image. User:Basotxerri, *Pastos en la niebla cerca de la Cruz del Gorbea. Álava, País Vasco, España*, 2017, commons.wikimedia.org
359 Original work by Karen Cantor.
361 Public domain image. Unknown, Franz Keller-Leuzinger, *Vom Amazonas und Madeira* (1874), p. 72, flickr.com/photos/britishlibrary
362 Public domain image. Unknown, Edward Laws, *The History of Little England beyond Wales, and the Non-Kymric Colony settled in Pembrokeshire* (1888), p. 48, flickr.com/photos/britishlibrary
369 Public domain image. Unknown, William Murray, *Daylight Land* (1888), p. 135, flickr.com/photos/britishlibrary
369 Public domain image. Unknown, Samuel Manning, *The Land of the Pharaohs* (1897), p. 132, flickr.com/photos/britishlibrary
369 Public domain image. Unknown, Robert Brown, *Our Earth and its Story* (1899), p. 186, flickr.com/photos/britishlibrary
369 Public domain image. Unknown, Alice Hart, *Picturesque Burma, Past and Present* (1897), p. 209, flickr.com/photos/britishlibrary
369 Public domain image. Unknown, Alexis M. G., *Le Congo Belge Illustre⊠* (1888), p. 231, flickr.com/photos/britishlibrary
371 Public domain image. Unknown, Jules Gourdault, *La France Pittoresque* (1893), p. 483, flickr.com/photos/britishlibrary
372 Original work by Michelle J. Felicetti.
376 Original work by Maureen Walrath.
379, 380 Public domain image. J Emslie, *Acoustics*, 1850, Wellcome Images, commons.wikimedia.org
382 Public domain image. Unknown, *Untitled*, Berkshire Community College Bioscience Image Library, commons.wikimedia.org
385 Original work by Deirdre O'Mahony.
387 Public domain image. Unknown, Lafayette Houghton Bunnell, *Discovery of the Yosemite* (1892), p. 384, flickr.com/photos/britishlibrary
388 Public domain image. Unknown, William Seward, *W. H. Seward's Travels around the World* (1873), p. 358, flickr.com/photos/britishlibrary
391 Public domain image. William Trost Richards, *Trees (from Sketchbook VII)*, 1886, Metropolitan Museum of Art, Purchase, Gifts in memory of Stephen D. Rubin, 1992, metmuseum.org
392, 393 Public domain image. Mori Yuzan; Yamada Geisodo, *Hamonshu v. 3* (1903), Internet Archive, archive.org
398 Public domain image. Unknown, Bertha Stoneman, *Plants and Their Ways in South Africa* (1915), p. 86, Internet Archive, archive.org
399 Public domain image. Unknown, Ralph Abercromby, *Seas and Skies in Many Latitudes* (1888), p. 77, flickr.com/photos/britishlibrary
400 Public domain image. Unknown, Jean Watson, *Round the Grange Farm* (1872), p. 310, flickr.com/photos/britishlibrary

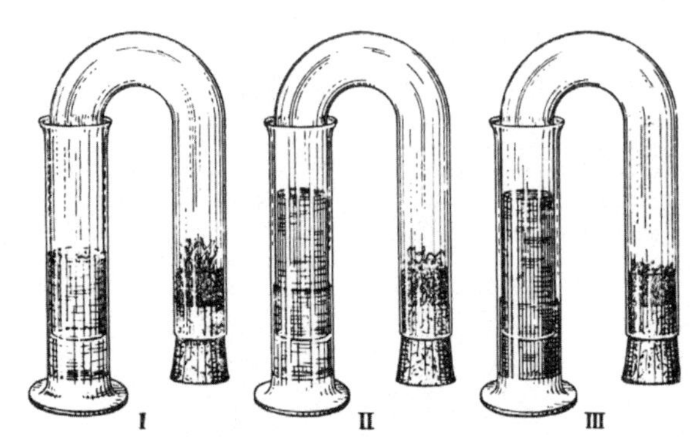

FIG. 79.—Peas germinating

ABOUT THE GREENHORNS

The Greenhorns is a ten-year-old grassroots organization with the mission to promote, support, and recruit an incoming generation of organic farmers and ranchers. Based in Downeast Maine, we produce diverse media and publications, including films, radio, guidebooks, and exhibits. Our various programs and projects address the practical and social concerns of those in their first years farming; we emphasize restorative land practices, skill building, networking, and dialogue.

The Greenhorns' new (permanent!) headquarters is an 1896 Odd Fellows Hall, lately christened "Reversing Hall." Visitors are welcome to our agrarian research library, seasonal workshops, and fun family camping on the saltwater farm down the road. We are nearly to Canada, with acres of wild Maine blueberries, mussels and clams, edible seaweeds, mossy forest in every direction, and dozens of grown-over grandmother orchards to restore and use for cider.

Stop by greenhorns.org and watch one of Up Up! Farm Film Fest's thirteen films about the future of farming, or dig into our archived podcasts, or the web films at OurLand.tv. Learn about the Greenhorns Sail Freight Project. Download a guidebook. Contribute to our blog. Or send us a note telling us how you'd like to collaborate!

Join our mailing list online to get news of naturalist trainings and adventures on land and sea.

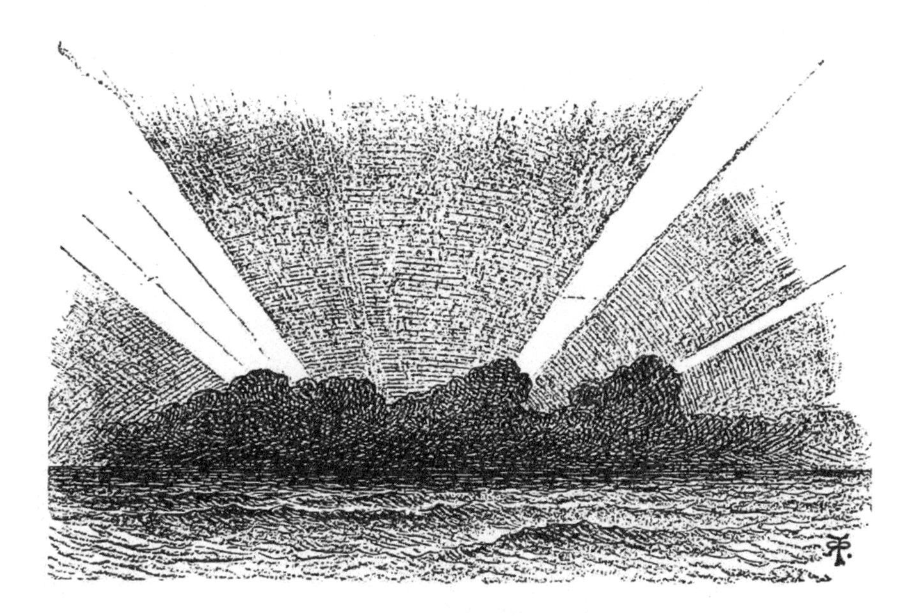

SUBMISSIONS

Was your voice missing from this volume? Remedy that by contributing to the next!

The theme for Volume V is restoration, with a focus on the social structures that hold the power to repair our ecosystems on a relevant local scale. How are people working together? Let's look at land like commoners, with the irrefutable magnetism of nature's changing pulse. Pushing like an overflowed river through chain-link fences, let's submerge the legacies of colonialism, slavery, and extraction with a fresh, fruitful layer of muck. All sorts of over-looked land can be the object of our devotional action. Mulching of railyards and vacant lots, native plantings of highway verges, negotiation with municipal keyholders. Imagine remediated peri-urban farmsteads where updated zoning laws make for bikeable food production. How are we approaching the project of land repair? How about a community potluck turned restoration hydrology project?

Please be in touch by March 2020 with inquiries and ideas for our 2021 edition. We're at almanac@greenhorns.org. We look forward to hearing from you.

THE END